THE APPLETON-CENTURY MATHEMATICS SERIES
Edited by Raymond W. Brink

SOLID ANALYTIC GEOMETRY

SOLID ANALYTIC
GEOMETRY

BY

JOHN M. H. OLMSTED, Ph.D.

PROFESSOR OF MATHEMATICS
AT THE UNIVERSITY OF MINNESOTA

NEW YORK
APPLETON-CENTURY-CROFTS, INC.

COPYRIGHT, 1947, BY
D. APPLETON-CENTURY COMPANY, INC.

All rights reserved. This book, or parts thereof, must not be reproduced in any form without permission of the publisher.

6123-8

PREFACE

There is wide variation in the content of courses in Solid Analytic Geometry. A very brief introductory course may place the principal emphasis on planes and lines, and treat only the simpler equations of the second degree. A standard first course probably includes the notion of the rank of a matrix, transformations of coördinates, and a more thorough study of equations of the second degree. A still more extended course may intensify the study of planes and lines, and make use of the elegant methods of matrix algebra in the treatment of transformations and the general equation of the second degree.

This book is designed to meet the needs of any of the courses just suggested. This flexibility is gained partly by the order of presentation of the material. Difficult topics are postponed as much as possible. For example, the first two chapters develop the geometry of the plane and the line, with determinants almost entirely restricted to those of the second order. Chapter III then lists the principal properties of determinants in general, introduces the concepts of a matrix and its rank, and applies these new ideas to a more elaborate discussion of planes and lines, including the application of the theory of systems of linear equations to a study of systems of planes. Again, the chapter on transformations (Chapter VII) is not presented until after the development of such topics in the theory of the general equation of the second degree as centers, diametral planes, and tangent planes, which do not require transformations of coördinates.

In those sections that would be included in any first course a special effort has been made to keep the discussion on as elementary a level as possible. On the other hand, although a class may not be formally exposed to certain related ideas, it is important that these ideas be available to those students who wish to explore beyond the confines of the standard course outline. It is partly for this reason that some sections, like those on higher dimensional Euclidean spaces (§ 62), complex space (§ 63), and symbolic notation for transformations (§ 137) have been included.

As an aid to adapting the book to courses of varied content, many sections that can be omitted without destroying the continuity of the course are marked with stars. Exercises that depend on these sections, as well as some of the other exercises, are similarly indicated. If all of these starred sections and exercises are omitted, the remaining portions of the book form a complete body of material suitable for a course of approximately thirty lessons. With this as a basis, alterations for a shorter or longer course can be made easily.

A consistent goal has been simplicity of presentation without sacrifice of completeness, with emphasis placed on logical reasoning and method rather than on rote and decree. Statements are formulated with precision and any exceptions are carefully pointed out. Careless statements, which hold only "in general" or whose validity depends on the existence of points and lines at infinity, are avoided. Numerous illustrative examples, some solved by more than one method, help to clarify new principles as they are introduced.

A feature of the book is the unusually extensive collection of exercises. Enough of them are given to permit their generous use for class demonstration without reducing their value as assigned problems. The exercises range from practice problems for developing technical facility, through problems that require some originality and independence from formulas, to problems that actually extend the subject matter of the text. Answers to all exercises, except those that ask for proofs or discussions, are given in the back of the book. Many of the harder exercises are followed by suggestions that either give a start toward a solution or furnish an outline of a method. Problems of this type are frequently marked with a star, and in some sections are assembled in a group labeled *Supplementary Exercises.* In some sets the hardest exercises are indicated by an asterisk. In nearly all exercises the arithmetic has been kept simple in order not to obscure the main purpose of the problem.

Matrices find many useful and simple applications in solid analytic geometry, and these applications are given unusually generous consideration in this book. The material has been arranged in such a way that the student can enjoy a large part of the benefits of matrix theory while knowing very little more about a matrix than its rank. One of these benefits is a simple method of identifying a quadric surface. For those who wish to minimize the use of matrices, the *applications* of rank are given first and are concentrated principally in Chapters III and VIII. An important theorem on invariants is stated in Chapter VIII. However, the proof of this theorem is

PREFACE

deferred to Chapter IX, since the proof requires much more knowledge of matrix theory than do the applications. In a brief course it is reasonable to include the statement and applications of the theorem while omitting Chapter IX, which supplies a proof.

Chapter IX contains a systematic elementary treatment of Matrix Algebra. It may well serve as a text in an introductory course in that subject. In that case the geometrical applications serve to illustrate and clarify the theory of matrices. In a substantial course in solid analytic geometry, on the other hand, the inclusion of Chapter IX supplies the proof of the theorem previously mentioned, completes the discussion of rotations and invariants, and gives formulas for the volume of a tetrahedron and the area of a triangle.

Little mathematical training is assumed on the part of the student beyond a course in plane analytic geometry. Although calculus is used only in the notation for partial derivatives, a first course in that subject is desirable, if only as a contribution to what is called "mathematical maturity." In fact, it has seemed desirable in this book to avoid the use of calculus and, instead, to place a heavy emphasis on algebraic methods. This emphasis is justified by the close ties existing between such concepts as *quadric surface* and *principal plane* on the one hand and *quadratic form* and *characteristic root* on the other. In line with this relationship, the language of algebra is used wherever it seems natural — as with *linear combinations* of planes and the *reducibility* of an algebraic surface.

Particular care is given to a discussion of surfaces and curves. Because of the impracticability of giving general definitions of these terms, the author simply warns the student of the difficulties, and then confines the treatment of the text to *algebraic* surfaces and curves. Definitions are fashioned to distinguish between such different surfaces as $xy^2 = 0$ and $x^2y = 0$. This permits an exact algebraic counterpart to the geometric statement that two algebraic surfaces are identical. The book explains in what sense one can speak of *the* equation of a given algebraic surface, and makes clear why it is necessary to include imaginary points in the discussion. The use of imaginary numbers, however, is not carried beyond the extent necessitated by this discussion, except for the section on complex space previously alluded to.

The attention of the reader is also called to the distinction that is made between *distance* and *directed distance*, the flexible treatment that is given to the normal equation of a plane, and the discussion that clarifies an important concept, namely that of *point transformations* as distinguished from *coördinate transformations*.

Homogeneous coördinates and vector notation have not been included since the former would decidedly alter the emphasis of the book and the latter would impose an unnecessary burden on many students.

The author wishes to express his deep appreciation of the very considerable aid given by Professor R. W. Brink of the University of Minnesota in the preparation of the manuscript. He is also indebted to others, including in particular Professor Kenneth Wegner of Carleton College, for their friendly and helpful counsel. The figures were drawn by Miss Dorothy Nori, a senior student at the University of Minnesota.

J. M. H. O.

CONTENTS

	PAGE
Preface	v

CHAPTER

I. Coördinates and Cosines

1. Distance and directed distance 1
2. Rectangular coördinates 1
3. Representing space figures by plane drawings 2
4. Symmetry 4
5. Projections 5
6. Traces and intercepts 5
7. Exercises 6
8. Explanation of certain terminology 7
9. Distance between two points 7
10. Angle between two directed lines 9
11. Projections on a directed line 9
12. Point of division 10
13. Exercises 12
14. Direction cosines 15
15. Exercises 16
16. Direction numbers 17
17. Exercises 18
18. Cosine of the angle between two directed lines 19
19. Perpendicular lines 20
20. Exercises 20
21. Sine of the angle between two lines 21
22. Exercises 22
23. Direction perpendicular to two directions 22
24. Exercises 23
25. Projection of a broken line segment 24
26. Exercises 24

II. Planes and Lines

27. Introduction 26
28. Equation of a plane 27
29. Exercises 29
30. Intercept form for the equation of a plane 30
31. Normal form for the equation of a plane 31
32. Directed distance from a plane to a point 32

CONTENTS

CHAPTER		PAGE
	33. Exercises	33
	34. Angle between two planes	35
	35. Direction of line of intersection of two planes	36
	36. Pencil of planes	37
	37. Exercises	39
	38. Equations of a line	41
	39. Parametric equations of a line	44
	40. Exercises	46
	★ 41. Another form for the parametric equations of a line	50
	★ 42. Parametric equations of a plane	50
	★ 43. Exercises	52
	★ 44. Distance between a point and a line	52
	★ 45. Distance between two skew lines	53
	★ 46. Exercises	54
III.	**Determinants and Matrices**	
	47. Introduction	56
	48. Some properties of determinants	56
	49. Matrix. Rank of a matrix	57
	50. Elementary transformations of a matrix	58
	51. Exercises	60
	52. Simultaneous linear equations	60
	53. Exercises	64
	54. Systems of planes	65
	55. Exercises	68
	★ 56. Linear dependence	68
	★ 57. Exercises	70
	★ 58. Families of points	71
	★ 59. Exercises	72
	60. Plane through three points	72
	61. Exercises	74
	★ 62. Space of four or more dimensions	77
	★ 63. Complex space	79
	★ 64. Exercises	80
IV.	**Surfaces and Curves**	
	65. Surfaces	83
	★ 66. Parametric equations of a surface	86
	67. Curves	87
	68. Sections, traces, and intercepts	89
	69. Cylinders	89
	70. Exercises	91
	★ 71. Projections of a curve. Elimination	92
	★ 72. Exercises	95
	73. Symmetry	95
	74. Exercises	96

★ Sections marked with a ★ may be omitted.

CONTENTS

CHAPTER		PAGE
	75. Symmetry for graphs	97
	76. Exercises	98
	★ 77. Symmetry for algebraic surfaces	98
	★ 78. Exercises	101
	79. Surfaces of revolution	101
	80. Exercises	103
	81. The sphere	104
	82. Sphere through four points	105
	83. The circle	106
	★ 84. Pencil of spheres	106
	85. Exercises	107
	86. Locus problems	110
	87. Exercises	112
	★ 88. Cylindrical and spherical coördinates	113
	★ 89. Exercises	115
V.	The Seventeen Quadric Surfaces	
	90. Introduction	117
	91. Real ellipsoid	117
	92. Exercises	118
	93. Imaginary ellipsoid	118
	94. Hyperboloid of one sheet	119
	95. Exercises	119
	96. Hyperboloid of two sheets	120
	97. Exercises	120
	98. Real quadric cone	121
	99. Exercises	121
	100. Imaginary quadric cone	122
	101. Elliptic paraboloid	122
	102. Exercises	123
	103. Hyperbolic paraboloid	123
	104. Exercises	123
	105. Real elliptic cylinder	124
	106. Imaginary elliptic cylinder	124
	107. Hyperbolic cylinder	124
	108. Real intersecting planes	125
	109. Imaginary intersecting planes	125
	110. Parabolic cylinder	125
	111. Quadric cylinders	126
	112. Real parallel planes	126
	113. Imaginary parallel planes	126
	114. Coincident planes	126
	115. Equations almost in canonical form	127
	116. Exercises	128

★ Sections marked with a ★ may be omitted.

CONTENTS

VI. The General Equation of the Second Degree

- 117. Introduction 131
- 118. Intersection of quadrics and lines 133
- 119. Degenerate sections of quadrics 134
- 120. Exercises 135
- 121. Diametral planes 135
- 122. Centers 136
- 123. Exercises 138
- 124. Singular points and regular points 140
- 125. Exercises 140
- 126. Tangent planes and normal lines 140
- 127. Normal planes and tangent lines 142
- 128. Exercises 143
- ★ 129. Poles and polar planes 146
- ★ 130. Exercises 148
- 131. Ruled surfaces 150
- ★ 132. Further discussion of ruled surfaces 154
- 133. Exercises 156

VII. Coördinate and Point Transformations

- 134. Introduction 159
- 135. Translations 161
- 136. Two aspects of a transformation 162
- ★ 137. Symbolic notation for transformations 162
- 138. Invariance 165
- 139. Simplification of equations by translations 166
- 140. Exercises 170
- 141. Rotations 172
- 142. Determinant of a rotation matrix 174
- ★ 143. Axis of a rotation 175
- ★ 144. Angle of a rotation 177
- 145. Invariance of degree 177
- 146. Plane sections of algebraic surfaces 178
- 147. Exercises 179

VIII. Analysis of the General Equation of the Second Degree

- 148. Introduction 184
- 149. Principal planes 184
- 150. Exercises 188
- 151. Reduction to canonical form 188
- 152. Equivalence of second degree equations 190
- 153. Invariants of a quadric surface 191
- 154. Exercises 193
- 155. Complete analysis of a quadric surface 194
- 156. Exercises 198

★ Sections marked with a ★ may be omitted.

CONTENTS

★ IX. Matrix Algebra

157. Introduction 207
158. Multiplication of matrices 207
159. Other operations on matrices 209
160. Transpose of a matrix 210
161. Zero and the identity 210
162. Multiplication of determinants 211
163. Rank of the product of two matrices 212
164. Exercises 213
165. Linear transformations 214
166. Inverse of a linear transformation 215
167. Inverse of a matrix 216
168. Orthogonal matrices 217
169. Exercises 220
170. Direct and inverse transformations 224
171. Quadratic forms 224
172. Proof of an earlier theorem 225
173. Exercises 227
174. Volume of a tetrahedron 228
175. Area of a triangle in space 229
176. Transforming a transformation 230
177. Exercises 232

Answers 235

Index 251

★ Sections marked with a ★ may be omitted.

CHAPTER I

COÖRDINATES AND COSINES

1. Distance and directed distance. In plane analytic geometry a correspondence is obtained between points and ordered pairs of real numbers. This is accomplished by means of a *rectangular coördinate system* based on two mutually perpendicular lines.† In a similar way a correspondence between points in space and ordered triads or triplets of real numbers can be set up with the aid of three mutually perpendicular planes. For a discussion of this question, the concepts of *distance* and *directed distance* are basic.

The single word *distance*, as used in this book, always refers to a non-negative quantity: for example, the distance between two points or the (perpendicular) distance between a plane and a point. On the other hand, *directed distance* carries the implication of direction, and may be represented by any real number, positive, zero, or negative. In any case, distance and directed distance are numerically equal. We shall consider two fundamental methods of defining directed distance:

(*i*) *Directed distance on a line.* If a direction is assigned to a line, the line is called a **directed line**. This direction is also called the **positive direction** of the line, the opposite direction being its **negative direction**. The **directed distance** from one point P on the line to another point Q on the line is then positive or negative according as Q lies in a positive or negative direction from P.

(*ii*) *Directed distance from a plane to a point.* In connection with any plane in space it is frequently convenient to assign a positive direction to the lines that are perpendicular to the plane. The **directed distance** from a plane to a point is then defined as the directed distance to the point from the foot of the perpendicular drawn from the plane to the point. A point not on the plane is said to lie on the *positive side* or on the *negative side* of the plane according as its directed distance from the plane is positive or negative.

2. Rectangular coördinates. The basis of a rectangular system of coördinates is three mutually perpendicular planes, called **coördinate**

† The word *line* is used consistently in this book to mean *straight line*.

1

planes, which intersect pairwise in three mutually perpendicular lines, called **coördinate axes**. The point common to all of these planes and lines is called the **origin**, and is usually designated by the letter O. The coördinate axes are usually designated as the x, y, and z axes, and the coördinate planes as the yz, xz, and xy planes corresponding to the axes that they contain. Let a positive direction be assigned to each coördinate axis. Each axis is then a directed line, and determines the positive side of the coördinate plane perpendicular to it. The **coördinates** (x, y, z) of any point are defined to be its directed distances from the coördinate planes: x is the directed distance from the yz plane, y is the directed distance from the xz plane, and z is the directed distance from the xy plane. The coördinates (x, y, z) are always to be written in that order and so constitute an *ordered triad* of real numbers. In this manner a *one-to-one correspondence* is established between points in space and ordered triads of real numbers. That is, corresponding to any point there is an ordered triad of real numbers, the coördinates of the point, and corresponding to any ordered triad of real numbers there is a point having these numbers as its coördinates. Because of this correspondence we can feel free to speak of *the point* (x, y, z) or *the point* $(2, -4, 1)$.

The three coördinate planes divide the points of space which are not on any coördinate plane into eight separate regions, called **octants**. The octant whose points have only positive coördinates is called the **first octant**. There will be no need to assign an order to the remaining octants.

3. Representing space figures by plane drawings. In order to represent points, lines, and other figures in space by means of drawings on paper (or blackboard), we must adopt certain conventions. One of the most convenient locations of a system of axes has one axis horizontal and directed to the right, one axis vertical and directed upward, and the other axis perpendicular to these two and directed toward the observer. We shall consider the axes to be so situated, and draw them on paper as shown in Fig. 1, with one line horizontal, one vertical, and the third making an angle of 135° with each of the

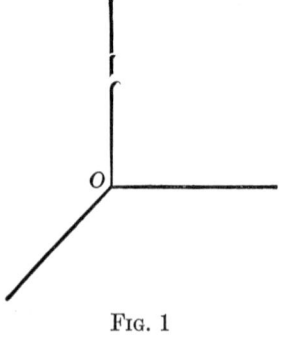

Fig. 1

first two. Only half of each axis is shown in order that we may say without ambiguity that the positive direction on each axis is away from the origin. We have still not labeled the axes. This can be done in six ways, shown in Fig. 2. The first three coördinate systems

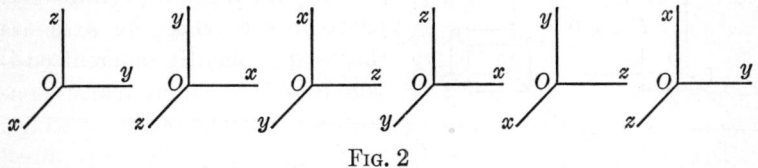

Fig. 2

of Fig. 2 are known as *right-handed systems*, and the last three as *left-handed systems*, according to the definition:

DEFINITION. *A rectangular coördinate system is **right-handed** if and only if the axes are arranged so that if the xy plane is viewed from a point on the positive half of the z axis, a counter-clockwise ninety-degree rotation in this plane about the origin carries points on the positive half of the x axis into points on the positive half of the y axis. The system is **left-handed** if and only if it is not right-handed.*

Using the concept of a right-handed screw (that is, one that is threaded in such a way that when it is turned about its axis in a clockwise sense it moves away from the observer), we can describe a right-handed rectangular coördinate system as one for which a right-handed screw whose axis coincides with the z axis advances in the positive z direction when turned through the right angle xOy.

The relationship between the axes expressed in this definition is a cyclic one, as shown in Fig. 3. The x, y, and z axes could be replaced (in the definition) by the y, z, and x axes, respectively, or by the z, x, and y axes, respectively, without changing the effect of the definition. A right-handed system is associated with a clockwise sense in Fig. 3, and a left-handed system with a counter-clockwise sense.

Because of its importance in applications, a right-handed system (the first of Fig. 2) has been chosen for most of the figures of this book.

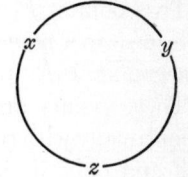

Fig. 3

Plane figures parallel to the yz plane are drawn without distortion. Parallel lines in space are represented by parallel lines, without any attempt being made to indicate perspective. Distances on lines parallel to the x axis are reduced by a constant factor to indicate foreshortening in this direction. This factor has been chosen as $1/\sqrt{2}$, partly because it gives approximately the right appearance to figures, and partly because it is convenient when used with rectangular coördinate graph paper. These conventions are admittedly a compromise, sacrificing photographic accuracy for the sake of simplicity. Later in the book the axes are chosen in a slightly different position, to obtain a more realistic appearance for some of the drawings.

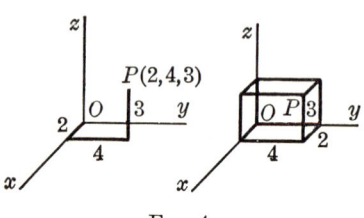

Fig. 4

As an illustrative example, Fig. 4 shows two methods of representing a point with given coördinates, in this case (2, 4, 3), one by means of a broken line segment and the other by means of a rectangular parallelopiped. Fig. 5 contains more illustrative representations by both methods.

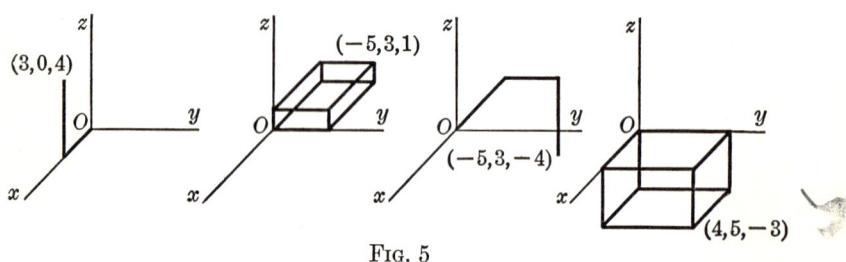

Fig. 5

4. Symmetry. The concept of *symmetry* is discussed in Chapter IV, but we shall introduce now some of the simpler aspects of the idea. Two points, P_1 and P_2, are said to be located symmetrically with respect to a *plane* or *line* if and only if this plane or line bisects the line segment P_1P_2 perpendicularly. These points are located symmetrically with respect to a *point* if and only if this point is the midpoint of the line segment P_1P_2. In each case each of the two points is said to be symmetrical to the other with respect to the plane, line, or point. In Fig. 6, $P(3, 4, 2)$ and $P_1(3, 4, -2)$ are located

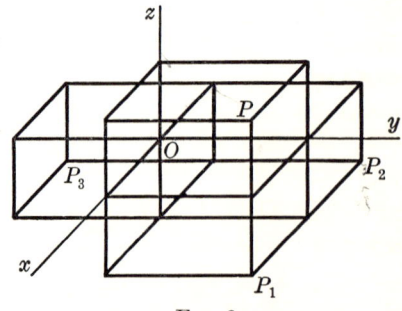

Fig. 6

symmetrically with respect to the xy plane, P and $P_2(-3, 4-2)$ are located symmetrically with respect to the y axis, and P and $P_3(-3, -4, -2)$ are located symmetrically with respect to the origin.

5. Projections. A point P in space may be projected on a plane Π or on a line Λ. The line through P perpendicular to Π intersects Π in a point called the *projection of P on Π*. The plane through P perpendicular to Λ intersects Λ in a point called the *projection of P on Λ*. The projection of any set† of points is defined to be the collection of all projections of the individual points of the set. For example, in Fig. 7, the projection of P_1 on the xy plane is F_1, and its projection on the x axis is A_1; the projection of the line segment D_1P_1 on the xz

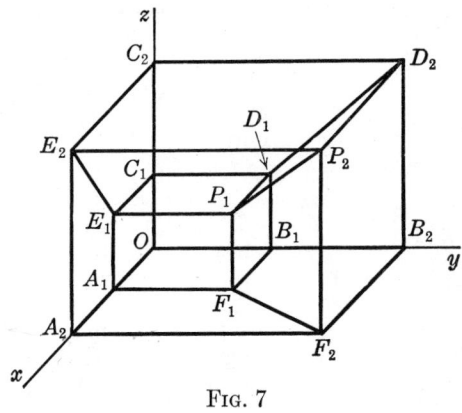

Fig. 7

plane is the line segment C_1E_1, and its projection on the y axis is the single point B_1; the projection of the line segment P_1P_2 on the xy plane is the line segment F_1F_2, and its projection on the x axis is the line segment A_1A_2.

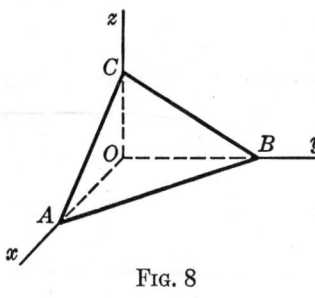

Fig. 8

6. Traces and intercepts. If a plane Π intersects a coördinate plane, the line of intersection is called the **trace** of the plane Π in that coördinate plane. A plane that is not parallel to any coördinate plane will therefore have three traces. If a plane intersects a coördinate axis, the point of intersection is called the **intercept** of the plane on that axis. It is also convenient to use this word in the following sense: Let Π be a plane with intercepts A, B, and C, as shown in Fig. 8. Then the x coördinate of A is called the x intercept of Π,

† The words *set*, *class*, and *collection* will be used synonymously, without any attempt being made to define this fundamental concept.

the y coördinate of B is called the y intercept of Π, and the z coördinate of C is called the z intercept of Π. The line AB (not to be confused with the line segment AB) is the trace of the plane Π in the xy plane, the other traces being the lines AC and BC.

7. Exercises.

1. Represent each of the following points by the two methods described in § 3, using a separate system of axes for each representation: $(3, 1, 4)$; $(3, -4, 1)$; $(2, 5, -1)$; $(2, -5, -1)$; $(-6, 3, 1)$; $(-6, -5, 4)$; $(-6, -5, -4)$; $(-4, 3, -2)$.

2. Represent the following points on one figure, using the same system of axes: $(0, 0, 0)$; $(2, 0, 0)$; $(0, -4, 0)$; $(0, 0, 6)$; $(0, -4, 6)$; $(2, 0, 6)$; $(2, -4, 0)$; $(2, -4, 6)$.

3. Represent the following points as the vertices of a rectangular parallelopiped, using one system of axes: $A(6, 2, 4)$; $B(6, 2, -4)$; $C(6, -2, 4)$; $D(-6, 2, 4)$; $E(6, -2, -4)$; $F(-6, 2, -4)$; $G(-6, -2, 4)$; $H(-6, -2, -4)$. Which pairs of points are located symmetrically with respect to the (a) yz plane? (b) xz plane? (c) xy plane? (d) x axis? (e) y axis? (f) z axis? (g) origin?

4. In Fig. 7, name the projection of (a) P_2 on the yz plane; (b) P_2 on the z axis; (c) the line segment E_2P_2 on the y axis; (d) the line segment E_2P_2 on the z axis; (e) the line segment P_1P_2 on the xz plane; (f) the line segment P_1P_2 on the y axis.

5. Draw a figure showing the traces of the plane whose intercepts are 3, 2, and 5.

6. A rectangular parallelopiped with faces parallel to the coördinate planes has opposite vertices at the origin and the point $(5, 4, -3)$. Find the other vertices and draw the figure.

7. A rectangular parallelopiped with faces parallel to the coördinate planes has opposite vertices at $(2, 3, 5)$ and $(8, 5, 7)$. Find the other vertices and draw the figure.

8. Find the coördinates of the point where the line through $(-3, 4, 1)$ parallel to the x axis meets the plane through $(7, 3, -1)$ parallel to the yz plane.

9. A plane parallel to one of the coördinate planes passes through three of the following points: $A(5, 4, -2)$, $B(5, 3, 6)$, $C(-2, 3, -2)$, $D(-2, 4, 6)$, $E(-5, 3, 2)$. Find the three points and name the parallel coördinate plane.

10. A plane with x intercept equal to 4 passes through the points $(2, 3, 0)$ and $(2, 0, 1)$. By means of a figure, find the other two intercepts.

11. Find the intercepts of the plane that passes through the three points $(2, 2, 0)$, $(3, 1, 0)$, and $(0, 2, 1)$. Draw the figure.

12. Give a rule stating conditions on their coördinates under which two points $P_1(x_1, y_1, z_1)$ and $P_2(x_2, y_2, z_2)$ are located symmetrically with respect to the (a) yz plane; (b) xz plane; (c) xy plane; (d) x axis; (e) y axis; (f) z axis; (g) origin.

13. Find the coördinates of the projection of the point $P(x, y, z)$ on the (a) yz plane; (b) xz plane; (c) xy plane; (d) x axis; (e) y axis; (f) z axis.

* 14. Describe what you would mean by an *oblique coördinate system*, based on three planes which intersect in one point, but which are not mutually perpendicular. The coördinates of a point should be related to the edges of a parallelopiped, three of whose faces lie in the coördinate planes. Define a *right-handed system* of oblique axes.

8. Explanation of certain terminology. In this book we shall consistently use the phrase "if P_1 and P_2 are two points" to mean "if P_1 is any point and if P_2 is any point." That is, P_1 and P_2 may be the same point. If it is necessary to indicate that they are not the same, we shall refer to them as *two different points* or *two distinct points*. A similar interpretation will be understood whenever a statement is made about two or more objects of one category.

We shall also say that a plane or a line is parallel to itself. Thus two parallel planes (lines) can be distinct or coincident, having no points or infinitely many points in common. This use of the word *parallel* is much more satisfactory in general than the one that excludes the possibility of identical parallel planes or lines. For example, we can say that if Π_1 and Π_2 are parallel planes and if Π_2 and Π_3 are parallel planes, then Π_1 and Π_3 are parallel planes; or that planes parallel to the same plane are parallel to each other. If identical planes were not to be called parallel, the above statement would be false, since Π_1 and Π_3 might be the same.

Two directed lines will be called parallel if and only if they are parallel without regard to direction, being either *parallel and similarly directed* or *parallel and oppositely directed*.

The points of any collection of points are said to be **collinear** if and only if they all lie on some line; otherwise they are called **non-collinear**. The points of any collection of points, or the lines of any collection of lines, are said to be **coplanar** if and only if they all lie in some plane; otherwise they are called **non-coplanar**. Two non-coplanar lines are also called **skew**.

The directed line segment from the origin to a point different from the origin is called the **radius vector** of that point.

The single word *segment* will henceforth mean *line segment*.

9. Distance between two points. Let $P_1(x_1, y_1, z_1)$ and $P_2(x_2, y_2, z_2)$ be any points in space. (See Fig. 9.) Then P_1 and P_2 are opposite vertices of a rectangular parallelopiped (which may collapse to a

---

* Sections and exercises marked with a * may be omitted without destroying the continuity of the remaining material.

rectangle, a line segment, or a point). The edges of this parallelopiped are parallel to the coördinate axes and equal in length to the projections of the line segment P_1P_2 on the axes, $|x_2 - x_1|$, $|y_2 - y_1|$, and $|z_2 - z_1|$,†

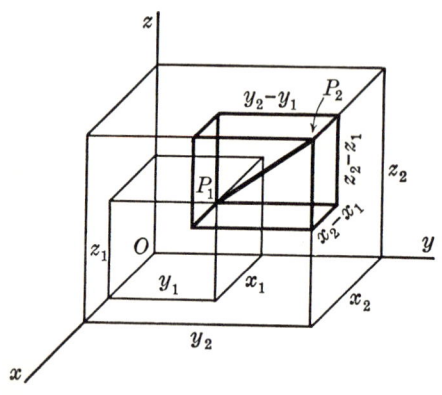

Fig. 9

as can be seen from Fig. 9. The problem of expressing the distance d between the points P_1 and P_2 is therefore reduced to that of finding the length of a diagonal of a box in terms of the edges. As in Fig. 10, let the edges be a, b, and c, the length of the diagonal be d, and the length of the diagonal of the base rectangle be e. A double application of the Pythagorean theorem gives the two equations:

$$d^2 = e^2 + c^2$$

and

$$e^2 = a^2 + b^2,$$

whence

$$d^2 = a^2 + b^2 + c^2.$$

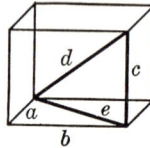 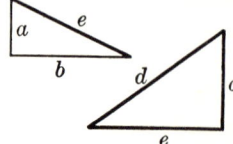

Fig. 10

Letting $a = |x_2 - x_1|$, $b = |y_2 - y_1|$, and $c = |z_2 - z_1|$, and examining the cases where some or all of the numbers a, b, and c are zero, we have the result:

The distance d between the points $P_1(x_1, y_1, z_1)$ and $P_2(x_2, y_2, z_2)$ is given by the formula

$$d = \sqrt{(x_2 - x_1)^2 + (y_2 - y_1)^2 + (z_2 - z_1)^2}.$$

EXAMPLE 1. Find the distance between the points $(5, -1, 3)$ and $(3, 4, 1)$.

Solution. The square of the distance is

$$(3 - 5)^2 + (4 + 1)^2 + (1 - 3)^2 = 4 + 25 + 4 = 33.$$

Therefore $d = \sqrt{33}$.

---

† The absolute value of a real number a, written $|a|$, is defined to be a if a is positive or zero, and $-a$ if a is negative. In other words, $|a| = \sqrt{a^2}$.

EXAMPLE 2. Write an equation which is satisfied by the coördinates of those points and only those points whose distance from the point (2, 5, −1) is 3.

Solution. Either of the following equations is such an equation (why?):

$$\sqrt{(x-2)^2 + (y-5)^2 + (z+1)^2} = 3,$$
$$(x-2)^2 + (y-5)^2 + (z+1)^2 = 9.$$

10. Angle between two directed lines. If two lines intersect, they lie in a plane and define certain angles. Unless the lines are perpendicular, one of the angles is acute and one is obtuse. If the intersecting lines are *directed lines*, an angle θ between 0° and 180° is uniquely determined, as indicated in Fig. 11, between their positive directions. This angle is called *the angle between the directed lines*, or *the angle that either directed line makes with the other*. If two directed lines are parallel we say that the angle between them is 0° if they are similarly directed and 180° if they are oppositely directed. The angle between two skew directed lines in space is defined to be the angle between two intersecting directed lines which are respectively parallel and similarly directed to the given two. It can be shown (Ex. **32**, § **13**) that this angle is independent of the point in which the two intersecting lines meet, so that this point can be chosen to be the origin. Therefore, between any two directed lines in space there is a unique angle θ satisfying the inequalities $0° \leq \theta \leq 180°$. For the case of two undirected lines we can define in a similar way a unique angle θ satisfying the inequalities $0° \leq \theta \leq 90°$, which we shall call *the angle between the lines*.

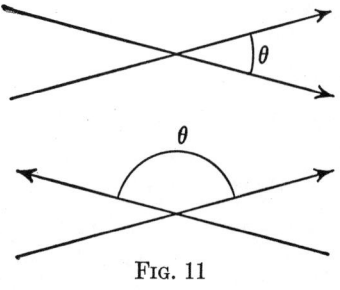

Fig. 11

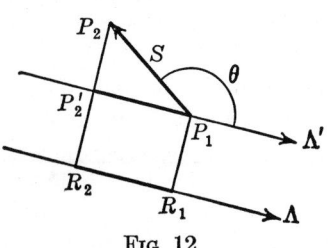

Fig. 12

11. Projections on a directed line. Let S be a directed line segment from the point $P_1(x_1, y_1, z_1)$ to $P_2(x_2, y_2, z_2)$, and let Λ be a directed line. If R_1 and R_2 are the projections on Λ of P_1 and P_2, respectively, then the directed line segment R_1R_2 is called the *projection on Λ of the directed line segment S*. (See Fig. 12. Here it is important to keep in mind the fact that S and Λ need not lie in a plane, in spite of the appearance of the drawing.) If Λ' is the line through P_1 parallel to Λ and similarly directed, and if P_2' is the projection of P_2 on Λ', the directed line segments P_1P_2' and R_1R_2 have the same length and sense. (Why?) It is our purpose now to determine this length and sense in terms of the length l of S and the angle θ between S and Λ. Since S and Λ' lie in a plane, it follows directly from the definition of the cosine of an angle that the length of P_1P_2' is the length of S multiplied by $|\cos\theta|$, the absolute value of $\cos\theta$. Taking account of the sign of $\cos\theta$ we can formulate this result as follows:

THEOREM. *The projection of a directed line segment S on a directed line Λ has directed length $l\cos\theta$, where l is the length of S and θ is the angle between S and Λ, and is directed similarly or oppositely to Λ, according as the sign of $l\cos\theta$ is positive or negative.*

Of special interest are the projections of the directed segment P_1P_2 on the coördinate axes. These projections have directed lengths $x_2 - x_1, y_2 - y_1,$ and $z_2 - z_1$. (Prove this.)

12. Point of division. Let P_1, P_2, and P be any three points on a directed line Λ and let R_1, R_2, and R be their corresponding projections on a directed line Λ'. (See Fig. 13.) Assuming that Λ and Λ' are not perpendicular, let us give an interpretation to the proportion:

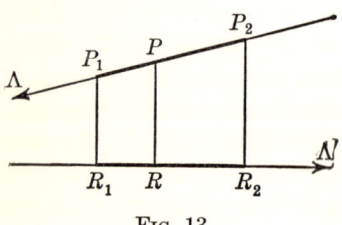

FIG. 13

(1) $\quad P_1P : PP_2 = R_1R : RR_2.$

This means two things: (*i*) The lengths of the line segments are proportional, a zero in any place implying a zero in the corresponding place on the other side of the proportion. (*ii*) The directions of the line segments on one side of the proportion are similar or opposite according as the directions of the line segments on the other side are similar or opposite. Knowing what the proportion (1) means, we can easily prove that it holds. The first part follows from the previous section, the factor of proportionality being $|\cos\theta|$, which is different from zero. The second part of the proof can safely be called trivial.

Assume now that $P_1(x_1, y_1, z_1)$ and $P_2(x_2, y_2, z_2)$ are any two distinct points in space, and let $P(x, y, z)$ be any point on the line through P_1

§ 12] POINT OF DIVISION 11

and P_2. Let r_1 and r_2 be any two real numbers, not both zero, whose absolute values are proportional to the lengths of the line segments P_1P and PP_2, having the same sign or opposite signs according as P_1P and PP_2 are similarly or oppositely directed. This relationship is written:

$$P_1P : PP_2 = r_1 : r_2 \quad \text{or} \quad \frac{P_1P}{PP_2} = \frac{r_1}{r_2},\dagger$$

and expressed by the statement

  P divides the line segment P_1P_2 in the ratio $r_1 : r_2$.

To illustrate this principle, Fig. 14 shows nine points, which (in alphabetical order) divide the line segment P_1P_2 in the ratios: $-\frac{1}{2}$, $-\frac{1}{3}$, 0, $\frac{1}{2}$, 1, 3, ∞, -3, -2.

A point between P_1 and P_2 is said to divide the segment P_1P_2 *internally*, and the ratio is positive. A point outside the segment P_1P_2 is said to divide P_1P_2 *externally*, and the ratio is negative.‡ If P is closer to P_1 than it is to P_2, $|r_1/r_2| < 1$, and if P is closer to P_2 than it is to P_1, $|r_1/r_2| > 1$. The midpoint is given by $r_1/r_2 = 1$. For no point is r_1/r_2 equal to -1. (Why?)

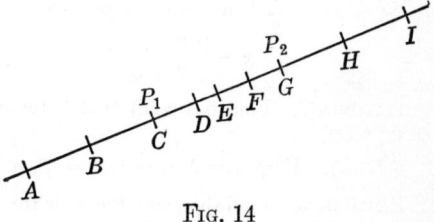

Fig. 14

Our problem now is to find the coördinates of a point that divides a given segment in a given ratio different from -1. If the line through P_1P_2 is not perpendicular to the x axis, the proportion (1), with Λ_1 as the x axis, becomes

$$P_1P : PP_2 = r_1 : r_2 = R_1R : RR_2 = (x - x_1) : (x_2 - x).$$

If $r_1 = 0$, $x = x_1$. If $r_2 = 0$, $x = x_2$. In any case

$$r_1(x_2 - x) = r_2(x - x_1), \text{ and therefore } x = \frac{r_2x_1 + r_1x_2}{r_1 + r_2}.$$

The student is asked to show that this last expression for x holds even if the line through P_1P_2 is perpendicular to the x axis, and that

---

† Here it is to be understood that $r_2 = 0$ does not mean division by zero, but only that the points P and P_2 are the same. In this case the ratio $r_1 : r_2$ can be represented by the symbol ∞.

‡ Occasionally one uses the word *externally* to take the place of the negative sign. For example, in Fig. 14, the point A could be said to divide the segment P_1P_2 externally in the ratio $1 : 2$.

the denominator can never vanish. By taking Λ' in turn as the y axis and the z axis, we obtain the y and z coördinates of P and the theorem:

Theorem. *If $P_1(x_1, y_1, z_1)$ and $P_2(x_2, y_2, z_2)$ are two distinct points and if $P(x, y, z)$ divides the line segment P_1P_2 in the ratio $r_1 : r_2$, then*

$$x = \frac{r_2 x_1 + r_1 x_2}{r_1 + r_2}, \quad y = \frac{r_2 y_1 + r_1 y_2}{r_1 + r_2}, \quad z = \frac{r_2 z_1 + r_1 z_2}{r_1 + r_2}.$$

In particular, if P is the midpoint of the segment, then

$$x = \tfrac{1}{2}(x_1 + x_2), \quad y = \tfrac{1}{2}(y_1 + y_2), \quad z = \tfrac{1}{2}(z_1 + z_2).$$

Example 1. Find the coördinates of the point that divides the segment from $(3, -1, 17)$ to $(8, 9, 2)$ in the ratio $2 : 3$.

Solution. Substitution in the point of division formulas gives

$$x = \frac{3 \cdot 3 + 2 \cdot 8}{2 + 3} = 5, \quad y = 3, \quad z = 11.$$

Example 2. Find the point that is two-thirds of the way from $(3, -1, 17)$ to $(8, 9, 2)$.

Solution. Here $r_1 = 2$, $r_2 = 1$, $x = \tfrac{19}{3}$, $y = \tfrac{17}{3}$, $z = 7$.

Example 3. Find the point P if P_2 is the midpoint of the segment P_1P, where $P_1 = (3, -1, 17)$ and $P_2 = (8, 9, 2)$.

Solution. In this case we can take $r_1 = 2$, $r_2 = -1$, and P is the point $(13, 19, -13)$.

For the student who (quite sensibly) wishes to minimize memorization of formulas, a "common sense" method can be applied to these problems. This method is that of finding on each axis the point that divides in the given ratio the projection on that axis of the given segment. To illustrate: For the first example, we wish to find the three numbers that are two-fifths of the way from (i) 3 to 8, (ii) -1 to 9, and (iii) 17 to 2. We obtain these by (i) increasing 3 by two-fifths of 5, or 2, (ii) increasing -1 by two-fifths of 10, or 4, and (iii) decreasing 17 by two-fifths of 15, or 6. In the second example, we (i) increase 3 by two-thirds of 5, (ii) increase -1 by two-thirds of 10, and (iii) decrease 17 by two-thirds of 15. Now apply this method to the third example.

13. Exercises.

In Exercises **1-6**, find the distance between the two given points.

1. $(6, 3, -1)$, $(4, 6, 5)$.
2. $(-1, 8, 1)$, $(6, 4, 5)$.
3. $(2, -5, -2)$, $(2, 7, -7)$.
4. $(7, 3, -5)$, $(2, 2, 1)$.
5. $(4, 0, 3)$, $(4, 4, 0)$.
6. $(7, 1, 2)$, $(7, 0, 2)$.

In Exercises **7-10**, find the distance of each point from the origin.

7. $(9, 2, -6)$.
8. $(4, -1, 8)$.
9. $(-3, 4, 2)$.
10. $(-2, 2, -1)$.

11. Show that the three points $(7, 2, -10)$, $(4, -6, 5)$, and $(3, -2, -3)$ are the vertices of an isosceles triangle.

12. Show that the three points $(4, 6, 10)$, $(8, 5, 1)$, and $(5, -3, 6)$ are the vertices of an equilateral triangle.

13. Show that the three points $(6, 1, 5)$, $(5, -1, 3)$, and $(5, 3, -1)$ are the vertices of a right triangle. Find its area.

14. Show that the origin and the three points $(1, 2, 2)$, $(2, 1, -2)$, and $(3, 3, 0)$ are the vertices of a square (and therefore lie in a plane).

15. Show that the origin and the three points $(3, 0, 3)$, $(-1, 1, 4)$, and $(0, -3, 3)$ are the vertices of a regular tetrahedron.

In Exercises **16-21**, write an equation that is satisfied by the coördinates of those points and only those points that satisfy the given condition.

16. The distance of each point from the origin is 4.

17. The distance of each point from the point $(5, -1, 3)$ is 2.

18. The distance of each point from the point $(2, -6, 3)$ is equal to its distance from the point $(4, 2, -5)$.

19. The points lie in the xz plane.

20. The points lie in a plane parallel to the xy plane and 2 units above it.

21. The points lie in the plane that passes through the z axis and bisects the first octant.

In Exercises **22-25**, find the midpoint of the line segment joining the two given points.

22. $(6, 1, 5)$, $(-2, 7, 3)$.   **23.** $(-3, 5, 2)$, $(5, 1, 2)$.
24. $(1, -4, 7)$, $(9, 0, -2)$.   **25.** $(3, 5, -3)$, $(-3, -5, 3)$.

In Exercises **26-29**, find the point that divides the segment from $(-2, 9, 7)$ to $(10, -3, 13)$ in the given ratio.

26. $1 : 3$.   **27.** $-5 : 2$.
28. $-1 : 4$.   **29.** -2.

30. Find the two points that trisect the segment from $(3, -2, 1)$ to $(6, 10, -5)$.

31. Find the four points that divide the segment from $(5, 11, -3)$ to $(17, -4, 17)$ in five equal parts.

32. Prove that the angle between two directed lines, as defined in § **10**, is uniquely determined.

* **33.** A **median** of a triangle is defined as a line segment drawn from a vertex to the midpoint of the opposite side. Prove that the three medians of a triangle meet in a point, and that if the vertices of the triangle are (x_1, y_1, z_1), (x_2, y_2, z_2), and (x_3, y_3, z_3), then the coördinates of the point of intersection of the medians are $\frac{1}{3}(x_1 + x_2 + x_3)$, $\frac{1}{3}(y_1 + y_2 + y_3)$, and $\frac{1}{3}(z_1 + z_2 + z_3)$. This point is called the **centroid** of the triangle. (Cf. almost any text on Integral Calculus.) *Suggestion:* Show that the point with coördinates $\frac{1}{3}(x_1 + x_2 + x_3)$, $\cdots$ divides each median in the ratio $2 : 1$.

---

* Exercises marked with an asterisk are somewhat more difficult than other exercises of the same section.

34. Find the centroid of the triangle whose vertices are $(5, 8, 1)$, $(-1, -2, 4)$, and $(2, 6, 4)$. (See Ex. **33**.)

* 35. Prove that the four line segments drawn from each vertex of a tetrahedron to the centroid of the opposite face meet in a point, that this point also lies on the three line segments joining midpoints of opposite edges of the tetrahedron, and that if the vertices of the tetrahedron are (x_1, y_1, z_1), (x_2, y_2, z_2), (x_3, y_3, z_3), and (x_4, y_4, z_4), the coördinates of this point of intersection are

$$\tfrac{1}{4}(x_1 + x_2 + x_3 + x_4), \quad \tfrac{1}{4}(y_1 + y_2 + y_3 + y_4), \quad \text{and} \quad \tfrac{1}{4}(z_1 + z_2 + z_3 + z_4).$$

This point is called the **centroid** of the tetrahedron. (Cf. Ex. **33**.) *Suggestion:* Show that the point with coördinates $\tfrac{1}{4}(x_1 + x_2 + x_3 + x_4)$, $\cdots$ divides each of the first four line segments mentioned above in the ratio $3 : 1$, and is the midpoint of each of the last three line segments.

* 36. It is a very common occurrence in mathematics for any object of some class to stand in a definite relation to certain other objects of the class. For example, if the class of objects is the set of real numbers, such a relation is that of being less than. Each number x is less than some numbers y, but there are numbers x and y for which the relation $x < y$ is false. Other examples of relations are given in the following list.

(*i*) $\Lambda_1 \parallel \Lambda_2$, or the line Λ_1 is parallel to the line Λ_2.
(*ii*) $\Lambda_1 \perp \Lambda_2$, or the line Λ_1 is perpendicular to the line Λ_2.
(*iii*) $\Delta_1 \cong \Delta_2$, or the triangle Δ_1 is congruent to the triangle Δ_2.
(*iv*) $\Delta_1 \sim \Delta_2$, or the triangle Δ_1 is similar to the triangle Δ_2.
(*v*) $x \leqq y$, or the real number x is less than or equal to the real number y.
(*vi*) $m \mid n$, or the positive integer m is a factor of the positive integer n.
(*vii*) The integral rational equation $f(x) = 0$ has the same roots as the integral rational equation $g(x) = 0$.
(*viii*) The variable x varies inversely as the variable y.
(*ix*) The variable x varies directly as the variable y.
(*x*) The line segment S_1 has the same length as the line segment S_2.

A particularly important type of relation R is one that satisfies the following three conditions:

(*i*) xRx. (The relation is *reflexive*.)
(*ii*) If xRy, then yRx. (The relation is *symmetric*.)
(*iii*) If xRy and yRz, then xRz. (The relation is *transitive*.)

Such a relation is called an **equivalence relation**. The study of equivalence relations occupies a fundamental position in mathematics. One could say that the underlying motivation for our defining a plane or a line to be parallel to itself was the desire that parallelism be an equivalence relation.

Determine which of the ten listed relations are equivalence relations, and for each relation that is not an equivalence relation state which of the three laws are violated.

14. Direction cosines. There is nothing in the geometry of three dimensions quite analogous to the slope of a line in plane analytic geometry. Instead of specifying the direction of a line by a trigonometric function evaluated for one angle, we use a trigonometric function evaluated for three angles.

The angles α, β, and γ that a directed line (or line segment) makes with the x, y, and z axes, respectively, are known as the **direction angles** of the line (or line segment), and the cosines of these angles are the **direction cosines** of the line (or line segment). The notation we shall use is

$$\lambda = \cos \alpha, \qquad \mu = \cos \beta, \qquad \nu = \cos \gamma.$$

If P_1P_2 is any directed line segment in space, with length d and direction cosines λ, μ, and ν, the theorem and final paragraph of § 11 imply immediately the relations:

(1) $\qquad x_2 - x_1 = \lambda d, \qquad y_2 - y_1 = \mu d, \qquad z_2 - z_1 = \nu d,$

or

(2) $\qquad \lambda = \dfrac{x_2 - x_1}{d}, \qquad \mu = \dfrac{y_2 - y_1}{d}, \qquad \nu = \dfrac{z_2 - z_1}{d}.$

Squaring these expressions for the direction cosines, adding, and using the formula for the distance d between two points, we have the following fundamental relation between direction cosines of a directed line segment:

$$\lambda^2 + \mu^2 + \nu^2 = 1.$$

Since any directed line contains a directed line segment, we have the theorem:

THEOREM. *The sum of the squares of the direction cosines of any directed line is equal to* 1.

If we choose the directed line segment from the origin O to the point $P(x, y, z)$ and let ρ be the distance between O and P, then the relations (2) above give formulas for the direction cosines of the radius vector of P:

$$\lambda = \frac{x}{\rho}, \qquad \mu = \frac{y}{\rho}, \qquad \nu = \frac{z}{\rho}.$$

In particular, if P is a point on the unit sphere,† then $\rho = 1$ and the coördinates of P are identical with the direction cosines of its radius

---

† The *unit sphere* is defined to be the set of all points that are at a unit distance from the origin.

vector. Conversely, if λ, μ, and ν are any set of direction cosines, the point with coördinates (λ, μ, ν) lies on the unit sphere. We have thus established a *one-to-one correspondence between the points on the unit sphere and all sets of direction cosines*, and proved that an ordered triad of real numbers is a set of direction cosines *if and only if* the sum of their squares is equal to 1.

Since the points (λ, μ, ν) and ($-\lambda$, $-\mu$, $-\nu$) are located symmetrically with respect to the origin, the two sets of direction cosines, λ, μ, and ν and $-\lambda$, $-\mu$, and $-\nu$, are direction cosines of the same line, directed in the two possible ways. Thus, although a directed line has just one set of direction cosines, an undirected line has two sets, each of which prescribes a direction on the line.

To achieve a certain simplification in terminology, we shall find it convenient to speak of the *direction* (λ, μ, ν), instead of using the longer phrase "the direction of a directed line with direction cosines λ, μ, and ν." For example, the statements "θ is the angle between the directions (λ_1, μ_1, ν_1) and (λ_2, μ_2, ν_2)" and "θ is the angle between two directed lines whose direction cosines are λ_1, μ_1, ν_1 and λ_2, μ_2, and ν_2" have the same meaning. We shall also speak of (λ, μ, ν) and ($-\lambda$, $-\mu$, $-\nu$) as *opposite directions*.

EXAMPLE. Find the direction cosines of the directed line segment from (13, -5, 0) to (11, 3, 16).

Solution. By (2), $\lambda = \dfrac{11-13}{18} = -\dfrac{2}{18} = -\dfrac{1}{9}$, $\mu = \dfrac{3+5}{18} = \dfrac{8}{18} = \dfrac{4}{9}$, $\nu = \dfrac{16-0}{18} = \dfrac{16}{18} = \dfrac{8}{9}$.

15. Exercises.

1. Of the following ordered triads of real numbers, choose those that are sets of direction cosines:

(a) $\frac{2}{3}, \frac{1}{3}, -\frac{2}{3}$;   (b) $1, -\frac{1}{2}, \frac{1}{2}$;   (c) $\frac{5}{8}, \frac{1}{3}, \frac{1}{2}$;
(d) $\frac{2}{7}, -\frac{6}{7}, \frac{3}{7}$;   (e) $\frac{20}{21}, -\frac{5}{21}, -\frac{4}{21}$;   (f) $0, 1, 0$.

2. Of the following ordered triads of angles, choose those that are sets of direction angles:

(a) 30°, 45°, 60°;   (b) 30°, 90°, 120°;   (c) 90°, 0°, 270°;
(d) 90°, 135°, 45°;   (e) 30°, 150°, θ;   (f) 120°, 135°, 60°.

3. If two direction angles of a directed line are 60° and 60°, find the third direction angle.

In Exercises **4-9**, find the direction cosines of the given directed line.

4. The x axis.   **5.** The y axis.   **6.** The z axis.

7. The line $x = y$ in the xy plane, directed from the origin into the third quadrant.

8. The line that makes equal angles with the coördinate axes, directed from the origin into the first octant.

9. The line of Exercise **8**, with the opposite direction.

10. Find to the nearest minute the angle that the line of Exercise **8** makes with each axis.

In Exercises **11-16**, find the direction cosines of the given line segment, directed from the first point to the second.

11. $(0, 0, 0)$, $(4, 7, -4)$;   **12.** $(0, 0, 0)$, $(1, 2, 3)$;
13. $(3, 8, 1)$, $(1, 2, 10)$;   **14.** $(0, -5, 13)$, $(6, 4, -5)$;
15. $(10, 11, 12)$, $(6, 3, 31)$;   **16.** $(6, -1, -2)$, $(4, 9, 9)$.

In Exercises **17-25**, explain the significance of the given condition on the direction cosines of a directed line.

17. $\nu > 0$.   **18.** $\nu < 0$.   **19.** $\nu = 0$.
20. $\lambda > 0$.   **21.** $\lambda = 0$.   **22.** $\mu < 0$.
23. $\lambda = 1$.   **24.** $\mu = -1$.   **25.** $\lambda = \mu = 0$.

In Exercises **26** and **27**, find the projections on the coördinate axes of the given directed line segment.

26. From $(5, 1, 1)$ to $(7, -3, 0)$.   **27.** From $(3, 8, 9)$ to $(1, 3, 10)$.

* **28.** Find the point in the first octant 28 units from the origin, whose radius vector has direction cosines satisfying the relations $\lambda = 2\mu = 3\nu$.

* **29.** (a) Define direction cosines for plane analytic geometry. (b) Prove that $\lambda^2 + \mu^2 = 1$. (c) To what trigonometric identity does this relation correspond? (d) Are direction cosines in the plane associated with directed lines or undirected lines? (e) What is the relation between the slope of a line and a set of direction cosines of the line? (f) Is slope associated with directed lines or undirected lines?

16. Direction numbers. If a line (or line segment) has direction cosines λ, μ, ν, any three real numbers l, m, n proportional to λ, μ, ν are called **direction numbers** of the line (or line segment). This relationship is expressed in any of the following three forms:

(1) $$l : m : n = \lambda : \mu : \nu,$$

(2) $$\frac{l}{\lambda} = \frac{m}{\mu} = \frac{n}{\nu},$$

(3) $$l = k\lambda, \quad m = k\mu, \quad n = k\nu,$$

where k is some number different from zero. It is to be understood that equations (2) are an expression of proportion, and that a zero denominator is to be interpreted not as "division by zero," but to mean that the corresponding numerator vanishes. For example, we can write

$$\frac{6}{9} = \frac{0}{0} = \frac{8}{12} \quad \text{or} \quad \frac{2}{1} = \frac{0}{0} = \frac{0}{0}.$$

but we shall *not* admit the "proportion"

$$\frac{0}{1} = \frac{0}{1} = \frac{0}{1}.$$

We see immediately that direction cosines are direction numbers (letting $k = 1$ or -1), but that direction numbers are direction cosines if and only if the sum of their squares is 1. However, *any* ordered triad of real numbers, not all zero, is a set of direction numbers of a family of parallel lines. For, if l, m, n is such a triad, then

(4) $$\lambda = \frac{l}{\pm\sqrt{l^2 + m^2 + n^2}}, \quad \mu = \frac{m}{\pm\sqrt{l^2 + m^2 + n^2}}$$

$$\nu = \frac{n}{\pm\sqrt{l^2 + m^2 + n^2}}$$

are direction cosines proportional to l, m, and n. The denominators must all carry the same sign, this sign depending on the direction assigned to the line. Notice that a set of direction numbers is associated with a family of *parallel lines*, and a set of direction cosines, with a family of *parallel and similarly directed lines*.

The formulas (1) of § **14** imply the theorem:

THEOREM. *If $P_1(x_1, y_1, z_1)$ and $P_2(x_2, y_2, z_2)$ are any two distinct points, then $x_2 - x_1$, $y_2 - y_1$, and $z_2 - z_1$ are a set of direction numbers of the line segment P_1P_2.*

EXAMPLE. Find direction cosines of a line with direction numbers 6, -2, 3.

Solution. $\pm \sqrt{36 + 4 + 9} = \pm 7$. Therefore, by (4), the two sets of direction cosines are $\frac{6}{7}, -\frac{2}{7}, \frac{3}{7}$, and $-\frac{6}{7}, \frac{2}{7}, -\frac{3}{7}$. The first set applies if the line is directed upward, making an acute angle with the z axis, and the second set applies if the line is directed downward.

17. Exercises.

In Exercises **1-6**, find a set of direction cosines for a line having the given direction numbers.

1. 9, -2, 6.
2. 11, 10, 2.
3. $-3, -5, -1$.
4. $-2, 1, -2$.
5. 0, 3, -4.
6. 0, 0, -5.

In Exercises **7-12**, find a set of direction numbers for the line segment joining the given points.

7. $(0, 0, 0)$, $(3, 5, -1)$. **8.** $(1, 2, 3)$, $(4, 5, 6)$.
9. $(6, -1, 2)$, $(4, 2, 6)$. **10.** $(8, 10, 7)$, $(3, 0, 2)$.
11. $(5, 1, 5)$, $(1, 5, 1)$. **12.** $(3, 4, 0)$, $(3, 0, 0)$.

* **13.** (a) Discuss direction numbers for lines in plane analytic geometry. (b) Give a set of direction numbers for the line $ax + by + c = 0$. (c) Give a set of direction numbers for the normals to this line. (d) Explain the relationship between slope and direction numbers of a line. (See Ex. **29**, § **15**.)

18. Cosine of the angle between two directed lines. Let θ be the angle between two directed lines with direction cosines λ_1, μ_1, ν_1 and λ_2, μ_2, ν_2. Since we may assume (§ **10**) that the lines pass through the origin, the angle θ is the angle between the radius vectors of the points $P_1(\lambda_1, \mu_1, \nu_1)$ and $P_2(\lambda_2, \mu_2, \nu_2)$, which are a unit distance from the origin. (See Fig. 15.)

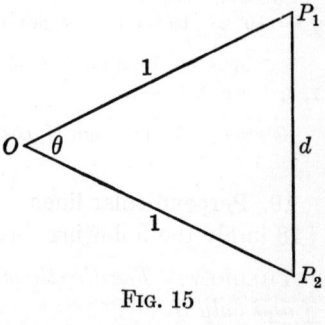

Fig. 15

By the law of cosines, if d is the distance between P_1 and P_2, $d^2 = 1^2 + 1^2 - 2 \cos \theta$, or

$$\cos \theta = \frac{2 - d^2}{2} = \frac{2 - [(\lambda_1 - \lambda_2)^2 + (\mu_1 - \mu_2)^2 + (\nu_1 - \nu_2)^2]}{2}$$

$$= \frac{2 - [(\lambda_1^2 + \mu_1^2 + \nu_1^2) + (\lambda_2^2 + \mu_2^2 + \nu_2^2) - 2(\lambda_1\lambda_2 + \mu_1\mu_2 + \nu_1\nu_2)]}{2}.$$

Simplifying this expression by using the fact that the sum of the squares of any set of direction cosines is 1, we have the result:

THEOREM I. *If θ is the angle between the directions $(\lambda_1, \mu_1, \nu_1)$ and $(\lambda_2, \mu_2, \nu_2)$ then*

$$\cos \theta = \lambda_1 \lambda_2 + \mu_1 \mu_2 + \nu_1 \nu_2.$$

Since θ is an angle between 0° and 180°, the value of $\cos \theta$ determines uniquely the value of θ.

In § **10** we stated what is meant by the angle between two undirected lines. From relations (4) of § **16**, the formula of Theorem I of this section, and the fact that the cosine of an angle between 0° and 90° is non-negative, we can conclude:

Theorem II. *If θ is the angle between two lines with direction numbers l_1, m_1, n_1 and l_2, m_2, n_2, then*

$$\cos\theta = \frac{|\,l_1 l_2 + m_1 m_2 + n_1 n_2\,|}{\sqrt{l_1^2 + m_1^2 + n_1^2}\ \sqrt{l_2^2 + m_2^2 + n_2^2}}.$$

As in Theorem I, the value of θ is uniquely determined by this formula.

Example 1. Find the angle between two directed lines with direction cosines $\frac{2}{3}, \frac{1}{3}, \frac{2}{3}$ and $-\frac{4}{9}, \frac{7}{9}, -\frac{4}{9}$.

Solution. $\cos\theta = -\frac{2}{3}\cdot\frac{4}{9} + \frac{1}{3}\cdot\frac{7}{9} - \frac{2}{3}\cdot\frac{4}{9} = -\frac{1}{3}$. Since arc $\cos\frac{1}{3} = 70°\,32'$ $\theta = 119°\,28'$, to the nearest minute.

Example 2. Find the (acute) angle between two lines with direction numbers $1, 1, 4$ and $0, 1, -1$.

Solution. By Theorem II, $\cos\theta = \dfrac{|\,0 + 1 - 4\,|}{\sqrt{18}\ \sqrt{2}} = \dfrac{1}{2}$. Therefore $\theta = 60°$.

19. Perpendicular lines. Since $\cos 90° = 0$, the two theorems of §18 imply the following theorem:

Theorem. *The directions $(\lambda_1, \mu_1, \nu_1)$ and $(\lambda_2, \mu_2, \nu_2)$ are perpendicular if and only if*

$$\lambda_1 \lambda_2 + \mu_1 \mu_2 + \nu_1 \nu_2 = 0.$$

Two lines with direction numbers l_1, m_1, n_1 and l_2, m_2, n_2 are perpendicular if and only if

$$l_1 l_2 + m_1 m_2 + n_1 n_2 = 0.$$

Example. Show that lines with direction numbers $3, 1, 4$ and $5, -3, -3$, are perpendicular.

Solution. $3\cdot 5 - 1\cdot 3 - 4\cdot 3 = 0$.

20. Exercises.

In Exercises **1-4**, find the cosine of the angle between the given directions.

1. $(\frac{3}{7}, -\frac{6}{7}, -\frac{2}{7})$, $(\frac{9}{11}, \frac{6}{11}, -\frac{2}{11})$. 2. $(\frac{1}{3}, \frac{2}{3}, -\frac{2}{3})$, $(\frac{8}{9}, \frac{1}{9}, \frac{4}{9})$.
3. $(\frac{1}{9}, -\frac{8}{9}, \frac{4}{9})$, $(,\frac{1}{19}, -\frac{18}{19}, \frac{6}{19})$. 4. $(0, \frac{3}{5}, -\frac{4}{5})$, $(\frac{12}{13}, \frac{5}{13}, 0)$.

In Exercises **5-8**, find the cosine of the (acute) angle between two lines having the given direction numbers.

5. $3, -2, 0;\ 4, 3, 1.$ 6. $0, 1, 2;\ 5, -2, -4.$
7. $5, 1, -4;\ 2, -1, 4.$ 8. $2, 3, 5;\ -3, 5, 2.$

In Exercises **9-12**, determine whether the two given sets of direction numbers are associated with perpendicular lines.

9. $2, 8, 3;\ 5, 1, -6.$ 10. $3, -1, 7;\ 5, 1, -2.$
11. $3, 2, -5;\ -2, 1, -1.$ 12. $-7, 5, 11;\ -3, -13, 4.$

§ 21] SINE OF ANGLE _length_ 21

In Exercises **13** and **14**, find the projection of the directed line segment, from the first point to the second point, on a line with the given set of direction cosines.

13. $(-1, 2, 4)$, $(5, 5, 1)$; $-\frac{6}{11}, \frac{2}{11}, \frac{9}{11}$.
14. $(7, -2, 8)$, $(1, 7, 5)$; $\frac{2}{3}, \frac{2}{3}, -\frac{1}{3}$.

15. Show that the segment joining $(5, -1, -2)$ and $(8, 4, 0)$ is perpendicular to the segment joining $(3, 11, -4)$ and $(9, 7, -3)$.

*** 16.** If θ is the angle between the directions $(\lambda_1, \mu_1, \nu_1)$ and $(\lambda_2, \mu_2, \nu_2)$, prove that

$$4 \sin^2 \frac{\theta}{2} = (\lambda_1 - \lambda_2)^2 + (\mu_1 - \mu_2)^2 + (\nu_1 - \nu_2)^2.$$

Suggestion: Examine the derivation of the formula for $\cos \theta$ given in §18.

*** 17.** (a) Derive a formula for $\cos \theta$, where θ is the angle between two directed lines in the plane, in terms of direction cosines. (b) Obtain a condition for perpendicularity of two lines, in terms of direction numbers. (c) Show that this result is consistent with the familiar perpendicularity condition involving slopes. (See Ex. **29**, § **15** and Ex. **13**, § **17**.)

21. Sine of the angle between two lines. The equation

(1)
$$\begin{vmatrix} \mu_1 & \nu_1 \\ \mu_2 & \nu_2 \end{vmatrix}^2 + \begin{vmatrix} \nu_1 & \lambda_1 \\ \nu_2 & \lambda_2 \end{vmatrix}^2 + \begin{vmatrix} \lambda_1 & \mu_1 \\ \lambda_2 & \mu_2 \end{vmatrix}^2$$
$$= (\lambda_1^2 + \mu_1^2 + \nu_1^2)(\lambda_2^2 + \mu_2^2 + \nu_2^2) - (\lambda_1 \lambda_2 + \mu_1 \mu_2 + \nu_1 \nu_2)^2$$

is an identity, holding for all values of the quantities involved. This can be established by expanding both members (Ex. **1**, § **22**).

In particular, if $(\lambda_1, \mu_1, \nu_1)$ and $(\lambda_2, \mu_2, \nu_2)$ are any two directions and if θ is the angle between these directions, then the right member of (1) is equal to $1 - \cos^2 \theta$. This fact establishes the theorem:

Theorem. *If θ is the angle between the directions $(\lambda_1, \mu_1, \nu_1)$ and $(\lambda_2, \mu_2, \nu_2)$, then*

$$\sin^2 \theta = \begin{vmatrix} \mu_1 & \nu_1 \\ \mu_2 & \nu_2 \end{vmatrix}^2 + \begin{vmatrix} \nu_1 & \lambda_1 \\ \nu_2 & \lambda_2 \end{vmatrix}^2 + \begin{vmatrix} \lambda_1 & \mu_1 \\ \lambda_2 & \mu_2 \end{vmatrix}^2.$$

Example. Find $\sin \theta$, where θ is the angle between the directions $(\frac{2}{3}, \frac{1}{3}, \frac{2}{3})$ and $(-\frac{4}{9}, \frac{7}{9}, -\frac{4}{9})$.

First solution. As shown in Example 1, § **18**, $\cos \theta = -\frac{1}{3}$, and therefore $\sin^2 \theta = 1 - \frac{1}{9} = \frac{8}{9}$, and $\sin \theta = \frac{2}{3} \sqrt{2}$.

Second solution. According to the theorem of this section,
$$\sin^2 \theta = (\tfrac{2}{3})^2 + (0)^2 + (\tfrac{2}{3})^2 = \tfrac{8}{9}; \ \sin \theta = \tfrac{2}{3} \sqrt{2}.$$

22. Exercises.

1. Actually carry through the work of proving the identity (1) § 21. Show that $\sin \theta$ is non-negative, and is therefore uniquely determined by the formula given in the theorem of § 21.

2. Find $\sin \theta$ in two ways, if θ is the angle between the directions $(\frac{2}{3}, \frac{2}{3}, \frac{1}{3})$ and $(\frac{3}{7}, \frac{6}{7}, \frac{2}{7})$.

23. Direction perpendicular to two directions.

The problem now proposed is that of finding a set of direction cosines λ, μ, ν for a line perpendicular to two given non-parallel lines with direction cosines λ_1, μ_1, ν_1 and λ_2, μ_2, ν_2. By § 19, these two perpendicularity conditions can be expressed:

$$(1) \qquad \begin{aligned} \lambda_1 \lambda + \mu_1 \mu + \nu_1 \nu &= 0, \\ \lambda_2 \lambda + \mu_2 \mu + \nu_2 \nu &= 0. \end{aligned}$$

This can be considered as a system of two homogeneous linear equations in the unknowns λ, μ, and ν.

Since the given lines are not parallel, the sine of the angle θ between them is different from zero, and therefore the three determinants appearing in the formula of § 21 for $\sin^2 \theta$ are not all zero. Let us assume for definiteness that the third determinant does not vanish. Subtracting $\nu_1 \nu$ from each member of the first equation above, and $\nu_2 \nu$ from each member of the second equation, we have a new system, regarded now as a system of two equations in the unknowns λ and μ, which are determined uniquely in terms of ν, since the determinant of their coefficients is not zero. The result is

$$\lambda = \frac{\begin{vmatrix} -\nu_1 \nu & \mu_1 \\ -\nu_2 \nu & \mu_2 \end{vmatrix}}{\begin{vmatrix} \lambda_1 & \mu_1 \\ \lambda_2 & \mu_2 \end{vmatrix}} = \begin{vmatrix} \mu_1 & \nu_1 \\ \mu_2 & \nu_2 \end{vmatrix} \cdot \frac{\nu}{\begin{vmatrix} \lambda_1 & \mu_1 \\ \lambda_2 & \mu_2 \end{vmatrix}},$$

$$\mu = \frac{\begin{vmatrix} \lambda_1 & -\nu_1 \nu \\ \lambda_2 & -\nu_2 \nu \end{vmatrix}}{\begin{vmatrix} \lambda_1 & \mu_1 \\ \lambda_2 & \mu_2 \end{vmatrix}} = \begin{vmatrix} \nu_1 & \lambda_1 \\ \nu_2 & \lambda_2 \end{vmatrix} \cdot \frac{\nu}{\begin{vmatrix} \lambda_1 & \mu_1 \\ \lambda_2 & \mu_2 \end{vmatrix}},$$

or

$$\frac{\lambda}{\begin{vmatrix} \mu_1 & \nu_1 \\ \mu_2 & \nu_2 \end{vmatrix}} = \frac{\mu}{\begin{vmatrix} \nu_1 & \lambda_1 \\ \nu_2 & \lambda_2 \end{vmatrix}} = \frac{\nu}{\begin{vmatrix} \lambda_1 & \mu_1 \\ \lambda_2 & \mu_2 \end{vmatrix}}.$$

This is a statement of proportionality, with the previously indicated interpretation of zero denominators.

A more general and useful expression of this principle is stated in the theorem:

THEOREM. *If l_1, m_1, n_1 and l_2, m_2, n_2 are direction numbers of non-parallel lines, then a set of direction numbers of a line perpendicular to these lines is*

$$\begin{vmatrix} m_1 & n_1 \\ m_2 & n_2 \end{vmatrix}, \quad \begin{vmatrix} n_1 & l_1 \\ n_2 & l_2 \end{vmatrix}, \quad \begin{vmatrix} l_1 & m_1 \\ l_2 & m_2 \end{vmatrix}.$$

Proof. Let $l_1 = k_1\lambda_1$, $m_1 = k_1\mu_1$, $n_1 = k_1\nu_1$, $l_2 = k_2\lambda_2$, $m_2 = k_2\mu_2$, $n_2 = k_2\nu_2$, where λ_1, μ_1, ν_1, and λ_2, μ_2, ν_2 are directon cosines of the original lines, and k_1 and k_2 are non-zero factors of proportionality. Substitution leads immediately to the fact that each determinant above is equal to k_1k_2 times the corresponding determinant with direction cosines as elements. The desired proportionality is thereby obtained.

The purely algebraic formulation of this essentially geometric theorem is discussed in Chapter III.

In practice one becomes adept at computing these three determinants from the 2 by 3 array or matrix

$$\begin{pmatrix} l_1 & m_1 & n_1 \\ l_2 & m_2 & n_2 \end{pmatrix}$$

However, a simple and easily remembered method should be mentioned, even though it requires slightly more work with pencil or chalk:

Write down the direction numbers twice and form the determinants provided by the middle three pairs of adjacent columns:

$$\begin{matrix} l_1 & m_1 & n_1 & l_1 & m_1 & n_1 \\ l_2 & m_2 & n_2 & l_2 & m_2 & n_2. \end{matrix}$$

EXAMPLE. Find a set of direction numbers for a line perpendicular to two lines having direction numbers 3, 4, −1 and 3, −4, 3.

Solution. From the array

$$\begin{matrix} 3 & 4 & -1 & 3 & 4 & -1 \\ 3 & -4 & 3 & 3 & -4 & 3 \end{matrix}$$

we obtain the direction numbers 8, −12, −24, or more simply, 2, −3, −6.

24. Exercises.

In Exercises **1-6,** find a set of direction numbers for a line perpendicular to two lines having the given direction numbers.

1. 1, 2, −3; 3, −7, 4.
2. 0, 5, 2; 0, 3, −7.
3. 3, −2, 8; −6, 4, 5.
4. 1, 7, 5; 1, 1, 1.
5. 3, −2, 1; −4, 1, 7.
6. 8, 6, −3; −6, 1, 5.

In Exercises **7** and **8,** find the directions perpendicular to the two given directions.

7. $(\frac{1}{3}, \frac{2}{3}, \frac{2}{3})$, $(\frac{2}{3}, \frac{1}{3}, -\frac{2}{3})$.   **8.** $(\frac{2}{7}, \frac{3}{7}, \frac{6}{7})$, $(\frac{3}{7}, -\frac{6}{7}, \frac{2}{7})$.

25. Projection of a broken line segment. Let $P_1P_2 \cdots P_n$ be a broken line segment, and let $R_1, R_2, \cdots, R_n$ be the projections of $P_1, P_2, \cdots, P_n$, respectively, on a directed line Λ. (See Fig. 16.) It is readily seen that the (algebraic) sum of the projections on Λ of the individual segments $P_1P_2, P_2P_3, \cdots, P_{n-1}P_n$ is R_1R_n, the projection of the closing segment P_1P_n. If we define, as in the case of vectors, the *sum* of the directed segments $P_1P_2, P_2P_3, \cdots, P_{n-1}P_n$ to be the single segment P_1P_n, we can conclude:

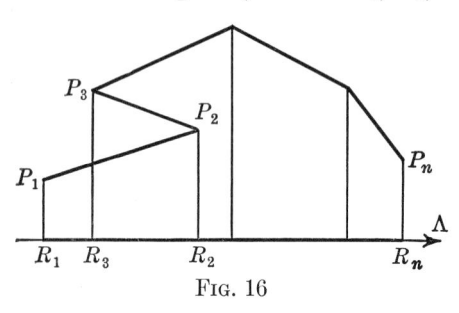

FIG. 16

The projection of the sum of the segments of a broken line segment is the sum of their projections.

This principle provides a means for deriving the formula for the cosine of the angle between two directed lines (Ex. **3, § 26**), and will be used in Chapter VII for the derivation of the formulas for a rotation of axes.

26. Exercises.

1. Let $P_1(3, 1, 9)$, $P_2(5, -3, 6)$, $P_3(1, -2, 7)$, and $P_4(8, 5, 2)$ be four given points. Show that the projection on each coördinate axis of P_1P_4 is equal to the sum of the projections on that axis of the segments P_1P_2, P_2P_3, and P_3P_4.

2. For the four points of Exercise **1**, show that the projection of P_1P_4 on a line with direction $(\frac{2}{3}, -\frac{2}{3}, \frac{1}{3})$ is equal to the sum of the projections on that line of P_1P_2, P_2P_3, and P_3P_4.

3. Derive the formula for $\cos \theta$, given in Theorem I, § **18**, as follows: Let Λ be a line with direction cosines λ_2, μ_2, ν_2, let P be the point $(\lambda_1, \mu_1, \nu_1)$, and let O be the origin. The projection of OP on Λ is $\cos \theta$. But OP is the sum of three segments of directed length $\lambda_1, \mu_1,$ and ν_1, parallel to the coördinate axes. Paying careful attention to signs, show that the sum of the projections on Λ of these three segments is $\lambda_1\lambda_2 + \mu_1\mu_2 + \nu_1\nu_2$.

4. Show that the three points $(5, 8, 6)$, $(8, 4, 1)$, and $(14, -4, -9)$ lie on a straight line.

5. Show that the four points $(3, -1, 4)$, $(7, 2, 1)$, $(11, 10, 3)$, and $(15, 13, 0)$ are the vertices of a parallelogram.

6. Show that the four points $(2, 9, 8)$, $(6, 4, -2)$, $(7, 15, 7)$, and $(11, 10, -3)$ are the vertices of a rectangle.

7. Using the formula of Theorem I, § **18**, for $\cos \theta$, find the angles of the triangle whose vertices are (3, 1, 2), (4, 5, 10), and (11, −3, 3). Check your answer by adding the angles.

8. Using the formula of Theorem I, § **18**, for cos θ, find the angles of the triangle whose vertices are (−3, 5, 6), (−2, 7, 9), and (2, 1, 7).

9. Using the formula of Theorem I, § **18**, for cos θ, find the angles of the triangle whose vertices are (2, 5, 4), (3, 6, 8), and (2, 10, −1).

10. What can you say about the direction numbers of a line
(a) parallel to the z axis?
(b) perpendicular to the z axis?

11. Determine k so that the segment joining the points (3, 2, k) and (k, 5, 0) is perpendicular to the segment joining the points (4, $2k$, 1) and (6, k, 2).

* **12.** Show that the three points (2, 1, 5), (3, 0, 1), and (5, −2, 4) lie in a vertical plane. *Suggestion:* Find direction numbers of the three segments joining pairs of the points.

* **13.** In what ratio does the point of intersection of the xy plane and the line segment from (3, 6, −2) to (5, 0, 2) divide this line segment? Find the coördinates of the point of intersection.

* **14.** Find the coördinates of the point of intersection of the line through the points (3, −2, 7) and (13, 3, −8) and the xz plane.

* **15.** Let $P_1(x_1, y_1, z_1)$ and $P_2(x_2, y_2, z_2)$ be any two distinct points equidistant from the origin O and not collinear with it (that is, O, P_1, and P_2 do not lie on one line). Prove that the line bisecting the angle P_1OP_2 passes through the midpoint of the segment P_1P_2 and therefore has direction numbers $x_1 + x_2$, $y_1 + y_2, z_1 + z_2$. What line through the origin has direction numbers $x_1 − x_2$, $y_1 − y_2, z_1 − z_2$?

* **16.** Let $P_1(x_1, y_1, z_1)$ and $P_2(x_2, y_2, z_2)$ be any two distinct points that are not collinear with the origin O, and let their distances from O be d_1 and d_2, respectively. Show that the bisector of the angle P_1OP_2 has direction numbers $d_2x_1 + d_1x_2, d_2y_1 + d_1y_2, d_2z_1 + d_1z_2$. (See Ex. **15**.)

In Exercises **17-20**, find a set of direction numbers of the bisector of the angle P_1OP_2, where O is the origin and P_1 and P_2 are the two given points. (See Exs. **15** and **16**.)

17. (9, 2, −6), (−9, 6, 2).   **18.** (4, −8, 1), (−7, 4, 4).
19. (14, −5, −2), (−2, 10, 11).   **20.** (2, 1, 2), (6, −2, 3).

CHAPTER II

PLANES AND LINES

27. Introduction. In plane analytic geometry an equation of the first degree corresponds to a straight line. In space this is not true. An equation of the first degree corresponds to a plane, and a system of *two* linear equations corresponds to a line. We first give some definitions:

DEFINITION I. *One equation is **equivalent** to another if and only if it can be obtained from the other by a sequence of the following two **elementary transformations:** (i) adding the same quantity to both members; (ii) multiplying both members by the same non-zero constant.* (Cf. Ex. **27**, § **29**.)

DEFINITION II. *A **linear equation,** or **equation of the first degree,** in the variables x, y, and z, is an equation of the type*

$$a_1x + b_1y + c_1z + d_1 = a_2x + b_2y + c_2z + d_2$$

*that is equivalent to an equation in the **standard** or **canonical** form*

$$ax + by + cz + d = 0,$$

where a, b, and c are not all zero.

DEFINITION III. *The **graph of an equation** in the variables x, y, z (whether any or all of these variables occur or not) is the set (collection, class, locus) of all points (x, y, z) whose coördinates satisfy the equation.*

DEFINITION IV. *The **graph of a function** of two variables, $f(x, y)$, is the graph of the equation $z = f(x, y)$.*

DEFINITION V. *The **graph of a system of equations** in the variables x, y, z is the set of all points (x, y, z) whose coördinates satisfy the system simultaneously.*

Except where an explicit statement to the contrary is made, *every linear equation $ax + by + cz + d = 0$ considered in this book is assumed to have real coefficients a, b, c, and d.*

28. Equation of a plane.

Theorem I. *The graph of any linear equation is a plane. Conversely, any plane is the graph of some linear equation.*

Proof. We shall start with any linear equation

$$ax + by + cz + d = 0.$$

Since the coefficients of x, y, and z are not all zero, there is a solution (x_0, y_0, z_0) of this equation. (Prove that there are infinitely many solutions.) From the definition of a solution of an equation, $d = -ax_0 - by_0 - cz_0$. Therefore the original equation can be written in the form

(1) $\qquad a(x - x_0) + b(y - y_0) + c(z - z_0) = 0.$

Let $P(x, y, z)$ be any point different from $P_0(x_0, y_0, z_0)$. Since $x - x_0$, $y - y_0$, $z - z_0$ are direction numbers of the line segment P_0P, this segment is perpendicular to a line with direction numbers a, b, c if and only if equation (1) holds. Therefore (since the coördinates of P_0 satisfy this equation) the graph of (1) consists precisely of those points that lie in the plane through P_0 perpendicular to a line with direction numbers a, b, c. Conversely, for any plane, let a, b, c be a set of direction numbers of the normals, and $P_0(x_0, y_0, z_0)$ be any point in the plane. Then, as above, this plane is the graph of the linear equation (1).

For convenience, we shall in the future speak of *the plane* $ax + by + cz + d = 0$, instead of saying "the plane that is the graph of the equation $ax + by + cz + d = 0$."

The proof of Theorem I also establishes the following theorem:

Theorem II. *The coefficients a, b, c in the equation*

$$ax + by + cz + d = 0$$

are direction numbers of the normals to the plane $ax + by + cz + d = 0$. The plane through the point (x_0, y_0, z_0) whose normals have direction numbers a, b, c is the graph of the equation

(1) $\qquad a(x - x_0) + b(y - y_0) + c(z - z_0) = 0.$

Equivalent equations have identical solutions and therefore identical graphs. The converse, however, is not true. The equations $x = 0$ and $xe^{y+z} = 0$, for example, are not equivalent, but the graph of each is the yz plane. However, it *is* true that two *linear* equations are equivalent if and only if their graphs are identical. With the linear equations in standard form this statement takes the form:

THEOREM III. *The planes*
$$a_1x + b_1y + c_1z + d_1 = 0$$
and
$$a_2x + b_2y + c_2z + d_2 = 0$$
are identical if and only if the two sets of numbers a_1, b_1, c_1, d_1 and a_2, b_2, c_2, d_2 are proportional.

Proof. As seen above, if these two sets of numbers are proportional the graphs of the two equations are the same. Assume now that one plane is the graph of each of the two equations. The coefficients of x, y, and z are proportional since they represent in each case direction numbers of the normals to the plane. Letting the constant of proportionality be k, we have $a_1 = ka_2$, $b_1 = kb_2$, $c_1 = kc_2$. If both members are multiplied by k, the second equation assumes the form
$$a_1x + b_1y + c_1z + kd_2 = 0.$$
Letting x_0, y_0, z_0 be the coördinates of any point in the plane, we have
$$a_1x_0 + b_1y_0 + c_1z_0 + d_1 = a_1x_0 + b_1y_0 + c_1z_0 + kd_2 = 0,$$
or $d_1 = kd_2$. This establishes the desired proportionality.

We have thus seen that any plane has infinitely many equations in standard form, each differing from any other by a non-zero factor in the left member. Properly, each of these equations should be called *an equation* of the plane, but for convenience we shall use the term *the equation* of a plane to mean any equation of the plane in standard form.

We now state three theorems, whose proofs are requested in the first three Exercises of the following section:

THEOREM IV. *A plane passes through the origin if and only if it has an equation of the form*
$$ax + by + cz = 0.$$

THEOREM V. *A plane is parallel to a coördinate plane if and only if its equation in standard form has two missing variables, the parallel coördinate plane being that of the missing variables.*

THEOREM VI. *A plane is parallel to just one coördinate axis if and only if its equation in standard form has just one missing variable, the parallel axis being that of the missing variable.*

EXAMPLE 1. Find the equation of the plane parallel to $3x - 2y + 4z + 3 = 0$ passing through the point $(1, 1, -2)$.

Solution. By Theorem II, the equation is $3(x - 1) - 2(y - 1) + 4(z + 2) = 0$ or $3x - 2y + 4z + 7 = 0$.

EXAMPLE 2. Find the equation of the plane through the three points $P_1(1, 12, 1)$, $P_2(2, 7, -1)$, and $P_3(-4, 5, 3)$.

First solution. The coefficients, a, b, c, in the equation of the plane are direction numbers of a line perpendicular to the plane and therefore to the segments P_1P_2 and P_1P_3, which have direction numbers 1, -5, -2 and 5, 7, -2, respectively. Therefore, as shown in § 23, $a : b : c = 24 : -8 : 32 = 3 : -1 : 4$. The equation thus has the form $3x - y + 4z + d = 0$. Substitution gives the value of d, and the equation is $3x - y + 4z + 5 = 0$.

Second solution. Let the equation of the plane be $ax + by + cz + d = 0$. Substitution of the coördinates of the three points gives the following system of three equations in the four unknowns, a, b, c, d:

$$a + 12b + c + d = 0$$
$$2a + 7b - c + d = 0$$
$$-4a + 5b + 3c + d = 0.$$

Subtraction of the left members of the first equation and the second, and of the first equation and the third, gives the two equations

$$a - 5b - 2c = 0$$
$$5a + 7b - 2c = 0.$$

As in the first solution, $a : b : c = 3 : -1 : 4$, and the work is completed in the same way. Notice that the algebraic work is practically the same by the two methods.

In Chapter III we shall discuss solutions of systems of linear equations and treat more completely the equation of the plane through three non-collinear points.

29. Exercises.

1. Prove Theorem IV, § 28.
2. Prove Theorem V, § 28.
3. Prove Theorem VI, § 28.
4. Find the equation of the plane through the point $(-3, 5, 1)$ perpendicular to a line with direction numbers 4, 1, -3.
5. Find the equation of the plane through the point $(3, -5, 1)$ perpendicular to the segment $(4, 2, 4)$ $(3, -5, 1)$.
6. Find the equation of the plane through the point $(1, 2, -4)$ perpendicular to the radius vector of the point.
7. Find the equation of the plane through the origin parallel to the plane $5x - y - 2z + 8 = 0$.
8. Find the equation of the plane through the point $(1, -1, 6)$ parallel to the plane $2x + 5y - z + 9 = 0$.

9. Find the equations of the parallel planes each containing one of the points (3, 1, 5) and (−4, 6, 1), both perpendicular to the segment joining these points.

10. Find the equation of the plane that is the perpendicular bisector of the segment (4, 3, −9)(8, −1, 5).

In Exercises **11-14**, find the equation of the plane through the point (5, 3, −1) parallel to the given plane.

11. The xy plane.
12. The yz plane.
13. The plane $y = 8$.
14. The plane $x = z$.

In Exercises **15** and **16**, find the equation of the plane through the points (4, −3, 2) and (1, 1, −5) and parallel to the given coördinate axis.

* **15.** The z axis. *Suggestion:* Find the equation of the line in the xy plane that passes through the points (4, −3) and (1, 1).

* **16.** The y axis.

17. Determine k if the planes $x + (k + 1)y - kz + 1 = 0$ and $(1 - 2k)x - 9y + 6z - 5 = 0$ are parallel.

In Exercises **18-23**, find the equation of the plane through the three given points.

18. (0, 0, 0), (1, 1, −1), (0, 2, 1).
19. (8, 0, 0), (0, 4, 0), (0, 0, −2).
20. (−3, 0, 0), (−6, −2, 0), (3, 0, 2).
21. (1, 2, 4), (5, 1, 1), (10, −5, 2).
22. (1, 1, 6), (2, −1, −1), (5, −2, 3).
23. (2, 1, 3), (6, 3, 5), (−2, 2, 8).

24. Show that the following four points lie in a plane and find its equation: (1, 4, 1), (2, 11, 0), (0, 5, −2), (3, 6, 5).

25. Determine the value of k if the following four points lie in a plane: (1, −1, 2), (0, 2, 4), (1, 0, 5), (2, −3, k). Find the equation of this plane.

26. Prove that the planes

$$a_1x + b_1y + c_1z + d_1 = 0 \quad \text{and} \quad a_2x + b_2y + c_2z + d_2 = 0$$

are parallel and distinct if and only if

$$\frac{a_1}{a_2} = \frac{b_1}{b_2} = \frac{c_1}{c_2} \neq \frac{d_1}{d_2}.$$

* **27.** Prove that the relation of equivalence of one equation to another is an equivalence relation. (See Ex. **36**, § **13**.)

30. Intercept form for the equation of a plane. If a plane has non-zero intercepts A, B, and C, its equation can be written

(1) $$\frac{x}{A} + \frac{y}{B} + \frac{z}{C} = 1.$$

This is true since the equation is linear and, as seen by substitution, is satisfied by the coördinates of each of the three points $(A, 0, 0)$, $(0, B, 0)$, and $(0, 0, C)$.

EXAMPLE 1. Find the equation of the plane with intercepts 3, -1, and 2.

Solution. By (1), the equation is $\frac{x}{3} + \frac{y}{-1} + \frac{z}{2} = 1$, or $2x - 6y + 3z - 6 = 0$.

EXAMPLE 2. Find the intercepts of the plane $3x - 12y - 8z + 24 = 0$.

First solution. To find the x intercept, we set $y = z = 0$. Then $3x + 24 = 0$, or $x = -8$. In a similar manner we find the y and z intercepts, which are 2 and 3, respectively.

Second solution. Subtract 24 from each side of the equation, and divide each side of the resulting equation by -24. The given equation can thus be written in intercept form:

$$\frac{x}{-8} + \frac{y}{2} + \frac{z}{3} = 1,$$

from which the intercepts are immediately seen to be those obtained in the first solution.

31. Normal form for the equation of a plane. Let (λ, μ, ν) be a direction normal to a given plane (that is, λ, μ, ν are a set of direction cosines for a line that is normal or perpendicular to the given plane). The equation of the plane can then be written in the form

$$\lambda x + \mu y + \nu z + d = 0.$$

This is known as a **normal** or **perpendicular form** for the equation of a plane. Any plane actually has two equations in normal form, since there are two normal directions. If the plane does not pass through the origin, one of these two forms is used somewhat more commonly than the other. We shall consider this first.

THEOREM I. *If a plane does not pass through the origin, its equation can be written*

$$\lambda x + \mu y + \nu z - p = 0,$$

where p is the (positive) distance between the plane and the origin, and (λ, μ, ν) is the normal direction from the origin to the plane.

Proof. The point $(\lambda p, \mu p, \nu p)$ lies in the plane, since, on substituting its coördinates in the given equation, we get

$$\lambda^2 p + \mu^2 p + \nu^2 p - p = (\lambda^2 + \mu^2 + \nu^2)p - p = 0.$$

It is at the distance p from the origin, and its radius vector has the direction (λ, μ, ν).

Any equation $ax + by + cz + d = 0$ can be reduced to normal form by division of both sides by $\pm\sqrt{a^2 + b^2 + c^2}$, since the coefficients become direction cosines. The sign to be chosen depends on the normal direction desired. For example, if the plane does not pass through the origin the sign can be chosen so that the constant term is negative, and the equation assumes the form used in the theorem above. Theorem I implies the following:

THEOREM II. *The origin is at a directed distance d from the plane $\lambda x + \mu y + \nu z + d = 0$, the direction being (λ, μ, ν). Equivalently, the plane is at a directed distance $-d$ from the origin.*

Proof. If d is negative and if $p = -d$, the conditions of Theorem I are satisfied. If $d = 0$, the plane passes through the origin. If d is positive, a reversal in direction brings a change in sign, establishing the truth of the theorem in this case.

THEOREM III. *The parallel planes $\lambda x + \mu y + \nu z + d_1 = 0$ and $\lambda x + \mu y + \nu z + d_2 = 0$ lie on the same side or opposite sides of the origin according as d_1 and d_2 have the same or opposite signs. In either case the directed distance from the first plane to the second is $d_1 - d_2$.*

Proof. By Theorem II, this is merely a matter of subtracting directed distances on a directed line that is normal to the two planes and passes through the origin.

EXAMPLE. Find the distance between the parallel planes
$4x - 7y - 4z + 5 = 0$ and $8x - 14y - 8z - 17 = 0$.

Solution. The equations in normal form are $\frac{4}{9}x - \frac{7}{9}y - \frac{4}{9}z + \frac{5}{9} = 0$ and $\frac{4}{9}x - \frac{7}{9}y - \frac{4}{9}z - \frac{17}{18} = 0$. The distance between them is $\frac{5}{9} + \frac{17}{18} = \frac{27}{18} = \frac{3}{2}$.

32. Directed distance from a plane to a point. We are now in a position to note the particular significance of the normal form for the equation of a plane, as disclosed in the following theorem (which can be considered as a generalization of Theorem II, § 31):

THEOREM. *The directed distance from the plane*
$$\lambda x + \mu y + \nu z + d = 0$$
to the point $P_0(x_0, y_0, z_0)$, the direction being (λ, μ, ν), is
$$\lambda x_0 + \mu y_0 + \nu z_0 + d.$$

Proof. The equation of the plane parallel to the given plane and passing through P_0 is (check by substitution):
$$\lambda x + \mu y + \nu z - (\lambda x_0 + \mu y_0 + \nu z_0) = 0.$$

Using the notation of Theorem III, § 31, we see that the directed distance sought is the directed distance from the first plane to the second, which is $d_1 - d_2$ or $d + (\lambda x_0 + \mu y_0 + \nu z_0)$.

According as $\lambda x_0 + \mu y_0 + \nu z_0 + d$ is *positive* or *negative*, (λ, μ, ν) is the direction from the plane *toward* the point or *away from* the point.

COROLLARY. *The distance between the plane* $ax + by + cz + d = 0$ *and the point* (x_0, y_0, z_0) *is the absolute value of*

(1) $$\frac{ax_0 + by_0 + cz_0 + d}{\pm\sqrt{a^2 + b^2 + c^2}}.$$

The direction of the normal is determined by the sign of the denominator of (1).

To clarify this question of sign, we present two examples.

EXAMPLE 1. Find the distance between the plane $2x - 2y + z + 1 = 0$ and the point $(1, 1, 5)$, and discuss their relative positions.

Solution. Substitute the coördinates of the point in the left member of the equation, and divide by 3, according to formula (1). This is equivalent to substituting the coördinates in the left member of the following normal form of the given equation:

(2) $$\tfrac{2}{3}x - \tfrac{2}{3}y + \tfrac{1}{3}z + \tfrac{1}{3} = 0.$$

The distance is thus found to be 2 (and is therefore directed positively toward the point). Since the constant term of (2) is positive, this distance is also directed positively toward the origin, and the point and the origin are on the same side of the plane. Since the coefficient of z is positive, the point lies above the plane. It also lies to the left of and in front of the plane.

EXAMPLE 2. Find the distance between the plane of Example 1 and the point $(1, 1, -10)$, and discuss their relative positions.

Solution. After the substitution of the coördinates, the left member of the given equation is negative. Although it is not necessary to do so, in order to have a *positive* directed distance we divide this time by -3. The distance is 3. Since the constant term after division by -3 is negative, this distance is directed away from the origin. It is also directed downward, to the right, and away from the observer. The point lies on the opposite side of the plane from the origin, and below, to the right of, and behind the plane.

33. Exercises.

1. Find the equation of the plane whose intercepts are 3, 1, -4.
2. Find the intercepts of the plane $5x - 3y + 2z + 15 = 0$.
3. Find the equation of the plane that has equal intercepts and passes through the point $(3, 7, -2)$.
4. Find the equation of the plane whose intercepts are proportional to $1 : 2 : 3$ and which passes through the point $(2, -5, 9)$.

5. Find the intercepts of the plane through the three points $(-2, 1, 1)$, $(2, 2, 2)$, and $(3, 4, -3)$.

6. What planes do not have equations in intercept form?

7. Write two equations in normal form for each of the following planes:

(a) $2x + 2y - z = 0$;
(b) $7x - 4y + 4z - 8 = 0$;
(c) $6x - 9y - 2z + 33 = 0$;
(d) $2x + 3y - 6z - 28 = 0$;
(e) $x + y + z + 6 = 0$;
(f) $8x - y - 4z - 72 = 0$.

8. Find the distance between each plane in Exercise **7** and the origin.

9. For each plane in Exercise **7**, except (a), give the normal direction from the plane to the origin. Name the cases where the origin lies above the plane.

10. Find the distance between each plane in Exercise **7** and the point $P(1, 2, 3)$. Name the cases where P lies above the plane. Name the cases where P and the origin lie on the same side of the plane.

11. Find the equations of the planes that are parallel to the plane $4x + 7y - 4z + 29 = 0$ and whose distance from the origin is 7.

12. Find the equations of the planes whose intercepts are proportional to $1 : 2 : 3$ and whose distance from the origin is 2.

13. Find the distance between each pair of parallel planes:

(a) $x - 2y - 2z + 1 = 0$, $x - 2y - 2z + 10 = 0$;
(b) $18x - y + 6z - 8 = 0$, $18x - y + 6z + 11 = 0$;
(c) $3x - 4y - 13 = 0$, $6x - 8y - 91 = 0$;
(d) $8x + 16y - 2z + 17 = 0$, $12x + 24y - 3z - 5 = 0$.

14. The point $(-2, 1, 1)$ lies between certain pairs of planes in Exercise **13**. Name these.

*15. Explain what one would mean by the graph of an inequality, $f(x, y, z) > 0$. Show that the graph of the inequality $ax + by + cz + d > 0$ consists of all points lying on one side of the plane $ax + by + cz + d = 0$ and that the graph of the inequality $ax + by + cz + d < 0$ consists of all points on the other side. Obtain thereby a simple test to determine whether two points lie on the same side or opposite sides of a given plane, and prove that the point (x_0, y_0, z_0) and the origin lie on the same side or opposite sides of the plane, according as $ax_0 + by_0 + cz_0 + d$ and d have the same sign or opposite signs.

16. State which of the following pairs of points lie on the same side and which lie on opposite sides of the plane $5x - y + 8z - 17 = 0$. (See Ex. **15** for one method.)

(a) $(1, 3, 1), (3, 1, 3)$;
(b) $(1, -3, 2), (-1, -2, 3)$;
(c) $(1, 2, -3), (-3, 2, -1)$;
(d) $(2, 1, -3), (2, 1, 3)$.

17. Find the equations of the planes parallel to the plane $x - 2y - 2z = 7$ and a distance 5 from it.

18. Find the equations of the planes at a distance 6 from the point $(-6, 2, -3)$ and parallel to the plane $9x + 2y - 6z + 3 = 0$.

19. Find the equation of the plane that is located half-way between the parallel planes $7x - 9y + 2z + 27 = 0$ and $7x - 9y + 2z - 31 = 0$.

20. Find the distance between the point $(1, 1, 1)$ and the plane of the points $(-5, 1, 1)$, $(-1, -1, -1)$, and $(1, 0, -1)$.

* **21.** Prove that the distance between the parallel planes
$$ax + by + cz + d_1 = 0 \text{ and } ax + by + cz + d_2 = 0,$$
containing the points (x_1, y_1, z_1) and (x_2, y_2, z_2), respectively, is equal to
$$\frac{\mid a(x_1 - x_2) + b(y_1 - y_2) + c(z_1 - z_2) \mid}{\sqrt{a^2 + b^2 + c^2}}.$$

* **22.** Prove that if A, B, and C are the intercepts of a plane and p its distance from the origin, then
$$\frac{1}{A^2} + \frac{1}{B^2} + \frac{1}{C^2} = \frac{1}{p^2}.$$

34. Angle between two planes. Let $\lambda_1 x + \mu_1 y + \nu_1 z + d_1 = 0$ and $\lambda_2 x + \mu_2 y + \nu_2 z + d_2 = 0$ be two non-parallel planes, Π_1 and Π_2, respectively. Fig. 17 indicates a section of the planes normal to their line of intersection. Directed normals of the planes at the point where this line of intersection meets the plane of the paper are represented by arrows. Unless the planes are perpendicular there are two distinct dihedral angles defined by them. If we wish to use directed normals, we can define *the* dihedral angle to be the angle between these directed normals. But this is rarely advantageous, what one generally wants to know being the *acute* angle defined by the planes. Since the two angles defined by the planes are the same as the two angles defined by the normals, the formulas obtained in the theorems of § **18** are again applicable. We now state the general result for two planes, parallel or not:

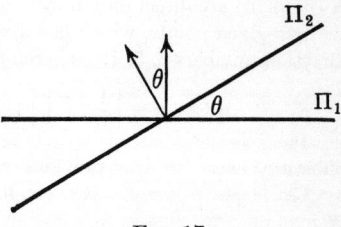

Fig. 17

THEOREM. *If θ is the dihedral angle between the planes*
$$\lambda_1 x + \mu_1 y + \nu_1 z + d_1 = 0 \quad and \quad \lambda_2 x + \mu_2 y + \nu_2 z + d_2 = 0$$
that satisfies the inequalities $0° \leq \theta \leq 90°$, *then*

(1) $$\cos \theta = \mid \lambda_1 \lambda_2 + \mu_1 \mu_2 + \nu_1 \nu_2 \mid.$$

If the equations of the planes are

(2) $\quad a_1 x + b_1 y + c_1 z + d_1 = 0 \quad and \quad a_2 x + b_2 y + c_2 z + d_2 = 0,$

then

(3) $$\cos\theta = \frac{|a_1a_2 + b_1b_2 + c_1c_2|}{\sqrt{a_1^2 + b_1^2 + c_1^2}\sqrt{a_2^2 + b_2^2 + c_2^2}}.$$

The two planes (2) *are perpendicular if and only if*

$$a_1a_2 + b_1b_2 + c_1c_2 = 0.$$

EXAMPLE 1. Find the acute angle between the planes $x + y = 0$ and $x + y + z = 0$.

Solution. By formula (3), $\cos\theta = \dfrac{1+1}{\sqrt{2}\sqrt{3}} = \dfrac{\sqrt{6}}{3}$. Therefore, to the nearest minute, $\theta = 35°\ 16'$.

EXAMPLE 2. Find the equation of the plane through the points (1, 4, 1) and (−1, 3, 2) and perpendicular to the plane $2x + 5y + z - 8 = 0$.

Solution. As with Example 2, § **28**, there are two methods. We present here the analogue of the first method given for that problem, and suggest that the student adapt the second to this example. The coefficients a, b, c in the equation of the plane are direction numbers of a line perpendicular to the segment joining the two given points, which has direction numbers 2, 1, −1, and to a line with direction numbers 2, 5, 1. Therefore

$$a : b : c = \begin{vmatrix} 1 & -1 \\ 5 & 1 \end{vmatrix} : \begin{vmatrix} -1 & 2 \\ 1 & 2 \end{vmatrix} : \begin{vmatrix} 2 & 1 \\ 2 & 5 \end{vmatrix} = 3 : -2 : 4.$$

Substitution as in the previous example gives $d = 1$, and the equation is $3x - 2y + 4z + 1 = 0$. (Cf. Ex. **6**, § **61**.)

35. Direction of line of intersection of two planes.

If two planes intersect, the line of intersection lies in each plane and is consequently perpendicular to the normals to each plane. We can therefore use the formula obtained in the first chapter for the direction numbers of a direction normal to two given directions, and state:

THEOREM. *A set of direction numbers of the line of intersection of the planes* $a_1x + b_1y + c_1z + d_1 = 0$ *and* $a_2x + b_2y + c_2z + d_2 = 0$ *is*

$$\begin{vmatrix} b_1 & c_1 \\ b_2 & c_2 \end{vmatrix}, \begin{vmatrix} c_1 & a_1 \\ c_2 & a_2 \end{vmatrix}, \begin{vmatrix} a_1 & b_1 \\ a_2 & b_2 \end{vmatrix}.$$

EXAMPLE. Find the direction cosines of the line of intersection of the planes $2x + y - 3z + 7 = 0$ and $4x + 2y - 3z - 1 = 0$, (taking it directed to the right).

Solution. The direction cosines are proportional to $1 : -2 : 0$, and are therefore $-\dfrac{1}{\sqrt{5}}, \dfrac{2}{\sqrt{5}}, 0$.

36. Pencil of planes. The set or family of all planes having a line in common is called a **pencil of planes.** The line common to the planes of a pencil of planes is called the **axis** of the pencil. A pencil of planes is completely determined by its axis, and therefore by any two distinct planes of the pencil. Our principal theorem is the following:

THEOREM. *If* $\Pi_1 = 0$ *and* $\Pi_2 = 0$, *where* $\Pi_1 = \Pi_1(x, y, z)$ *and* $\Pi_2 = \Pi_2(x, y, z)$ *are two linear expressions in* x, y, *and* z, *are two intersecting planes, then the family of planes*

(1) $$k_1\Pi_1 + k_2\Pi_2 = 0,$$

where k_1 and k_2 are parameters or arbitrary constants, not both zero, is the pencil of planes passing through the line of intersection of the planes $\Pi_1 = 0$ and $\Pi_2 = 0$.

Proof. There are two parts to this proof. (*i*) We must first show that for any pair of values of k_1 and k_2, not both zero, equation (1) is the equation of a plane passing through the line of intersection of the two given planes. For any values of k_1 and k_2, not both zero, equation (1) is the equation of a plane. For it is of at most the first degree, and if non-zero values of k_1 and k_2 existed for which this equation were of degree zero, the coefficients in Π_1 and Π_2 would be proportional and the two given planes would be parallel. Furthermore, for any values of the parameters, the plane (1) contains the line of intersection of the given planes, since the coördinates of any point on this line satisfy simultaneously the equations $\Pi_1 = 0$ and $\Pi_2 = 0$, and hence also the equation (1). (*ii*) We must now establish the fact that any plane Π_0 of the pencil is the graph of some equation of the form (1). Let $P_0(x_0, y_0, z_0)$ be any point in the plane Π_0 that is not on the axis of the pencil. We can now determine values of k_1 and k_2, not both zero, so that x_0, y_0, and z_0 will satisfy equation (1). For on substitution, we get

(2) $$k_1\Pi_1(x_0, y_0, z_0) + k_2\Pi_2(x_0, y_0, z_0) = 0.$$

Since P_0 does not lie both on $\Pi_1 = 0$ and on $\Pi_2 = 0$, the coefficients of k_1 and k_2 in this equation are not both zero. We can therefore determine values of k_1 and k_2, not both zero, that satisfy equation (2). Finally, with these values of k_1 and k_2 equation (1) is an equation of the plane of the pencil that passes through the point P_0, and hence of the plane Π_0.

For any given values of k_1 and k_2, the expression $k_1\Pi_1 + k_2\Pi_2$ is called a **linear combination** of the expressions Π_1 and Π_2, and the plane (1) is called a linear combination of the planes $\Pi_1 = 0$ and $\Pi_2 = 0$. More generally, if $\Pi_1 = 0$, $\Pi_2 = 0$, $\cdots$, $\Pi_n = 0$ are any n planes and if $k_1, k_2, \cdots, k_n$ are any n constants, the plane

$$k_1\Pi_1 + k_2\Pi_2 + \cdots + k_n\Pi_n = 0$$

is called a linear combination of these n planes.

Example 1. Find the equation of the plane passing through the line of intersection of the planes $3x + 4y - z + 5 = 0$ and $2x + y + z + 10 = 0$ and through the point $(-2, 5, 1)$.

Solution. The equation has the form

(3) $\qquad k_1(2x + y + z + 10) + k_2(3x + 4y - z + 5) = 0.$

Substitution of the coördinates $-2, 5, 1$ gives $12k_1 + 18k_2 = 0$, or $2k_1 + 3k_2 = 0$. A simple solution of this equation is $k_1 = 3$, $k_2 = -2$. Substitution of these values of k_1 and k_2 gives the equation $-5y + 5z + 20 = 0$, which reduces to $y - z = 4$.

Example 2. Find the equation of the plane through the line of intersection of the planes of Example 1 perpendicular to the plane $6x + y + 2z = 5$.

Solution. In this case we write equation (3) in the form

$$(2k_1 + 3k_2)x + (k_1 + 4k_2)y + (k_1 - k_2)z + 10k_1 + 5k_2 = 0.$$

The condition that this plane be perpendicular to the plane $6x + y + 2z = 5$ is

$$6(2k_1 + 3k_2) + (k_1 + 4k_2) + 2(k_1 - k_2) = 0,$$

which reduces to $3k_1 + 4k_2 = 0$. Let $k_1 = -4$, $k_2 = 3$. Then the desired equation is $x + 8y - 7z - 25 = 0$.

Example 3. Find the equation of the plane through the line of intersection of the planes of Example 1 perpendicular to the plane $x = y + z$.

Solution. We start as in Example 2, and write the condition for perpendicularity:

$$(2k_1 + 3k_2) - (k_1 + 4k_2) - (k_1 - k_2) = 0.$$

But this equation in k_1 and k_2 is an identity. Therefore *every* plane of the pencil is perpendicular to $x = y + z$. On further investigation we see that $1, -1, -1$ is a set of direction numbers of the line of intersection.

The process called *elimination by addition or subtraction*, used in solving simultaneous systems of linear equations, is similar to what we have been doing in forming equations of planes of a pencil. Let us consider now the significance of eliminating variables. We have already observed (Theorems V and VI, § **28**) that any plane is parallel to the axis of any variable that is missing from its equation. This gives the result:

The graph of an equation obtained by eliminating a variable by the method of addition or subtraction from the equations of two intersecting planes is a plane through the line of intersection of the two planes parallel to the axis of the variable eliminated.

These **projecting planes** are discussed again in § **38**, and in Chapter IV a more complete treatment of elimination of variables is given.

EXAMPLE 4. Find equations of the planes, each of which is parallel to a coördinate axis, that pass through the line of intersection of the planes

$$3x + y - 2z - 1 = 0 \text{ and } 2x - 5y + 4z - 6 = 0.$$

Solution. Multiplying both sides of the equations by 2 and -3, respectively, and adding the left members, we eliminate x and obtain $17y - 16z + 16 = 0$. Similarly, elimination of y and elimination of z give

$$17x - 6z - 11 = 0 \text{ and } 8x - 3y - 8 = 0,$$

respectively.

EXAMPLE 5. Find equations of the planes, each of which is parallel to a coördinate axis, that pass through the line of intersection of the planes

$$4x - y + z + 5 = 0 \text{ and } 3x + 2y - 2z - 7 = 0.$$

Solution. Elimination of x gives $11y - 11z - 43 = 0$. Elimination of y automatically eliminates z to give $11x + 3 = 0$. The line of intersection is parallel to the yz plane.

EXAMPLE 6. Find equations of the planes, each of which is parallel to a coördinate axis, that pass through the line of intersection of the planes

$$3x - y - 5 = 0 \text{ and } x + 2y + 3 = 0.$$

Solution. Elimination of x and elimination of y give

$$y = -2, \ x = 1.$$

The line is parallel to the z axis and *every* plane through the line is parallel to the z axis.

37. Exercises.

In Exercises **1-4**, find the cosine of the acute angle between the two planes.

1. $x + 2y + 2z - 17 = 0, 6x + 2y - 3z + 1 = 0.$
2. $x + 2y + 2z - 17 = 0, 4x - 4y - 7z + 5 = 0.$
3. $3x - 6y - 2z + 10 = 0, 3y - 4z - 6 = 0.$
4. $3x - 4z - 11 = 0, 3x + 5y - 4z = 0.$

In Exercises **5** and **6**, show that the two planes are perpendicular.

5. $8x + 7y - 10z + 3 = 0, 2x + 2y + 3z + 13 = 0.$
6. $5x - y - 13z - 1 = 0, 2x - 3y + z + 2 = 0.$

In Exercises **7-10**, (*a*) find a set of direction numbers for the line of intersection of the two planes; (*b*) find the equation of the plane through the origin perpendicular to the two planes.

7. $x + 2y - z + 8 = 0, 2x = z.$
8. $x - y + z + 2 = 0, x + y - 5z - 8 = 0.$
9. $x + 5y + z + 1 = 0, 5x - 3y - 3z - 3 = 0.$
10. $3x - 7y - 6z = 18, x + y + 2z = 10.$

In Exercises **11-18**, find the equation of the plane that passes through the line of intersection of the planes

$$x - y + 2z + 5 = 0 \text{ and } 2x + 3y - z - 1 = 0$$

and that satisfies the given condition.

11. It passes through the origin.
12. It passes through the point $(1, -1, 4)$.
13. It is parallel to the z axis.
14. It is perpendicular to the plane $x + 2y - 2z = 0$.
15. It is parallel to the segment $(1, 1, -1)$ $(3, 5, -3)$.
16. It has equal y and z intercepts.
17. It bisects a dihedral angle between the given planes.
* **18.** It bisects the acute dihedral angle between the given planes.

19. Find the equation of the plane through the point $(3, -2, -1)$ and perpendicular to the line of intersection of the planes $2x - 2y + z + 5 = 0$ and $9x - 7y + 2z - 11 = 0$.

In Exercises **20** and **21**, determine whether the origin lies in the acute angle or the obtuse angle of the two planes.

* **20.** $3x + 7y - z + 10 = 0$, $2x - 3y - 5z + 17 = 0$.
* **21.** $2x + 8y - 3z + 5 = 0$, $4x - y + 6z - 16 = 0$.

In Exercises **22** and **23**, find the equation of the plane through the points $(2, 1, -4)$ and $(5, -3, 2)$ and parallel to the given line.

22. The x axis.
23. A line that makes equal angles with the coördinate axes.

In Exercises **24** and **25**, find the equation of the plane through the points $(1, 0, 4)$ and $(5, -1, 6)$ and perpendicular to the given plane.

24. $4x - 3y - 2z + 7 = 0$.   **25.** $8x - 2y + 13z - 28 = 0$.

In Exercises **26-30**, Λ designates the trace of the plane

$$3x - 2y + 4z - 24 = 0$$

in the xy plane. Find the equation of the plane that satisfies the given conditions.

26. It passes through the origin and is perpendicular to Λ.
27. It bisects perpendicularly the segment of Λ cut by the x and y axes.
28. It contains Λ and the point $(1, -2, 1)$.
29. It contains the origin and the point $(3, -3, 1)$ and is parallel to Λ.
30. It contains the origin and is parallel to the z axis and Λ.

In Exercises **31** and **32**, find the equations of the planes through the line of intersection of the planes $x + y + z - 4 = 0$ and $y + z - 2 = 0$ that are at the given distance from the specified point.

31. Distance 2 from the origin.
32. Distance 3 from the point $(5, -3, 7)$.

In Exercises **33-36**, write the equations of the family of planes containing the two given points.

* **33.** (3, −1, 1), (3, 2, 1).   * **34.** (5, −2, 1), (1, 6, 1).
* **35.** (0, 0, 0), (1, 1, 1).   * **36.** (4, −1, 2), (3, 1, 4).

37. Prove that through a line that is not parallel to a second line there passes exactly one plane parallel to the second line. Thus prove that through any line that is not parallel to any coördinate axis there pass at most three planes each of which is parallel to (at least) one coördinate axis. Under what circumstances will there be just three planes having this property? Just two planes?

38. Prove that if $\Pi_1 = 0$ and $\Pi_2 = 0$ are intersecting planes and k is a parameter, then the family of planes $\Pi_1 + k\Pi_2 = 0$ is the pencil of planes discussed in § **36**, with one exception. What is this exception?

* **39.** Prove that if $\Pi_1 = 0$ and $\Pi_2 = 0$ are distinct parallel planes, then the family of planes $k_1\Pi_1 + k_2\Pi_2 = 0$, where k_1 and k_2 are parameters, not both zero, is the family of planes parallel to $\Pi_1 = 0$ and $\Pi_2 = 0$, except for values of k_1 and k_2 having a certain ratio. What is the significance of these exceptional values?

38. Equations of a line. A line is determined in three principal ways: (*i*) by the intersection of two planes, (*ii*) by two points on the line, and (*iii*) by a point and a set of direction numbers.

The first method has already been discussed. A system of two linear equations is a set of equations of a line if and only if their graphs are planes intersecting in the line. Any line obviously has infinitely many systems of equations, but for convenience we shall henceforth feel free to speak of *the line* $\Pi_1 = 0, \Pi_2 = 0$.

Now let $P_1(x_1, y_1, z_1)$ and $P_2(x_2, y_2, z_2)$ be any two distinct points on a line Λ, and let $P(x, y, z)$ be any point on the line. Then $x_2 - x_1$, $y_2 - y_1$, $z_2 - z_1$ is a set of direction numbers of the line, and unless P is the point P_1, $x - x_1$, $y - y_1$, $z - z_1$ is also a set of direction numbers of the line. In any case, whether P is P_1 or not, we have the system of equations given below satisfied by the coördinates of those points and only those points that lie on the line Λ:

A system of equations of the line Λ *through the points* $P_1(x_1, y_1, z_1)$ *and* $P_2(x_2, y_2, z_2)$ *is*

(1) $$\frac{x - x_1}{x_2 - x_1} = \frac{y - y_1}{y_2 - y_1} = \frac{z - z_1}{z_2 - z_1}.$$

By exactly analogous reasoning we have another **form for** the equations of a line:

A system of equations of the line Λ through the point $P_1(x_1, y_1, z_1)$ with direction numbers l, m, n is

(2) $$\frac{x - x_1}{l} = \frac{y - y_1}{m} = \frac{z - z_1}{n}.$$

Equations (2) are called equations of the line in **symmetric form**. It is to be understood again that a vanishing denominator implies the vanishing of the corresponding numerator.

Let us assume first that l, m, and n are all different from zero. Equations (2) can then be regarded as a system of three equations

(3) $$\frac{y - y_1}{m} = \frac{z - z_1}{n}, \quad \frac{z - z_1}{n} = \frac{x - x_1}{l}, \quad \frac{x - x_1}{l} = \frac{y - y_1}{m}.$$

These equations are not independent, since any one can be obtained from the other two. They are the equations of the three planes through the line Λ each of which is parallel to one of the three coördinate axes. (Prove this.) These three planes are called the **projecting planes** of the line. (See Fig. 18.) (Why is this terminology suitable?)

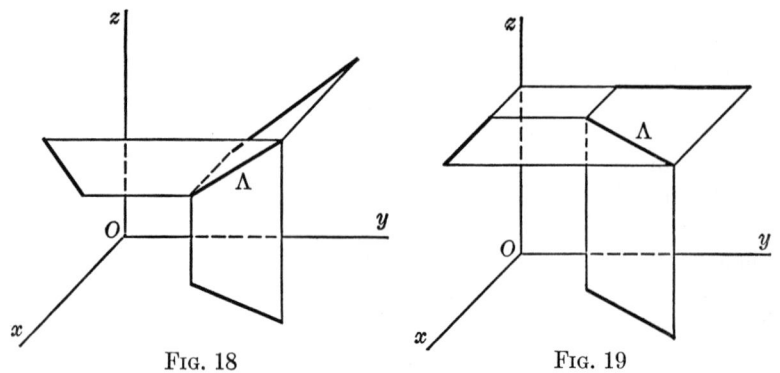

Fig. 18

Fig. 19

Let us consider now the case where just one of the direction numbers, say n, vanishes. In this case the system (2) is equivalent to the following system:

(4) $$\frac{x - x_1}{l} = \frac{y - y_1}{m}, \quad z = z_1.$$

The line Λ is parallel to the xy plane, but not parallel to a coördinate axis. The **projecting planes** defined by equations (4) are the two planes through Λ, each parallel to at least one of the coördinate axes. (See Fig. 19.)

§ 38] EQUATIONS OF A LINE

Finally, if two direction numbers, say l and m, are zero, the system (2) reduces to the two equations

(5) $$x = x_1,\ y = y_1,$$

which are the equations of the planes through Λ each of which is parallel to a coördinate plane. Since Λ is parallel to the z axis, every plane through Λ is parallel to the z axis. The line is said to have two **projecting planes,** given by equations (5). (See Fig. 20.)

EXAMPLE 1. Find equations of the line through each pair of points:

(a) $(3, 1, -4), (5, -6, 1)$; (b) $(4, 5, -3)$, $(6, -1, -3)$; (c) $(6, 1, 7), (1, 1, 7)$.

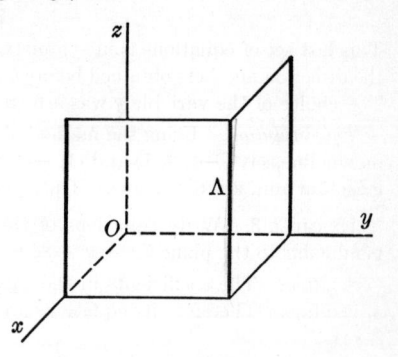

FIG. 20

Solution. Direct substitution in equations (1) gives for the first pair

$$\frac{x-3}{2} = \frac{y-1}{-7} = \frac{z+4}{5}$$

for the second pair

$$\frac{x-4}{2} = \frac{y-5}{-6} = \frac{z+3}{0}$$

or

$$\frac{x-4}{1} = \frac{y-5}{-3} \text{ and } z = -3$$

and for the third pair

$$\frac{x-6}{-5} = \frac{y-1}{0} = \frac{z-7}{0},$$

or

$$y = 1 \text{ and } z = 7.$$

EXAMPLE 2. Find equations in symmetric form for the line

$$x - 4y + 2z + 7 = 0,\ 3x + 3y - z - 2 = 0.$$

First solution. First, by the method of § **35**, we find $\lambda : \mu : \nu = -2 : 7 : 15$. To find a point on the line we can eliminate z to obtain $7x + 2y + 3 = 0$. A solution of this equation is easily found by giving either x or y a value. A convenient solution is $x = -1, y = 2$. The corresponding value of z is 1, and the equations are

$$\frac{x+1}{-2} = \frac{y-2}{7} = \frac{z-1}{15}.$$

Second solution. As in the first solution, eliminate z. Then solve for y in terms of x. In a similar way, eliminate x and solve for y in terms of z. The two resulting equations can be combined as follows:

or
$$\frac{-7x-3}{2} = y = \frac{7z+23}{15},$$

$$\frac{x+\frac{3}{7}}{-2} = \frac{y}{7} = \frac{z+\frac{23}{7}}{15}.$$

This last set of equations is in symmetric form and shows that the given line has the direction numbers obtained before and passes through the point $(-\frac{3}{7}, 0, -\frac{23}{7})$. The choice of the variable y was arbitrary.

Third solution. Using the method of the first solution, we find any two points on the line, say $(-1, 2, 1)$ and $(1, -5, -14)$, and from their coördinates, a set of direction numbers of the line. The equations can then be written down as before.

EXAMPLE 3. Write equations of the line through the point $(1, 5, -2)$ perpendicular to the plane $7x - y + 8z + 11 = 0$.

Solution. The coefficients in the equation of the plane are direction numbers of the line. Therefore its equations are

$$\frac{x-1}{7} = \frac{y-5}{-1} = \frac{z+2}{8}.$$

EXAMPLE 4. Write an equation of the plane through $(1, 5, -2)$ perpendicular to the line

$$\frac{x+9}{7} = \frac{y-3}{-1} = \frac{z}{8}.$$

Solution. The denominators can be used as coefficients in the equation of the plane. Determining the constant term by substitution, we have the equation: $7x - y + 8z + 14 = 0$. Alternatively, we could substitute in equation (1), § 28.

EXAMPLE 5. Show that the plane $x + 4y - z - 1 = 0$ contains the line
$$\frac{x-3}{2} = \frac{y+1}{1} = \frac{z+2}{6}.$$

Solution. Since $2 \cdot 1 + 1 \cdot 4 - 6 \cdot 1 = 0$, the line is perpendicular to the normals to the plane and is therefore parallel to the plane. Furthermore, the point $(3, -1, -2)$ lies on the line and in the plane. Therefore the line lies entirely in the plane.

39. Parametric equations of a line.
For any point $P(x, y, z)$ on the line Λ with equations

(1) $$\frac{x-x_1}{l} = \frac{y-y_1}{m} = \frac{z-z_1}{n},$$

the three fractions (or at least those without zero denominators) are equal. If we denote by t this common value, the equations can be written

(2) $$\begin{aligned} x &= x_1 + lt, \\ y &= y_1 + mt, \\ z &= z_1 + nt. \end{aligned}$$

An important fact to notice is that the equations (2) define the line Λ even when the direction numbers are not all different from zero. That is, the following two statements hold in every case; (*i*) Corresponding to any value of t, equations (2) give the coördinates of a point on the line Λ. (*ii*) Corresponding to any point $P(x, y, z)$ on the line Λ, there is a value of t such that x, y, z, and t satisfy equations (2). Equations (2) are called **parametric equations** of the line, the parameter being t.

Since, as the parameter t varies over all real values equations (2) give the coördinates of all of the points and only the points of the line Λ, these equations establish a one-to-one correspondence between the points of Λ and the real numbers. Positive values of t correspond to points on one side of $P_1(x_1, y_1, z_1)$, negative values of t to points on the other side, and the value zero to the point P_1.

If λ, μ, and ν are direction cosines of a directed line Λ, then the equations

(3)
$$x = x_1 + \lambda t,$$
$$y = y_1 + \mu t,$$
$$z = z_1 + \nu t,$$

are again parametric equations of the line Λ. In this case, however, the parameter t has special significance, being the actual directed distance from P_1 to P along the line. To show this, assume first that (λ, μ, ν) is the direction from P_1 to P, the (positive) distance between the points being d. Then, as in § **14**, we can see that

$$\frac{x - x_1}{d} = \lambda, \quad \frac{y - y_1}{d} = \mu, \quad \frac{z - z_1}{d} = \nu,$$

from which the result comes immediately, with $t = d$. On the other hand, a change of direction and a change of sign lead again to equations (3), with $t = -d$.

One simple but useful application of the parametric equations of a line is finding the point where a line meets a plane, as illustrated in the following examples:

EXAMPLE 1. Find the coördinates of the point where the line
$$\frac{x+3}{2} = \frac{y+1}{-3} = \frac{z-5}{-1}$$
meets the plane $5x - y + 4z + 3 = 0$.

Solution. Substitute in the equation of the plane $x = -3 + 2t$, $y = -1 - 3t$, $z = 5 - t$. The result is $9t + 9 = 0$. Therefore $t = -1$ and $x = -5$, $y = 2$, $z = 6$.

EXAMPLE 2. Find the projection of the point (10, 13, −6) on the plane $4x + 7y - z - 5 = 0$.

Solution. The point desired is the point where the normal to the plane through the point (10, 13, −6) meets the plane. Accordingly, we substitute in the equation of the plane: $x = 10 + 4t$, $y = 13 + 7t$, $z = -6 - t$. This gives $132 + 66t = 0$, or $t = -2$. Therefore the projection is $(2, -1, -4)$.

40. Exercises.

In Exercises **1-4**, find the angle between the two given lines, (a) assuming that each line is directed upward; (b) assuming that each line is directed to the right.

1. $\dfrac{x+5}{3} = \dfrac{y+1}{-3} = \dfrac{z}{-6}$; $\dfrac{x}{5} = \dfrac{y-2}{-5} = \dfrac{z-7}{2}$.

2. $\dfrac{x+1}{0} = \dfrac{y-3}{1} = \dfrac{z-1}{-1}$; $\dfrac{x-1}{1} = \dfrac{y}{1} = \dfrac{z+5}{4}$.

3. $\dfrac{2x-1}{4} = \dfrac{2-y}{3} = \dfrac{z}{5}$; $6x = 6y + 1 = 6 - z$.

4. $3x + y - 7z + 4 = 0$, $x - y + 7z - 2 = 0$;
   $x + 2y + z - 1 = 0$, $2x + y - 2z + 8 = 0$.

5. Show that the following lines are parallel:

   $x + y - 2z + 5 = 0$, $2x + 5y + 2z - 6 = 0$;
   $2x + y - 6z - 1 = 0$, $3x + 4y - 4z = 0$.

6. Show that the following lines are perpendicular:

   $2x + y - 4z + 10 = 0$, $x + 5y + 7z - 6 = 0$;
   $5x - y - 3z + 18 = 0$, $3x - 4y + 5z + 1 = 0$.

In Exercises **7-9**, show that the two lines are identical:

7. $\dfrac{x+2}{5} = \dfrac{y-2}{4} = \dfrac{z}{1}$; $\dfrac{x+7}{5} = \dfrac{y+2}{4} = \dfrac{z+1}{1}$.

8. $x = 2 - 4t$, $y = 3 + t$, $z = 7 + 2t$;
   $x = 10 - 4t$, $y = 1 + t$, $z = 3 + 2t$.

9. $x - y - z + 1 = 0$, $2x + 3y + 2z - 7 = 0$;
   $6x - y - 2z - 3 = 0$, $x + 4y + 3z - 8 = 0$.

10. Show that the following three lines are identical:

    $\dfrac{x-1}{1} = \dfrac{y+9}{-2} = \dfrac{z-28}{5}$; $x = -4 + t$, $y = 1 - 2t$, $z = 3 + 5t$;
    $7x + y - z + 30 = 0$, $4x - 3y - 2z + 25 = 0$.

In Exercises **11-14**, write equations in symmetric form for the line through the given point having the given direction numbers.

11. (1, 3, 1); 2, 1, 6.
12. (1, −2, −4); 3, −2, 1.
13. (5, 1, 0); 5, 0, −3.
14. (−2, 0, 7); 0, 1, 0.

In Exercises **15-18**, write equations in symmetric form for the line through the given points.

15. (5, 1, 8), (7, 4, 5). **16.** (−3, 2, 0), (1, 1, 1).
17. (3, 4, 1), (3, 7, 7). **18.** (7, −9, −2), (20, −9, −2).

19. Write equations in the parametric form (2), § **39**, for each line of Exercises **11-14**.

In Exercises **20-23**, find the coördinates of every point where the given line meets a coördinate plane.

20. $\dfrac{x+6}{3} = \dfrac{y-7}{1} = \dfrac{z-8}{-2}$.   **21.** $\dfrac{x-14}{2} = \dfrac{y-18}{-3} = \dfrac{z+10}{5}$.

22. $\dfrac{x+3}{3} = \dfrac{y-11}{0} = \dfrac{z-12}{-4}$.   **23.** $3x + y - z + 26 = 0$,
$\phantom{\mathbf{23.}\ }3x - y - 2z + 16 = 0$.

In Exercises **24-27**, write a set of equations in symmetric form for the line through the given point perpendicular to the given plane.

24. (3, 5, 6); $5x + 8y - z + 4 = 0$.
25. (−2, 1, 0); $x - 3y - 2z - 5 = 0$.
26. (0, 5, 1); $6x + 2y - 7 = 0$.
27. (1, −3, −4); $3y = 10$.

In Exercises **28-33**, write a set of equations in symmetric form for the given line.

28. $\dfrac{5x}{3} = \dfrac{2-y}{4} = \dfrac{2z-5}{-2}$.

29. $x = 2 + 3t, y = -5 + t, z = -4t$.
30. $x + 5y + 5z + 6 = 0, 7x - 10y + 5z + 57 = 0$.
31. $x + 2y - 3z + 14 = 0, 3x + 4y - 6z + 25 = 0$.
32. The line through the origin that makes equal angles with the axes.
33. The line through the point (−4, −7, 2) parallel to the x axis.

In Exercises **34** and **35**, write equations in symmetric form for the line through the given point parallel to the given planes.

34. (−5, 1, 0); $2x + y - z + 11 = 0, 5x - y - 2z + 13 = 0$.
35. (0, 0, 0); $2x + 3y + 2z - 5 = 0, 2x + 6y - z - 7 = 0$.

In Exercises **36-39**, draw a figure, showing the segment joining the given points, and indicate the projecting planes of each line.

36. (1, 2, 1), (3, 7, 4).   **37.** (3, 2, 4), (1, 7, 1).
38. (1, 2, 4), (3, 7, 4).   **39.** (1, 2, 1), (1, 2, 4).

In Exercises **40-42**, write the equation of the plane through the point (5, −1, 0) perpendicular to the given line.

40. $\dfrac{x-2}{3} = \dfrac{y}{-2} = \dfrac{z+5}{-6}$.   **41.** $\dfrac{x+3}{4} = \dfrac{y+7}{0} = \dfrac{z-1}{-5}$.

42. $3x - y - 3z - 16 = 0, 3x + y - 2z - 5 = 0$.

In Exercises **43-46**, write the equations of the projecting planes of the given line.

43. $\dfrac{x-3}{-5} = \dfrac{y+2}{6} = \dfrac{z}{1}$.

44. $\dfrac{x-2}{0} = \dfrac{y+7}{2} = \dfrac{z+1}{-3}$.

45. $x = 4 + 3t,\ y = -5 - 2t,\ z = 1 + 5t$.

46. $6x - y + 2z - 8 = 0,\ 3x + 2y - z - 5 = 0$.

47. Assuming that l and m are not both zero, write in symmetric form the equations of the projection in the xy plane of the line

$$\dfrac{x-x_1}{l} = \dfrac{y-y_1}{m} = \dfrac{z-z_1}{n}.$$

In Exercises **48-51**, show that the two lines intersect, and find the point of intersection.

48. $\dfrac{x+2}{5} = \dfrac{y-2}{3} = \dfrac{z+2}{0};\ \dfrac{x+3}{2} = \dfrac{y-5}{0} = \dfrac{z-7}{-3}$.

49. $\dfrac{x+11}{1} = \dfrac{y+1}{0} = \dfrac{z-4}{0};\ \dfrac{x+1}{2} = \dfrac{y-1}{-1} = \dfrac{z+8}{6}$.

50. $x = 3 + 5t,\ y = 5 + 8t,\ z = -t;\ x = 10 + 6t,\ y = -7 - 2t,\ z = 7 + 3t$.

51. $3x + 5y + 2z - 8 = 0,\ 2x - y - z - 13 = 0;$
$4x - y + 2z = 0,\ x - 4y + z + 11 = 0$.

In Exercises **52-55**, find the point where the given line intersects the given plane.

52. $x = -9 + 5t,\ y = -17 + 8t,\ z = 12 - 6t;\ 5x - 3y + 6z - 8 = 0$.

53. $x = 3,\ y = 16 + 5t,\ z = 2 - t;\ x - 2y - 4z + 19 = 0$.

54. $\dfrac{x+8}{2} = \dfrac{y-8}{-4} = \dfrac{z+2}{1};\ 4x + y + z + 11 = 0$.

55. $x + y + z - 8 = 0,\ 5x - y + 2z - 4 = 0;\ 2x - y + 10z + 21 = 0$.

In Exercises **56** and **57**, find the projection of the given point on the given plane.

56. $(13, 4, -27);\ 3x + y - 7z + 4 = 0$.

57. $(-10, 13, 3);\ 6x - 5y - 2z + 1 = 0$.

In Exercises **58** and **59**, find the angle between the given line and the given plane.

58. $\dfrac{x+3}{4} = \dfrac{y+7}{-1} = \dfrac{z-6}{-4};\ 2x - 5y - 2z - 13 = 0$.

59. $x = 5 + 3t,\ y = 3 + 4t,\ z = t;\ x - 3y - 4z + 10 = 0$.

In Exercises **60** and **61**, show that the two lines intersect, and find the equation of the plane containing them.

60. $\dfrac{x-1}{2} = \dfrac{y+4}{5} = \dfrac{z-3}{-3}; \dfrac{x+3}{3} = \dfrac{y-9}{-4} = \dfrac{z+14}{7}.$

61. $2x + 2y + 3z + 3 = 0,\ x + 2y + 2z + 1 = 0;$
$x - y + z + 3 = 0,\ 2x + y + 8z + 9 = 0.$

* **62.** Find the equation of the plane containing the parallel lines of Exercise **5**.

In Exercises **63** and **64**, write the equation of the plane containing the first line and parallel to the second.

* **63.** $x = -y = z;\ \dfrac{x-5}{1} = \dfrac{y+2}{11} = \dfrac{z+5}{-3}.$

* **64.** $\dfrac{x-4}{1} = \dfrac{y-5}{2} = \dfrac{z+1}{3};\ \dfrac{x+13}{2} = \dfrac{y+10}{4} = \dfrac{z+3}{-5}.$

* **65.** Find equations in symmetric form of the line of intersection of the plane $5x + 2y + z + 2 = 0$ and the plane containing the line

$$\dfrac{x+5}{3} = \dfrac{y}{-1} = \dfrac{z+13}{5}$$

that intersects the given plane in a line perpendicular to the given line.

* **66.** The two lines

$$\dfrac{x-1}{1} = \dfrac{y-2}{1} = \dfrac{z-3}{-2} \quad \text{and} \quad \dfrac{x-1}{1} = \dfrac{y-2}{2} = \dfrac{z-3}{-1}$$

meet in a point P and lie in a plane Π. Find equations in symmetric form of the following three lines, and show that they are mutually perpendicular:

(a) The line through P in Π that bisects the acute angle between the given lines.

(b) The line through P in Π that bisects the obtuse angle between the given lines.

(c) The line through P perpendicular to Π.

In Exercises **67-70**, write equations of the indicated line in parametric form, where the parameter is distance from the given point, directed upward.

67. Through the point $(5, 1, -6)$, direction numbers $4, -1, 8$.
68. Through the origin, direction numbers $1, -2, -2$.
69. Through the point $(1, -3, 0)$, perpendicular to the plane

$$6x + 2y - 3z + 10 = 0.$$

70. Through the origin, perpendicular to the plane $3x = 4z$.

★ **41. Another form for the parametric equations of a line.** Let $P_1(x_1, y_1, z_1)$ and $P_2(x_2, y_2, z_2)$ be any two points on a line Λ. The parametric equations of Λ can be written

(1) $$\begin{aligned} x &= x_1 + (x_2 - x_1)t, \\ y &= y_1 + (y_2 - y_1)t, \\ z &= z_1 + (z_2 - z_1)t; \end{aligned} \quad \text{or} \quad \begin{aligned} x &= (1 - t)x_1 + tx_2, \\ y &= (1 - t)y_1 + ty_2, \\ z &= (1 - t)z_1 + tz_2. \end{aligned}$$

The point P_1 corresponds to $t = 0$ and P_2 to $t = 1$.

More generally, the significance of t can be seen in this way: Write the first equation of (1) in the form $x = x_1 + \dfrac{x_2 - x_1}{d} \cdot dt$, where d is the distance P_1P_2, similar equations holding for y and z. Then $(x_2 - x_1)/d$, $(y_2 - y_1)/d$, and $(z_2 - z_1)/d$ are the direction cosines of the line Λ, with the direction taken from P_1 to P_2. Therefore, by equations (3), § **39**, dt is the directed distance from P_1 to P. The values of t are therefore proportional to the directed distances of P from P_1. We could say that t measures directed distances on the line adjusted so that the distance P_1P_2 is 1. Similarly, $1 - t$ measures distances of P from P_2 directed oppositely, with the same scale adjustment.

We formulate some of these results in a slightly different form:

Theorem. *A necessary and sufficient condition for a point $P(x, y, z)$ to lie on the line through the points $P_1(x_1, y_1, z_1)$ and $P_2(x_2, y_2, z_2)$ is that its coördinates be representable in the form*

(2) $$x = t_1x_1 + t_2x_2, \quad y = t_1y_1 + t_2y_2, \quad z = t_1z_1 + t_2z_2,$$

where $t_1 + t_2 = 1$. The numbers t_1 and t_2 can be interpreted as directed distances with the scale so adjusted that P_1 and P_2 are a unit distance apart, t_1 being the directed distance from P_2 to P directed toward P_1, and t_2 being the directed distance from P_1 to P directed toward P_2.

Notice that the representation of x, y, and z in the form (2) is merely a repetition of the point of division formulas, with $t_1 = \dfrac{r_2}{r_1 + r_2}$, $t_2 = \dfrac{r_1}{r_1 + r_2}$. With the new notation the ratio $r_1 : r_2$ becomes $t_2 : t_1$.

★ **42. Parametric equations of a plane.** The following theorem corresponds closely to the theorem of § **41**:

Theorem. *A necessary and sufficient condition for a point $P(x, y, z)$ to lie in the plane through the three non-collinear points $P_1(x_1, y_1, z_1)$,*

$P_2(x_2, y_2, z_2)$, and $P_3(x_3, y_3, z_3)$ *is that its coördinates be representable in the form*

(1)
$$x = t_1 x_1 + t_2 x_2 + t_3 x_3,$$
$$y = t_1 y_1 + t_2 y_2 + t_3 y_3,$$
$$z = t_1 z_1 + t_2 z_2 + t_3 z_3,$$

where $t_1 + t_2 + t_3 = 1$.

Proof. For the first part of the proof, let P be any point in the plane $P_1 P_2 P_3$. If P lies on the line through two of the points, say P_1 and P_2, then, by § **41**, it is representable in the form above with $t_3 = 0$. Assume now that P does not lie on any one of the three lines $P_1 P_2$, $P_1 P_3$, or $P_2 P_3$, and furthermore assume that $P_3 P$ is not parallel to $P_1 P_2$. (There is no loss of generality in this assumption, since any one of the three points could be labeled P_3, and there is at least one labeling of the points for which $P_3 P$ is not parallel to $P_1 P_2$.) If $P_4(x_4, y_4, z_4)$ is the point where the lines $P_1 P_2$ and $P_3 P$ meet (see Fig. 21), then (since it lies on the line $P_1 P_2$) its coördinates are representable in the form

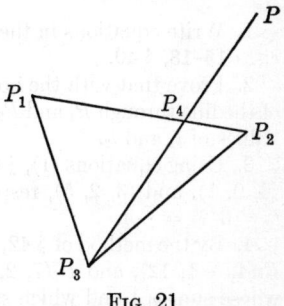

Fig. 21

(2)
$$x_4 = s_1 x_1 + s_2 x_2,$$
$$y_4 = s_1 y_1 + s_2 y_2,$$
$$z_4 = s_1 z_1 + s_2 z_2,$$

where $s_1 + s_2 = 1$. Similarly, the coördinates of P are representable in the form

(3) $\qquad x = s_3 x_3 + s_4 x_4, \qquad y = s_3 y_3 + s_4 y_4, \qquad z = s_3 z_3 + s_4 z_4,$

where $s_3 + s_4 = 1$. Combining these formulas, we have the coördinates of P represented in the form (1), where $t_1 = s_1 s_4$, $t_2 = s_2 s_4$, and $t_3 = s_3$.

Conversely, let $P(x, y, z)$ be a point whose coördinates are represented in the form (1). The t's cannot all be equal to 1. Assume for definiteness that t_3 is different from 1. The steps of the first part of the proof can now be retraced, with

$$s_1 = \frac{t_1}{1 - t_3}, \qquad s_2 = \frac{t_2}{1 - t_3}, \qquad s_3 = t_3, \qquad s_4 = 1 - t_3.$$

That is, let P_4 be the point whose coördinates are given by (2). Then, since $s_1 + s_2 = 1$, P_4 lies on the line P_1P_2. Finally, the coördinates of P are given by (3) and, since $s_3 + s_4 = 1$, P lies on the line P_3P_4, and therefore in the plane $P_1P_2P_3$. This completes the proof.

A plane through three points is discussed again in Chapter III. The methods of that chapter furnish another means of deriving equations (1). (Cf. Ex. **24**, § **61**.)

* **43. Exercises.**

1. Write equations in the parametric form (2), § **41**, for each line of Exercises **15-18**, § **40**.

2. Prove that with the notation of § **41**, those points and only those points of the line through P_1 and P_2 that are *between* P_1 and P_2 correspond to positive values of t_1 and t_2.

3. Using equations (1), § **42**, letting P_1, P_2, and P_3 be the points $(1, 3, -2)$, $(4, 0, 1)$, and $(3, 2, 5)$, respectively, find the point corresponding to $t_1 = 2$, $t_2 = 3$, $t_3 = -4$.

4. By the method of § **42**, show that the four points $P(-4, 9, 4)$, $P_1(1, 5, 2)$, $P_2(4, -1, 12)$, and $P_3(7, 2, -6)$ lie in a plane by finding numbers t_1, t_2, t_3 whose sum is 1 and which satisfy equations (1) of that section.

* **5.** Prove that if the three points P_1, P_2, P_3 are collinear but not identical, equations (1), § **42**, give coördinates of all points on the line through these points.

* **6.** Prove that the points *inside* the triangle $P_1P_2P_3$, and only those points, correspond to positive values of t_1, t_2, t_3 in equations (1), § **42**. To what values does the centroid correspond? (See Ex. **33**, § **13**. Cf. Ex. **2** of this section.)

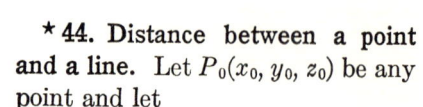

Fig. 22

* **44. Distance between a point and a line.** Let $P_0(x_0, y_0, z_0)$ be any point and let

$$\frac{x - x_1}{\lambda} = \frac{y - y_1}{\mu} = \frac{z - z_1}{\nu}$$

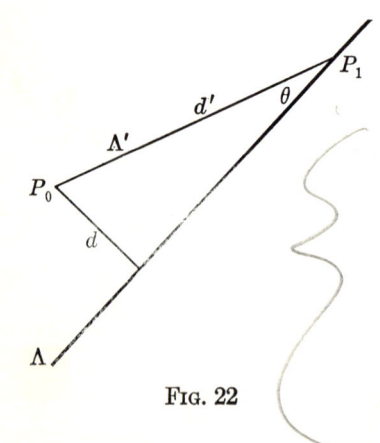

be any line Λ. Denote by d the (perpendicular) distance between P_0 and Λ. (See Fig. 22.) Our problem is to determine d in terms of the coördinates of P_0 and $P_1(x_1, y_1, z_1)$ and the direction cosines λ, μ, ν. Let Λ' be the line through P_0 and P_1 (assuming momentarily that P_0 and P_1 are distinct), let θ be the angle ($0° \leq \theta \leq 90°$) between Λ and Λ', and let d' be the distance P_0P_1. Then a set of direction cosines of Λ' is

$$\frac{x_0 - x_1}{d'}, \quad \frac{y_0 - y_1}{d'}, \quad \frac{z_0 - z_1}{d'}.$$

Therefore, since $d^2 = d'^2 \sin^2 \theta$ and since $\sin^2 \theta$ can be represented as the sum of the squares of three second order determinants, (§ 21), if we multiply each of these determinants by d' we have

$$d^2 = \begin{vmatrix} y_0 - y_1 & z_0 - z_1 \\ \mu & \nu \end{vmatrix}^2 + \begin{vmatrix} z_0 - z_1 & x_0 - x_1 \\ \nu & \lambda \end{vmatrix}^2 + \begin{vmatrix} x_0 - x_1 & y_0 - y_1 \\ \lambda & \mu \end{vmatrix}^2.$$

This formula obviously holds if P_0 and P_1 are identical.

In practice it is usually simpler to use essentially the method of derivation rather than the formula itself. This also has the advantage of impressing the method on one's memory and developing freedom from a list of formulas. The following example is illustrative.

EXAMPLE. Find the distance between the point $(2, -1, 7)$ and the line $\frac{x-5}{1} = \frac{y-2}{-2} = \frac{z+1}{1}$.

Solution. Using the notation of this section, we have $d'^2 = 3^2 + 3^2 + 8^2 = 82$ and $\cos \theta = \frac{11}{\sqrt{82} \cdot \sqrt{6}}$. Therefore

$$\sin^2 \theta = \frac{492 - 121}{492} = \frac{371}{492}, \text{ and } d = d' \sin \theta = \sqrt{\frac{371}{6}}.$$

* 45. Distance between two skew lines. The problem proposed in this section is to find the (perpendicular) distance between two skew lines:

$$\Lambda_1 : \frac{x - x_1}{\lambda_1} = \frac{y - y_1}{\mu_1} = \frac{z - z_1}{\nu_1}; \quad \Lambda_2 : \frac{x - x_2}{\lambda_2} = \frac{y - y_2}{\mu_2} = \frac{z - z_2}{\nu_2}.$$

Let Π_1 and Π_2 be the parallel planes containing the lines Λ_1 and Λ_2, respectively. (Prove that these planes exist and are unique and that the distance between Λ_1 and Λ_2 is the distance between Π_1 and Π_2.) The normals to each plane are perpendicular to both lines, and therefore have direction numbers

$$l = \begin{vmatrix} \mu_1 & \nu_1 \\ \mu_2 & \nu_2 \end{vmatrix}, \quad m = \begin{vmatrix} \nu_1 & \lambda_1 \\ \nu_2 & \lambda_2 \end{vmatrix}, \quad n = \begin{vmatrix} \lambda_1 & \mu_1 \\ \lambda_2 & \mu_2 \end{vmatrix}.$$

The equations of the planes are

$$lx + my + nz - (lx_1 + my_1 + nz_1) = 0$$

and

$$lx + my + nz - (lx_2 + my_2 + nz_2) = 0.$$

Therefore the required distance is (Theorem III, § 31):

$$\frac{|\, l(x_2 - x_1) + m(y_2 - y_1) + n(z_2 - z_1)\,|}{\sqrt{l^2 + m^2 + n^2}}.$$

This can also be expressed as the absolute value of

$$\frac{1}{\sin \theta} \begin{vmatrix} x_2 - x_1 & y_2 - y_1 & z_2 - z_1 \\ \lambda_1 & \mu_1 & \nu_1 \\ \lambda_2 & \mu_2 & \nu_2 \end{vmatrix}.$$

The distance between two parallel lines can be found by the method of the preceding section.

Again, the student is advised not to substitute mechanically in a formula, but to use the method of derivation.

EXAMPLE. Find the distance between the lines

$$\frac{x+5}{3} = \frac{y-3}{2} = \frac{z}{4} \quad \text{and} \quad \frac{x-1}{1} = \frac{y+2}{0} = \frac{z+3}{4}.$$

Solution. If $ax + by + cz + d = 0$ is a plane parallel to both lines, then $a:b:c = 4:-4:-1$. Therefore the parallel planes, each containing one of the given lines are

$$4x - 4y - z + 32 = 0 \quad \text{and} \quad 4x - 4y - z - 15 = 0.$$

The distance between these planes is $\dfrac{32 + 15}{\sqrt{33}} = \dfrac{47}{\sqrt{33}}$.

★ **46. Exercises.**

In Exercises **1-6,** find the distance between the given point and the given line.

1. $(6, 1, 8);\ \dfrac{x+3}{2} = \dfrac{y-4}{1} = \dfrac{z-2}{-2}.$
2. $(6, 3, -10);\ \dfrac{x-6}{4} = \dfrac{y+2}{3} = \dfrac{z+5}{-5}.$
3. $(7, 1, -6);\ \dfrac{x+5}{3} = \dfrac{y-6}{-4} = \dfrac{z+2}{1}.$
4. $(7, 4, 9);\ \dfrac{x+7}{6} = \dfrac{y-13}{-6} = \dfrac{z-1}{1}.$
5. $(1, -1, 2);\ \dfrac{x+1}{5} = \dfrac{y-1}{-2} = \dfrac{z-3}{4}.$
6. $(1, 1, 1);\ x + y + z = 0,\ 3x - 2y + 4z = 0.$

EXERCISES

In Exercises **7-12,** find the distance between the two skew lines.

7. $\dfrac{x-4}{3} = \dfrac{y-7}{1} = \dfrac{z-1}{-4}$; $\dfrac{x-8}{2} = \dfrac{y-1}{-1} = \dfrac{z-5}{-6}$.

8. $\dfrac{x-2}{0} = \dfrac{y-2}{2} = \dfrac{z-1}{1}$; $\dfrac{x-1}{3} = \dfrac{y-6}{-4} = \dfrac{z+2}{-1}$.

9. $\dfrac{x-1}{1} = \dfrac{y+2}{4} = \dfrac{z+3}{-1}$; $\dfrac{x+1}{1} = \dfrac{y-4}{0} = \dfrac{z+2}{-2}$.

10. $\dfrac{x-1}{1} = \dfrac{y-1}{1} = \dfrac{z-1}{1}$; $\dfrac{x+1}{1} = \dfrac{y+6}{-8} = \dfrac{z-2}{-2}$.

11. $2x + y - z = 0$, $x - y + 2z - 3 = 0$;
$x + 2y - 3z = 4$, $2x - 3y + 4z = 5$.

12. $\dfrac{x-1}{2} = \dfrac{y-2}{4} = \dfrac{z-3}{1}$; $y - 3x = z = 0$.

13. Find the distance between a vertex of a cube, with edges of length a, and a diagonal not passing through it.

14. Find the distance between a diagonal of a cube, with edges of length a, and an edge not meeting it.

15. Lines Λ_1 and Λ_2 are non-intersecting diagonals of adjacent faces of a cube with edges of length a. Find

(a) the angle between Λ_1 and Λ_2;
(b) the distance between Λ_1 and Λ_2.

In Exercises **16-19,** find the projection of the given point on the given line.

16. $(6, 9, -2)$; $\dfrac{x+10}{4} = \dfrac{y-8}{-1} = \dfrac{z-3}{-3}$.

17. $(1, 0, 0)$; $x = y = z$.

18. $(7, 4, -5)$; $x = -13 + 5t$, $y = 8 - 3t$, $z = -6 + 2t$.

19. $(2, 0, k)$; $x = y$, $z = 0$.

20. Write equations for the "projecting line" for each of Exercises **16-19**.

In Exercises **21** and **22,** find equations of the projection of the given line on the given plane.

* 21. $\dfrac{x}{0} = \dfrac{y}{1} = \dfrac{z}{1}$; $x = y$.

* 22. $x - 2y + 9 = 0$, $3x - 2z = 7$; $5x - y + 2z + 4 = 0$.

* 23. Find a point on each of the lines

$$\dfrac{x-1}{1} = \dfrac{y-2}{1} = \dfrac{z+6}{1} \quad \text{and} \quad \dfrac{x-1}{2} = \dfrac{y+2}{-3} = \dfrac{z-10}{1}$$

such that the distance between these points has the smallest value possible. What is this value?

CHAPTER III

DETERMINANTS AND MATRICES

47. Introduction. Problems of analytic geometry frequently involve solutions of systems of linear equations. For example, the point of intersection of three planes in space can be found by solving simultaneously three linear equations in three variables. In order to treat these questions with some degree of completeness we first recall some fundamental properties of determinants, and then introduce the notion of *matrix*.†

48. Some properties of determinants.

I. *The value of a determinant is not changed if the rows and columns are interchanged.*

II. *If the elements of one row (or column) of a determinant are all zero, the value of the determinant is zero.*

III. *If the elements of one row (or column) of a determinant are multiplied by the same constant factor, the value of the determinant is multiplied by this factor.*

IV. *If one determinant is obtained from another by interchanging any two rows (or columns), the value of either is the negative of the value of the other.*

V. *If two rows (or columns) of a determinant are identical, the value of the determinant is zero.*

VI. *If two determinants are identical except for one row (or column), the sum of their values is given by a single determinant obtained by adding corresponding elements of the dissimilar rows (or columns) and leaving unchanged the remaining elements.*

VII. *The value of a determinant is not changed if to the elements of any row (or column) are added a constant multiple of the corresponding elements of any other row (or column).*

---

† For a more complete treatment see Maxime Bôcher, *Introduction to Higher Algebra* (The Macmillan Company, 1935), or Garrett Birkhoff and Saunders MacLane, *A Survey of Modern Algebra* (The Macmillan Company, 1944).

VIII. *The value of a determinant is equal to the sum of the products obtained by multiplying each element of any row (or column) by its cofactor.*

IX. *If each element of any row (or column) of a determinant is multiplied by the cofactor of the corresponding element of a different row (or column), the sum of these products is zero.*

49. Matrix. Rank of a matrix.

DEFINITION I. *A rectangular array of mn quantities, arranged in m rows and n columns,*

$$(1) \quad \begin{pmatrix} a_{11} & a_{12} & \cdots & a_{1n} \\ a_{21} & a_{22} & \cdots & a_{2n} \\ \cdot & \cdot & \cdot & \cdot \\ \cdot & \cdot & \cdot & \cdot \\ a_{m1} & a_{m2} & \cdots & a_{mn} \end{pmatrix},$$

*is called a **matrix**.* These mn quantities are called the **elements** of the matrix. If $m = n$, the matrix is said to be **square** and of **order** n.

The element of a matrix that is in the ith row and jth column, where i may have any value from 1 to m and j may have any value from 1 to n, is called the **general element** of the matrix. Either notation (a_{ij}) or A is often used to represent the matrix (1) whose general element is a_{ij}.

A matrix, even if it is square, does not have a numerical value, as a determinant does. However, if a matrix is square, a determinant can be formed which has the same elements as the matrix. This will be called *the determinant of the matrix.* Other determinants which can be formed from elements of a matrix play an important rôle in the theory of systems of linear equations. Such a determinant is formed by choosing a certain number of rows and an equal number of columns. If every element of the matrix that is *not* in one of these rows *and* in one of these columns is removed, the remaining elements define what is called *a determinant of the matrix.* Any matrix consisting of more than one element will therefore have more than one determinant. Some of these determinants may vanish (that is, have the value zero) and some may not. Consideration of these determinants leads to the definition:

DEFINITION II. *A matrix has **rank** r if and only if it has a non-vanishing determinant of order r and no non-vanishing determinant of order greater than r.*

An equivalent definition of rank of a matrix is obtained by replacing in the definition the phrase "of order greater than r" by "of order $r + 1$." Proof, by mathematical induction, of this equivalence would involve showing that if all determinants of order k vanish, then all determinants of order $k + 1$ vanish. This last fact is an immediate consequence of the eighth property of determinants given in the preceding section. (Complete the proof.)

If the determinant of a square matrix is zero (that is, if the rank is less than the order) the matrix is called **singular**; otherwise it is called **non-singular**.

The concept of rank is an important one, but if one were forced to determine rank directly by means of the definition the task would in some cases be enormous. For example, to show that the rank of a 4 by 5 matrix is 2 it would be necessary to evaluate forty third order determinants.

50. Elementary transformations of a matrix. Fortunately, the determination of the rank of a matrix is facilitated by the following **elementary transformations** of a matrix:

(*i*) *Interchanging any two rows (or columns).*

(*ii*) *Multiplying the elements of any row (or column) by the same non-zero constant.*

(*iii*) *Adding to the elements of any row (or column) the corresponding elements of any other row (or column).*

For the sake of brevity we shall find it convenient, when referring to elementary transformations of types (*ii*) or (*iii*), to speak simply of multiplying a row (or column) by a non-zero constant and adding one row (or column) to another.

DEFINITION. *A matrix A is **equivalent** to a matrix B, written $A \sim B$, if and only if A can be transformed into B by means of a finite sequence of elementary transformations.*

The fundamental fact that is of immediate importance is expressed by the theorem:

THEOREM. *Two matrices with the same number of rows and the same number of columns are equivalent if and only if they have the same rank.*

Proof. Only an outline of the proof, which consists of two parts, will be given. For the first part we wish to show that any elementary transformation leaves the rank unchanged. Equivalently, since any elementary transformation can be reversed by means of elementary transformations, it is sufficient to show that the rank cannot be increased. Only the third elementary transformation presents any difficulty. Let A be a matrix of rank r, and let B be a matrix obtained from A by an elementary transformation of the third type. Any determinant of order $r + 1$ contained in B either is a determinant of A (by the seventh property of determinants, § **48**), or can be represented as a determinant of A plus (or minus) another determinant of A (by the sixth property of determinants, § **48**). Since every $(r + 1)$th order determinant of A vanishes, any $(r + 1)$th order determinant of B also vanishes, and the rank of B is less than or equal to r. Equivalent matrices therefore have the same rank.

To prove that two matrices that have the same number of rows, the same number of columns, and the same rank are equivalent, one shows that both can be reduced by a chain of elementary transformations to the same standard form.

This procedure will be exemplified by the simplification of the particular matrix:

$$\begin{pmatrix} 5 & 17 & 8 & 26 & 1 \\ 2 & 9 & 1 & 17 & -4 \\ 3 & 9 & 6 & 12 & 3 \\ 1 & 4 & 1 & 7 & -1 \end{pmatrix}.$$

If we subtract twice the fourth row from the second row and if we subtract the third row and twice the fourth row from the first row, the result is the matrix

$$\begin{pmatrix} 0 & 0 & 0 & 0 & 0 \\ 0 & 1 & -1 & 3 & -2 \\ 3 & 9 & 6 & 12 & 3 \\ 1 & 4 & 1 & 7 & -1 \end{pmatrix}.$$

Dividing the third row by 3 and subtracting the resulting row and the second row from the fourth row yield the matrix

$$\begin{pmatrix} 0 & 0 & 0 & 0 & 0 \\ 0 & 1 & -1 & 3 & -2 \\ 1 & 3 & 2 & 4 & 1 \\ 0 & 0 & 0 & 0 & 0 \end{pmatrix}.$$

Similar operations on the columns and interchanging the first and third rows complete the reduction of the original matrix to the following standard form:

$$\begin{pmatrix} 1 & 0 & 0 & 0 & 0 \\ 0 & 1 & 0 & 0 & 0 \\ 0 & 0 & 0 & 0 & 0 \\ 0 & 0 & 0 & 0 & 0 \end{pmatrix}.$$

The rank is obviously 2.

51. Exercises.

In Exercises **1-5**, evaluate the determinant.

1. $\begin{vmatrix} 3 & -2 & 0 \\ 4 & 1 & 5 \\ -1 & 0 & 6 \end{vmatrix}$.
2. $\begin{vmatrix} 10 & 20 & 30 \\ 5 & 15 & 25 \\ 6 & 16 & 26 \end{vmatrix}$.
3. $\begin{vmatrix} 5 & 2 & 3 & -1 \\ -4 & 1 & 1 & 2 \\ 3 & -1 & -2 & 1 \\ 1 & 6 & 7 & -5 \end{vmatrix}$.

4. $\begin{vmatrix} 0 & 1 & 1 & 1 \\ 1 & 0 & 2 & 3 \\ 1 & 2 & 0 & 4 \\ 1 & 3 & 4 & 0 \end{vmatrix}$.
5. $\begin{vmatrix} 1 & 1 & 1 & 1 & 1 \\ 1 & 2 & 2 & 2 & 2 \\ 1 & 2 & 3 & 3 & 3 \\ 1 & 2 & 3 & 4 & 4 \\ 1 & 2 & 3 & 4 & 5 \end{vmatrix}$.

In Exercises **6-9**, determine the rank of the matrix.

6. $\begin{pmatrix} 3 & -2 & 7 \\ -15 & 10 & -35 \\ 6 & -4 & 14 \end{pmatrix}$.
7. $\begin{pmatrix} 1 & 2 & -4 & 3 & 5 \\ 1 & 1 & -9 & 2 & 3 \\ 2 & 4 & -7 & 13 & 5 \end{pmatrix}$.

8. $\begin{pmatrix} 4 & 7 & -2 \\ -3 & 1 & -11 \\ 0 & -2 & 4 \\ 8 & 5 & 14 \end{pmatrix}$.
9. $\begin{pmatrix} 2 & -3 & 4 & 1 & -11 \\ -1 & 4 & 2 & 5 & 6 \\ 3 & -2 & 10 & 7 & -16 \\ 1 & 6 & 14 & 17 & -4 \end{pmatrix}$.

* **10.** Prove that the relation $A \sim B$ as defined in § **50**, is an equivalence relation. (See Ex. 36, § 13.)

52. Simultaneous linear equations. Before investigating systems of planes by means of the corresponding systems of linear equations, we shall state the general problem of linear equations and prove two basic theorems.

A system of m linear equations in n variables or unknowns has the form

(1)
$$\begin{aligned} a_{11}x_1 + a_{12}x_2 + \cdots + a_{1n}x_n + k_1 &= 0, \\ a_{21}x_1 + a_{22}x_2 + \cdots + a_{2n}x_n + k_2 &= 0, \\ &\cdots \\ a_{m1}x_1 + a_{m2}x_2 + \cdots + a_{mn}x_n + k_m &= 0. \end{aligned}$$

A set of numbers $x_1, x_2, \ldots, x_n$ that satisfies all m equations is called a **solution**. Two systems of linear equations will be called **equivalent** if and only if they have the same solutions. A system of equations that has at least one solution is said to be **consistent**; otherwise it is called **inconsistent**. If $k_1 = \ldots = k_m = 0$, the system (1) and its equations are called **homogeneous**. Obviously any homogeneous system is consistent, since it possesses at least the so-called **trivial** solution

$$x_1 = x_2 = \ldots = x_n = 0.$$

Any solution of a system of homogeneous linear equations in which at least one of the variables has a value different from zero is called a **non-trivial** solution.

If $m = n$ and the determinant of the coefficients is different from zero, Cramer's Rule states that the system (1) is consistent, guarantees that there is only one solution, and gives a formulation for this solution. We state Cramer's Rule, assuming for simplicity that the constants appear on the right sides of the equations instead of on the left.

CRAMER'S RULE. *If the determinant of the coefficients of a system of n linear equations in n variables is different from zero, the system has one and only one solution. In this solution the value of each variable is equal to the quotient of two determinants; the denominator is the determinant of the coefficients, and the numerator is obtained from it by replacing the column of coefficients of the variable in question by the column of constants appearing as the right members in the system of equations.*

Associated with the system (1) are two matrices, the **coefficient matrix** A and the **augmented matrix** K:

$$A = \begin{pmatrix} a_{11} & \cdots & a_{1n} \\ \cdot & \cdot & \cdot \\ a_{m1} & \cdots & a_{mn} \end{pmatrix}, \quad K = \begin{pmatrix} a_{11} & \cdots & a_{1n} & k_1 \\ \cdot & \cdot & \cdot & \cdot \\ a_{m1} & \cdots & a_{mn} & k_m \end{pmatrix}.$$

Let the ranks of A and K be denoted by r and R, respectively. Then either $R = r$ or $R = r + 1$. (If there were a non-vanishing determinant of K of order greater than $r + 1$, it would necessarily contain a column of k's, and since the cofactor of each of these k's would vanish, expansion of this determinant with respect to the k's would lead to a contradiction.)

The fundamental theorem on consistency of a system of linear equations is the following:

THEOREM I. *A system of linear equations is consistent if and only if the ranks of the coefficient matrix and the augmented matrix are equal.*

Proof. We shall limit the proof to the special case in which we are particularly interested, a system of m linear equations in three variables, or unknowns,

x, y, and z. The same method will establish the theorem in general. We shall assume, furthermore, that each variable actually occurs in at least one equation. Otherwise the system is essentially one involving only two variables, or possibly only one. In either case the problem is simplified and can be resolved in a similar manner.

The general plan will be to write down a sequence of successively simpler systems, each of which is equivalent to the preceding, with the goal of obtaining a system so simple that the truth of the theorem for that system will be obvious. Simultaneously, as we change from one system of equations to the next, we shall observe the effect of the change upon the coefficient and augmented matrices, and notice in particular that the transformations of the equations have their counterparts in elementary transformations of the coefficient and augmented matrices. As we write each system of equations, we shall also write down after it the augmented matrix (and therefore the elements of the coefficient matrix also).

Accordingly, let the original system and its augmented matrix be

$$(2) \quad \begin{matrix} a_1x + b_1y + c_1z + d_1 = 0, \\ \cdots\cdots\cdots\cdots \\ a_mx + b_my + c_mz + d_m = 0, \end{matrix} \quad \text{and} \quad \begin{pmatrix} a_1 & b_1 & c_1 & d_1 \\ \cdot & \cdot & \cdot & \cdot \\ a_m & b_m & c_m & d_m \end{pmatrix}.$$

Let the equations be arranged so that the coefficient of x in the first equation is not zero (this corresponds to rearranging rows of the coefficient and augmented matrices). We can now eliminate x from all but the first equation. If we "multiply" the first equation by a_2/a_1 and "subtract" the "product" from the second equation, multiply the first by a_3/a_1 and subtract the product from the third, and so on, and finally "divide" the first equation by a_1, we reduce the system of equations to an equivalent system of the form

$$(3) \quad \begin{matrix} x + e_1y + f_1z + g_1 = 0, \\ e_2y + f_2z + g_2 = 0, \\ \cdots\cdots\cdots\cdots \\ e_my + f_mz + g_m = 0. \end{matrix} \quad \begin{pmatrix} 1 & e_1 & f_1 & g_1 \\ 0 & e_2 & f_2 & g_2 \\ \cdot & \cdot & \cdot & \cdot \\ 0 & e_m & f_m & g_m \end{pmatrix}$$

The coefficient and augmented matrices of system (3) have been obtained from those of system (2) by elementary transformations on the rows, and therefore have the ranks r and R, respectively.

If $e_2 = \ldots = e_m = f_2 = \ldots = f_m = 0$, the equations (3) clearly are satisfied and the system is consistent if and only if $g_2 = \ldots = g_m = 0$; that is, if and only if $r = R = 1$. If, on the other hand, one of these numbers, e_2 to e_m or f_2 to f_m, is different from zero, we can assume by rearranging the equations (rows of the coefficient and augmented matrices), and possibly by interchanging the order of y and z (interchanging the second and third columns of the coefficient and augmented matrices), that e_2 is different from zero. Again by transformations of the same type as those used in reducing system (2) to system (3), we can reduce the latter system (and simultaneously the coefficient and augmented matrices) to the form

$$(4) \quad \begin{matrix} x + h_1z + i_1 = 0, \\ y + h_2z + i_2 = 0, \\ h_3z + i_3 = 0, \\ \cdots\cdots\cdots \\ h_mz + i_m = 0. \end{matrix} \quad \begin{pmatrix} 1 & 0 & h_1 & i_1 \\ 0 & 1 & h_2 & i_2 \\ 0 & 0 & h_3 & i_3 \\ \cdot & \cdot & \cdot & \cdot \\ 0 & 0 & h_m & i_m \end{pmatrix}$$

§ 52] LINEAR EQUATIONS 63

If, now, $h_3 = \ldots = h_m = 0$, the system (4) is obviously consistent if and only if $i_3 = \ldots = i_m = 0$; that is, if and only if $r = R = 2$. On the other hand, if these numbers, h_3 to h_m, are not all zero, we can proceed as before and obtain the system

(5)
$$\begin{aligned} x + k_1 &= 0, \\ y + k_2 &= 0, \\ z + k_3 &= 0, \\ &\cdots \\ k_m &= 0. \end{aligned} \qquad \begin{pmatrix} 1 & 0 & 0 & k_1 \\ 0 & 1 & 0 & k_2 \\ 0 & 0 & 1 & k_3 \\ \cdot & \cdot & \cdot & \cdot \\ 0 & 0 & 0 & k_m \end{pmatrix}$$

In this final case the system is consistent if and only if $k_4 = \ldots = k_m = 0$; that is, if and only if $r = R = 3$. This completes the proof.

The method of the proof just completed indicates that any (consistent) system of linear equations can be solved by applying elementary transformations to the *rows* of the augmented matrix. This process is illustrated in the example given below. If one is interested only in the ranks of the coefficient and augmented matrices of a given system of linear equations, and not in the solutions, these two ranks can be obtained simultaneously with comparative ease by a sequence of elementary transformations applied to both rows and columns of the augmented matrix if the following two transformations are excluded: (*i*) interchanging the last column and any other; (*ii*) adding the last column to any other.

If a system of m linear equations in n variables is consistent and if $r < n$, then we can solve for r of the variables in terms of the remaining $n - r$, which can be given arbitrary values. This can be shown by ordering the equations and the variables in such a way that the determinant made up of elements in the first r rows and the first r columns of the coefficient matrix is different from zero. Then the original system is equivalent to the system consisting of only these first r equations, since elementary transformations of the type used in the proof of Theorem I, applied to rows of the augmented matrix will reduce this matrix to one in which all elements except those in the first r rows are zero. Cramer's Rule finally permits us to solve for the first r variables in terms of the remaining $n - r$. It is now readily seen that a system of linear equations has either (*i*) infinitely many solutions, or (*ii*) one solution, or (*iii*) no solution.†

We conclude with a theorem that is implied by the remarks of the preceding paragraph.

---

† Since zero is only one of infinitely many numbers, the vanishing of a determinant can be thought of as an exceptional occurrence and the non-vanishing of a determinant as customary and expected. In this sense, in the absence of other information a system of m linear equations in n variables can be expected to be consistent with infinitely many solutions if $m < n$, consistent with one solution if $m = n$, and inconsistent if $m > n$.

64   DETERMINANTS AND MATRICES   [Ch. III

Theorem II. (*i*) *A system of homogeneous linear equations has non-trivial solutions if and only if the rank of the coefficient matrix is less than the number of variables.* (*ii*) *A system of n homogeneous linear equations in n variables has non-trivial solutions if and only if the determinant of its coefficients vanishes.*

Example. Find the general solution of the system of equations

$$5x + 10y + 2z - 2u + 3v - 7 = 0,$$
$$10x + 20y - 2z + 5u + 3v - 2 = 0,$$
$$5x + 10y - 10z + 16u - 3v + 17 = 0.$$

Solution. Elementary transformations on the rows of the augmented matrix can be used to simplify the given system of equations. For example, if we add three times the first row to the third and subtract twice the second row from the third, the third row is reduced to a row of zeros. Subtracting twice the first row from the second, dividing the new second row by 3, and adding the resulting second row to the first, yield the matrix

$$\begin{pmatrix} 5 & 10 & 0 & 1 & 2 & -3 \\ 0 & 0 & -2 & 3 & -1 & 4 \\ 0 & 0 & 0 & 0 & 0 & 0 \end{pmatrix}.$$

It is now evident that the ranks of the coefficient and augmented matrices are 2 and that the given system is consistent, with solutions given by the simplified system

$$5x + 10y + u + 2v - 3 = 0,$$
$$-2z + 3u - v + 4 = 0.$$

We should be able to solve this system for two of the variables in terms of the others. These two variables cannot be x and y, since their coefficient matrix is singular. However, we can solve for x and z, and express the general solution in the form

$$x = \frac{-10y - u - 2v + 3}{5}, \qquad z = \frac{3u - v + 4}{2},$$

where y, u, and v can be assigned arbitrary values.

53. Exercises.

In Exercises **1-4**, determine whether the system is consistent or not. If it is consistent, obtain the general solution.

1. $x + 3y + 2z + 2 = 0,$
   $x + 8y + 7z + 2 = 0,$
   $2x + y - z + 4 = 0.$
2. $x + 4y + 3z - 8w = 3,$
   $x + 2y - 3w = 2,$
   $x + 4y + 6z - 12w = 5.$
3. $2x - 3y + z + 3 = 0,$
   $x + y + 3z + 4 = 0,$
   $x - y + z - 8 = 0.$
4. $2x + y - 4z + 3w = 7,$
   $x + 2y - 5z + 3w = 8,$
   $x + y - 3z + 2w = 5.$

* **5.** A definition of what is meant by saying that two systems of linear equations are equivalent is given in § **52**. Prove that this relation is an equivalence relation. (See Ex. **36**, § **13**.)

§ 54] SYSTEMS OF PLANES 65

* **6.** Let a matrix A be defined to be **row-equivalent** to a matrix B if and only if A can be transformed into B by means of a finite sequence of elementary transformations applied to rows only. (a) Prove that this relation of *row-equivalence* is an equivalence relation. (b) Prove that two consistent systems of m linear equations in n variables are equivalent if and only if their augmented matrices are row-equivalent. (See Ex. **36**, § **13**.)

54. Systems of planes. The theory of systems of linear equations is immediately applicable to systems of planes. We shall assume the equation of any plane considered in this section to be in standard form, and, as in § 52, denote the ranks of the coefficient and augmented matrices of any system of linear equations by r and R, respectively.

The simplest case is that of two planes:

(1) $$\begin{aligned} a_1x + b_1y + c_1z + d_1 &= 0, \\ a_2x + b_2y + c_2z + d_2 &= 0. \end{aligned}$$

If these planes are parallel, the coefficients of x, y, and z in their equations are proportional, and $r = 1$. In this case, the system (1) is obviously consistent or inconsistent according as the planes are identical or distinct. Consequently, if the planes are identical, $r = R = 1$, and if the planes are parallel and distinct, $r = 1$ and $R = 2$. If the planes (1) are not parallel, the coefficients are not proportional and hence form sets of direction numbers of non-parallel lines. Therefore, by the Theorem of § **23**, the three second order determinants of the coefficient matrix are not all zero, and $r = R = 2$. We thus see that the relationship of the planes (1) is completely determined by the values of r and R.

Consider now a system of three planes:

(2) $$\begin{aligned} a_1x + b_1y + c_1z + d_1 &= 0, \\ a_2x + b_2y + c_2z + d_2 &= 0, \\ a_3x + b_3y + c_3z + d_3 &= 0. \end{aligned}$$

If these planes are parallel, then $r = 1$, and $R = 1$ or 2 according as the planes are identical or distinct. Conversely, if $r = 1$, the planes are parallel. (These results follow directly from the preceding paragraph.) If the planes have exactly one line in common, then the system (2) is consistent and $r = R$. In fact, $r = R = 2$, since, by Cramer's Rule, if r were equal to 3 the system (2) would have exactly one solution, and the planes would have exactly one point in common. Suppose now that two of the planes intersect in a line that is parallel to but not contained in the third plane. Then the system (2) is inconsistent, and $R = r + 1$. Since the planes are not all parallel, r cannot be equal to 1, and therefore $r = 2$ and $R = 3$. Finally, we must show that if the planes have exactly one point in common, then $r = R = 3$. Since the various methods of establishing this fact are good illustrations of the ideas of this chapter, we restate what remains to be proved as a lemma, and give three proofs.

LEMMA. *If the planes* (2) *have exactly one point in common, then* $r = R = 3$.

First proof. Since the system (2) is assumed to be consistent, $r = R$. A set of direction numbers for the line of intersection of the second and third planes is

$$\begin{vmatrix} b_2 & c_2 \\ b_3 & c_3 \end{vmatrix}, \quad \begin{vmatrix} c_2 & a_2 \\ c_3 & a_3 \end{vmatrix}, \quad \begin{vmatrix} a_2 & b_2 \\ a_3 & b_3 \end{vmatrix}.$$

From the assumption of the lemma, we know that the normals to the first plane are not perpendicular to this line of intersection, and therefore that

$$a_1 \begin{vmatrix} b_2 & c_2 \\ b_3 & c_3 \end{vmatrix} + b_1 \begin{vmatrix} c_2 & a_2 \\ c_3 & a_3 \end{vmatrix} + c_1 \begin{vmatrix} a_2 & b_2 \\ a_3 & b_3 \end{vmatrix}$$

is different from zero. But this expression is the expansion of the coefficient determinant of system (2) with respect to the elements of the first row. Therefore $r = R = 3$.

Second proof. As in the first proof, we see that $r = R$. Assume that $r = R = 2$. By the arguments used in establishing Theorem II, § 52, we see that elementary transformations applied to the rows of the augmented matrix will eliminate one equation completely. Therefore system (2) is equivalent to one consisting of two linear equations. But two planes cannot have exactly one point in common. This fact contradicts the assumption that $r = R = 2$. Therefore $r = R = 3$.

Third proof. Assume that $r = R = 2$, and consider the system of homogeneous linear equations

(3)
$$\begin{aligned} k_1 a_1 + k_2 a_2 + k_3 a_3 &= 0, \\ k_1 b_1 + k_2 b_2 + k_3 b_3 &= 0, \\ k_1 c_1 + k_2 c_2 + k_3 c_3 &= 0, \\ k_1 d_1 + k_2 d_2 + k_3 d_3 &= 0, \end{aligned}$$

in the unknowns k_1, k_2, and k_3. The coefficient matrix of this system is obtained by interchanging the rows and columns of the augmented matrix of system (2), and therefore has the rank 2. Hence, by Theorem II, § **52**, system (3) has nontrivial solutions. Assuming that $k_3 \neq 0$, we can solve these equations to obtain

$$\begin{aligned} a_3 &= k_1' a_1 + k_2' a_2, \\ b_3 &= k_1' b_1 + k_2' b_2, \\ c_3 &= k_1' c_1 + k_2' c_2, \\ d_3 &= k_1' d_1 + k_2' d_2. \end{aligned}$$

But this means that the third plane belongs to the pencil of planes defined by the first two (see § **36**). This provides the desired contradiction of the assumption that $r = R = 2$.

As in the case of a system of two planes, we see that the general relationship of three planes is determined by the values of r and R.

We now catalog a classification of systems of planes according to the ranks r and R. We shall do this for systems of two, three and m planes, where $m > 3$. The student should supply the verification for the case of more than three planes.

Classification of Systems of Planes
Two Planes

$r = 2, R = 2$. The planes intersect in a line.
$r = 1, R = 2$. The planes are parallel and distinct.
$r = 1, R = 1$. The planes coincide.

Three Planes

$r = 3, R = 3$. The planes intersect in just one point.
$r = 2, R = 3$. The planes have no point in common. One plane is parallel to the line of intersection of the other two. Two of the planes may be parallel.
$r = 2, R = 2$. The planes have a line in common, but are not coincident. Two of the planes may coincide.
$r = 1, R = 2$. The planes are parallel and not coincident. Two of the planes may coincide.
$r = 1, R = 1$. The planes coincide.

More than Three Planes

$r = 3, R = 4$. The planes have no point in common. Three of the planes meet in a point.
$r = 3, R = 3$. The planes have just one point in common.
$r = 2, R = 3$. Two of the planes meet in a line parallel to all of the remaining planes, but not lying in all of them.
$r = 2, R = 2$. The planes have a line in common.
$r = 1, R = 2$. The planes are parallel and not coincident.
$r = 1, R = 1$. The planes coincide.

In the sense of the footnote on page 63, in the absence of other information one can expect two planes to have just one line in common, three planes to intersect in just one point, and four or more planes to have no point in common.

EXAMPLE. Determine the relationship between the three planes:
$$2x - y + 8z - 3 = 0,$$
$$x - 3y - z + 1 = 0,$$
$$2x - 3y + 4z + 7 = 0.$$

Solution. The ranks of the coefficient and augmented matrices are 2 and 3, respectively. The planes form a "prism." Any two intersect in a line that is parallel to but not contained in the third plane.

55. Exercises.

In Exercises **1-8**, determine the relationship of the planes. Find the ranks r and R. If the planes have just one point in common, find its coördinates. If the planes have a line in common, write its equations in symmetric form.

1. $x + y = 2,$
   $y + z = 2,$
   $x + z = 2.$

2. $x - y = 2,$
   $y - z = 2,$
   $x - z = 2.$

3. $x - y = 1,$
   $y - z = 1,$
   $x - z = 2.$

4. $x + y - 2z = -1,$
   $2x - y + z = 4,$
   $4x + y - 3z = 5.$

5. $x + y - 2z = -1,$
   $2x - y + z = 4,$
   $x + 4y - 7z = -7.$

6. $x + y - 2z = -1,$
   $2x - y + z = 4,$
   $x - z = 1.$

7. $x - y + z - 1 = 0,$
   $x + y + 2z + 2 = 0,$
   $3x + y + 5z + 3 = 0,$
   $4x - 2y + 5z - 1 = 0.$

8. $x + y = 1,$
   $x + z = 2,$
   $y + z = 3,$
   $x + y + z = 4.$

* **9.** Carry through an analysis, similar to that of § **54,** for systems of lines in plane analytic geometry.

* **56. Linear dependence.** In § **50** we spoke casually of adding rows (or columns) of a matrix, and of multiplying a row (or column) by a constant. In this section we shall consider these and similar concepts on a more formal basis.

DEFINITION. *The **sum** of two rows*

$$a_1 \quad b_1 \cdots r_1$$

and

$$a_2 \quad b_2 \cdots r_2$$

of a matrix is the sequence of numbers

$$a_1 + a_2 \quad b_1 + b_2 \cdots r_1 + r_2,$$

obtained by adding corresponding elements of the two given rows. The **product** *of a constant k and a row*

$$a \quad b \cdots r$$

of a matrix is the sequence of numbers

$$ka \quad kb \cdots kr,$$

obtained by multiplying each element of the given row by the constant k.

§ 56]   LINEAR DEPENDENCE   69

*A **linear combination** of a collection of rows,*

$$a_1 \quad b_1 \quad \cdots \quad r_1,$$
$$\cdot \quad \cdot \quad \cdot \quad \cdot \quad \cdot \quad \cdot$$
$$a_p \quad b_p \quad \cdots \quad r_p,$$

of a matrix is a sum of multiples of these rows,

$$(k_1 a_1 + \cdots + k_p a_p) \quad \cdots \quad (k_1 r_1 + \cdots + k_p r_p).$$

*The rows of a matrix are **linearly dependent** if and only if some linear combination of them, not all of the constant factors being zero, consists entirely of zeros. Otherwise they are **linearly independent**.*

A similar definition applies to the columns of a matrix.

Consider the matrix $A = (a_{ij})$ with m rows and n columns. A necessary and sufficient condition that the rows of A be linearly dependent is that there exist m constant factors, $k_1, \cdots, k_m$, not all zero, such that the sum of the products of these constants and the corresponding rows consists entirely of zeros; that is, that the system of homogeneous linear equations

(1)
$$a_{11}k_1 + \cdots + a_{m1}k_m = 0,$$
$$\cdots \cdots \cdots \cdots \cdots$$
$$a_{1n}k_1 + \cdots + a_{mn}k_m = 0$$

have non-trivial solutions. (Notice that the coefficient matrix of this system is obtained by interchanging the rows and columns of A. Compare the third proof of the Lemma of § **54**.) Combining the first part of Theorem II, § **52**, and these remarks, we have established the first part of the theorem:

THEOREM. *Each of the following is a necessary and sufficient condition for the rows (or columns) of a matrix to be linearly dependent:*
(i) the rank of the matrix is less than the number of rows (or columns);
(ii) some row (or column) is a linear combination of the others.

The second part of this theorem can be proved with the aid of system (1). For example, if the rows of A are linearly dependent, equations (1) have a non-trivial solution where k_i, say, is different from zero. Hence we can solve for $a_{i1}, a_{i2}, \cdots, a_{in}$ in a form that represents the ith row as a linear combination of the others. (Complete the proof.)

Consider now a system of two homogeneous linear equations in three variables

(2)
$$a_1 x + b_1 y + c_1 z = 0,$$
$$a_2 x + b_2 y + c_2 z = 0,$$

neither row of whose coefficient matrix consists entirely of zeros. As a result of the Theorem of this section it is apparent that the statement that the coefficients of the two equations of system (2) are not proportional is equivalent to the statement that the rows of the coefficient matrix are linearly independent, and that this is equivalent to the statement that the rank of this matrix is equal to 2. Therefore the solution given in § 23 is justified on purely algebraic grounds. The general solution of the system (2), with rank equal to 2, is

$$x = k\begin{vmatrix} b_1 & c_1 \\ b_2 & c_2 \end{vmatrix}, \quad y = k\begin{vmatrix} c_1 & a_1 \\ c_2 & a_2 \end{vmatrix}, \quad z = k\begin{vmatrix} a_1 & b_1 \\ a_2 & b_2 \end{vmatrix},$$

where k is an arbitrary number.

EXAMPLE. Express the first row of the matrix

$$\begin{pmatrix} 3 & 2 & -1 & 5 \\ 5 & 1 & 4 & -2 \\ 1 & -4 & 11 & -19 \end{pmatrix}$$

as a linear combination of the other two.

SOLUTION. Consider the system of homogeneous linear equations

$$3a + 5b + c = 0,$$
$$2a + b - 4c = 0,$$
$$-a + 4b + 11c = 0,$$
$$5a - 2b - 19c = 0.$$

Since the rank of the coefficient matrix is 2 and since no two rows are proportional, the general solution is that of any two equations, say the first two. Therefore the non-trivial solutions are given by the proportion $a : b : c = 3 : -2 : 1$. That is, for the original matrix, 3 times the first row minus twice the second row plus the third row is a row of zeros. Therefore the first row is two-thirds of the second row minus one-third of the third row.

* **57. Exercises.**

In Exercises **1-3**, determine the rank of the given matrix, and express the first row as a linear combination of the others.

1. $\begin{pmatrix} 6 & 15 & -3 & 18 \\ 10 & 25 & -5 & 30 \end{pmatrix}.$

2. $\begin{pmatrix} 2 & 3 & 3 \\ 1 & 2 & 4 \\ 1 & 1 & 1 \\ 2 & 1 & -2 \end{pmatrix}.$

3. $\begin{pmatrix} 9 & -7 & 1 & 11 \\ 3 & 2 & 1 & 2 \\ 1 & 5 & 1 & -1 \end{pmatrix}.$

4. Express the third plane of Exercise **3**, § **55**, as a linear combination of the other two. (See § **36**.)

5. Express the third plane of Exercise **5**, § **55**, as a linear combination of the other two. (See § **36**.)

6. Express the third plane and the fourth plane of Exercise **7**, § **55**, as a linear combination of the first two. (See § **36**.)

* **7.** Determine the rank of the following matrix, and express the first row

§ 58] FAMILIES OF POINTS 71

as a linear combination of the second and third; of the third and fourth; of the second, third, and fourth:

$$\begin{pmatrix} 1 & 1 & 5 \\ 2 & -1 & -2 \\ 1 & -1 & -3 \\ 1 & 0 & 1 \end{pmatrix}.$$

8. Prove that the rows (or columns) of a square matrix are dependent if and only if the matrix is singular.

* 58. Families of points. A set of points can be subjected to an analysis similar to that used for a system of planes. With each finite collection of points $P_1(x_1, y_1, z_1), \cdots, P_m(x_m, y_m, z_m)$ will be associated the matrix

(1) $$\begin{pmatrix} x_1 & y_1 & z_1 & 1 \\ \cdots & \cdots & \cdots & \cdots \\ x_m & y_m & z_m & 1 \end{pmatrix}.$$

We shall consider in turn the possible values of the rank r of this matrix.

If $r = 1$, any two rows are proportional, and since the last column consists entirely of 1's, the constant of proportionality is 1. Therefore all of the points are identical. Conversely, if all of the points are identical, $r = 1$.

If $r = 2$, there are two rows that are not proportional (assume they are the first two), and any other row (say the kth) is a linear combination of the first two. That is, the four equations

(2) $$\begin{aligned} x_k &= t_1 x_1 + t_2 x_2, \\ y_k &= t_1 y_1 + t_2 y_2, \\ z_k &= t_1 z_1 + t_2 z_2, \\ 1 &= t_1 + t_2 \end{aligned}$$

are simultaneously satisfied. As proved in § 41, the point P_k is on the line through P_1 and P_2. Therefore all of the points lie on a line, but are not all identical. Conversely, if all of the points lie on a line determined by any two, the coördinates of any point of the set can be represented in the form (2). This means that all but two rows of the matrix (1) can be reduced to zeros by elementary row operations, and that therefore the rank is 2.

In a similar manner, the rank can be shown to be 3 if and only if the coördinates of any point of the family can be represented in terms of three of the points by a set of equations of the form

$$\begin{aligned} x_k &= t_1 x_1 + t_2 x_2 + t_3 x_3, \\ y_k &= t_1 y_1 + t_2 y_2 + t_3 y_3, \\ z_k &= t_1 z_1 + t_2 z_2 + t_3 z_3, \\ 1 &= t_1 + t_2 + t_3, \end{aligned} \tag{3}$$

where the three points P_1, P_2, and P_3 do not lie on a line. In other words, the rank is 3 if and only if the points have a plane but no line in common.

The following table summarizes these results, and includes the case $r = 4$:

$r = 1$. *The points are identical.*
$r = 2$. *The points are collinear but not identical.*
$r = 3$. *The points are coplanar but not collinear.*
$r = 4$. *The points are not coplanar.*

* 59. Exercises.

In Exercises **1-6**, determine whether the given points are collinear, coplanar but not collinear, or non-coplanar. If they are collinear, find equations of the line containing them. If they are coplanar but not collinear, find the equation of the plane containing them.

1. $(5, -1, -2)$, $(9, -11, 0)$, $(1, 9, -4)$, $(-1, 14, -5)$.
2. $(2, 0, 0)$, $(0, 2, 0)$, $(0, 0, 2)$, $(1, 1, 1)$.
3. $(1, 1, 1)$, $(3, 8, -1)$, $(2, 0, 0)$, $(6, 5, -4)$.
4. $(0, 1, 0)$, $(0, 0, -1)$, $(2, 2, -3)$, $(1, 4, 1)$, $(-3, -1, 4)$.
5. $(-3, -2, -1)$, $(-2, 0, 2)$, $(-1, 2, 5)$, $(1, 6, 11)$, $(2, 8, 14)$.
6. $(2, 0, 0)$, $(0, 2, 3)$, $(1, 1, 0)$, $(0, 0, 0)$, $(2, 0, 1)$, $(1, 1, 1)$.

7. Discuss the following statement: One can expect two points to be distinct, three points to be non-collinear, and four or more points to be non-coplanar. (See the footnote on page 63.)

8. Carry through an analysis, similar to that of § **58**, for families of points in plane analytic geometry.

60. Plane through three points.

THEOREM I. *The equation of the plane through three non-collinear points $P_1(x_1, y_1, z_1)$, $P_2(x_2, y_2, z_2)$, and $P_3(x_3, y_3, z_3)$ can be written in the form*

$$\begin{vmatrix} x & y & z & 1 \\ x_1 & y_1 & z_1 & 1 \\ x_2 & y_2 & z_2 & 1 \\ x_3 & y_3 & z_3 & 1 \end{vmatrix} = 0. \tag{1}$$

Proof. Assume first that $P(x, y, z)$ is any point in the plane $P_1P_2P_3$, and that the equation of this plane is

(2) $$ax + by + cz + d = 0.$$

Since the three given points lie in this plane, we can obtain three more equations by substituting their coördinates in (2). The resulting system,

(3)
$$\begin{aligned} ax + by + cz + d &= 0, \\ ax_1 + by_1 + cz_1 + d &= 0, \\ ax_2 + by_2 + cz_2 + d &= 0, \\ ax_3 + by_3 + cz_3 + d &= 0, \end{aligned}$$

considered as a system of four homogeneous linear equations in the unknowns a, b, c, d, is assumed to have non-trivial solutions, and therefore (Theorem II, § **52**) its coefficient determinant vanishes. That is, the coördinates of P satisfy equation (1). Conversely (Theorem II, § **52**), if the coördinates of P satisfy (1), non-trivial solutions, a, b, c, d, of the corresponding system (3) of homogeneous linear equations exist. But this means that there is a first degree equation that is satisfied by the coördinates of P as well as by those of P_1, P_2, and P_3, and that therefore P lies in the same plane as the three given points. Thus (1) is a necessary and sufficient condition for the point P to lie in the plane through the three given points. Since (1) is an equation of at most the first degree (as can be seen by expanding the determinant with respect to the elements of the first row), it must therefore be the equation of this plane.

To indicate an alternative method of attacking this type of problem, we present a second proof:

* *Alternative proof.* As in the first proof, we see by expanding the determinant with respect to the elements of the first row that equation (1) is of at most the first degree. We shall show next that it is of *exactly* the first degree. Since the three given points do not lie on a line, the rank of the matrix obtained by omitting the first row of the determinant (see § **58**) is 3. Therefore, if the coefficients of x, y, and z vanished, the cofactor of the element 1 in the first row would *not* vanish. But this would mean that the equation had no solutions, which would be a contradiction of the obvious fact that $x = x_1$, $y = y_1$, $z = z_1$ is a solution. In fact, equation (1) is satisfied by the coördinates of each of the three given points (since substitution in each case gives two identical rows), and therefore the plane defined by (1) is the plane through the three given points.

A consequence of what has just been proved, either in this section or in § **58** is the theorem:

Theorem II. *A necessary and sufficient condition for four points $P_1(x_1, y_1, z_1), \cdots, P_4(x_4, y_4, z_4)$ to be coplanar is that the determinant*

$$\begin{vmatrix} x_1 & y_1 & z_1 & 1 \\ x_2 & y_2 & z_2 & 1 \\ x_3 & y_3 & z_3 & 1 \\ x_4 & y_4 & z_4 & 1 \end{vmatrix} = 0$$

vanish.

It is shown in Chapter IX that if this determinant does not vanish it has a simple relation to the volume of the tetrahedron having the given points as vertices.

61. Exercises.

In Exercises **1** and **2**, write an equation in determinant form for the plane through the given points.

1. $(1, 0, 0), (0, 1, 0), (0, 0, 1)$. **2.** $(3, 8, 1), (0, -3, 2), (2, 4, -5)$.

3. Let $P_1(x_1, y_1, z_1)$ and $P_2(x_2, y_2, z_2)$ be any two points that are not collinear with the origin O. Prove that the equation of the plane OP_1P_2 can be written in the form

$$\begin{vmatrix} x & y & z \\ x_1 & y_1 & z_1 \\ x_2 & y_2 & z_2 \end{vmatrix} = 0.$$

4. Prove that a necessary and sufficient condition for three points $P_1(x_1, y_1, z_1)$, $P_2(x_2, y_2, z_2)$, and $P_3(x_3, y_3, z_3)$ to lie on a plane through the origin is

$$\begin{vmatrix} x_1 & y_1 & z_1 \\ x_2 & y_2 & z_2 \\ x_3 & y_3 & z_3 \end{vmatrix} = 0.$$

Show that this condition is automatically satisfied if the three points are collinear.

5. Prove that the equation of the plane through three non-collinear points $P_1(x_1, y_1, z_1)$, $P_2(x_2, y_2, z_2)$, and $P_3(x_3, y_3, z_3)$ can be written in the form

$$\begin{vmatrix} x - x_1 & y - y_1 & z - z_1 \\ x_2 - x_1 & y_2 - y_1 & z_2 - z_1 \\ x_3 - x_1 & y_3 - y_1 & z_3 - z_1 \end{vmatrix} = 0.$$

6. A plane is drawn through two distinct points $P_1(x_1, y_1, z_1)$ and $P_2(x_2, y_2, z_2)$ parallel to a line with direction numbers l, m, n (perpendicular to the plane $lx + my + nz + d = 0$). Assuming that the segment P_1P_2 is not parallel to this line (perpendicular to this plane), prove that the equation of the plane can be written in the form

$$\begin{vmatrix} x & y & z & 1 \\ x_1 & y_1 & z_1 & 1 \\ x_2 & y_2 & z_2 & 1 \\ l & m & n & 0 \end{vmatrix} = 0 \quad \text{or} \quad \begin{vmatrix} x - x_1 & y - y_1 & z - z_1 \\ x_2 - x_1 & y_2 - y_1 & z_2 - z_1 \\ l & m & n \end{vmatrix} = 0.$$

7. A plane is drawn through the point $P_1(x_1, y_1, z_1)$ parallel to two non-parallel lines with direction numbers l_1, m_1, n_1 and l_2, m_2, n_2 (or parallel to a line with direction numbers l_1, m_1, n_1 and perpendicular to the plane $l_2x + m_2y + n_2z + d_2 = 0$, or perpendicular to the planes $l_1x + m_1y + n_1z + d_1 = 0$ and $l_2x + m_2y + n_2z + d_2 = 0$). Prove that the equation of the plane can be written in the form

$$\begin{vmatrix} x & y & z & 1 \\ x_1 & y_1 & z_1 & 1 \\ l_1 & m_1 & n_1 & 0 \\ l_2 & m_2 & n_2 & 0 \end{vmatrix} = 0 \quad \text{or} \quad \begin{vmatrix} x - x_1 & y - y_1 & z - z_1 \\ l_1 & m_1 & n_1 \\ l_2 & m_2 & n_2 \end{vmatrix} = 0.$$

* Supplementary Exercises

8. Prove that if $P_1(x_1, y_1, z_1)$, $P_2(x_2, y_2, z_2)$, $P_3(x_3, y_3, z_3)$, and $P_4(x_4, y_4, z_4)$ are the vertices of a tetrahedron, then the six planes that are the perpendicular bisectors of the edges of the tetrahedron meet in a point. Is this point the centroid? (See Ex. **35**, § **13**.) *Suggestion:* Determine the ranks of a coefficient and an augmented matrix, the latter consisting of 6 rows and 4 columns.

9. Prove that the lines

$$\frac{x - x_1}{l_1} = \frac{y - y_1}{m_1} = \frac{z - z_1}{n_1} \quad \text{and} \quad \frac{x - x_2}{l_2} = \frac{y - y_2}{m_2} = \frac{z - z_2}{n_2}$$

lie in a plane if and only if

$$\begin{vmatrix} x_2 - x_1 & y_2 - y_1 & z_2 - z_1 \\ l_1 & m_1 & n_1 \\ l_2 & m_2 & n_2 \end{vmatrix} = 0.$$

10. Letting

$$l = \begin{vmatrix} m_1 & n_1 \\ m_2 & n_2 \end{vmatrix}, \quad m = \begin{vmatrix} n_1 & l_1 \\ n_2 & l_2 \end{vmatrix}, \quad n = \begin{vmatrix} l_1 & m_1 \\ l_2 & m_2 \end{vmatrix},$$

where

$$\frac{x - x_1}{l_1} = \frac{y - y_1}{m_1} = \frac{z - z_1}{n_1} \quad \text{and} \quad \frac{x - x_2}{l_2} = \frac{y - y_2}{m_2} = \frac{z - z_2}{n_2}$$

are any two skew lines, prove that the equations of the line containing the shortest line segment from a point on one line to a point on the other can be written in the form

$$\begin{vmatrix} x - x_1 & y - y_1 & z - z_1 \\ l_1 & m_1 & n_1 \\ l & m & n \end{vmatrix} = 0, \quad \begin{vmatrix} x - x_2 & y - y_2 & z - z_2 \\ l_2 & m_2 & n_2 \\ l & m & n \end{vmatrix} = 0.$$

Suggestion: Show that each of these planes contains one of the given lines and is perpendicular to any plane that is parallel to both. (Cf. Ex. **7**.)

11. Prove that the lines

$$\frac{x}{l_1} = \frac{y}{m_1} = \frac{z}{n_1}, \quad \frac{x}{l_2} = \frac{y}{m_2} = \frac{z}{n_2}, \quad \frac{x}{l_3} = \frac{y}{m_3} = \frac{z}{n_3}$$

lie in a plane if and only if
$$\begin{vmatrix} l_1 & m_1 & n_1 \\ l_2 & m_2 & n_2 \\ l_3 & m_3 & n_3 \end{vmatrix} = 0.$$

12. Let $(l_1, m_1, n_1), \cdots, (l_k, m_k, n_k)$ be direction numbers of a collection of lines, and let r be the rank of the matrix

$$\begin{pmatrix} l_1 & m_1 & n_1 \\ l_2 & m_2 & n_2 \\ . & . & . \\ l_k & m_k & n_k \end{pmatrix}.$$

Prove that the lines are parallel if and only if $r = 1$, and that they are non-parallel, but parallel to some plane (or perpendicular to some line) if and only if $r = 2$. (Cf. Ex. **11**.)

13. If l_1, m_1, n_1 and l_2, m_2, n_2 are direction numbers of two lines, parallel or not, prove that a line with direction numbers l, m, n lies in a plane parallel to the two given lines if and only if there exist numbers t_1 and t_2 such that
$$l = t_1 l_1 + t_2 l_2, \quad m = t_1 m_1 + t_2 m_2, \quad n = t_1 n_1 + t_2 n_2.$$

14. If (l_1, m_1, n_1), (l_2, m_2, n_2), and (l_3, m_3, n_3) are direction numbers of three lines that are not parallel to any plane, prove that any set of direction numbers l, m, n can be expressed in the form
$$l = t_1 l_1 + t_2 l_2 + t_3 l_3,$$
$$m = t_1 m_1 + t_2 m_2 + t_3 m_3,$$
$$n = t_1 n_1 + t_2 n_2 + t_3 n_3.$$

15. Prove that the parametric representation (2), § **41**, for the coördinates of a point on the line through two given distinct points is unique. That is, show that corresponding to any point on the line there is precisely one appropriate set of values for t_1 and t_2.

16. Let $P_1(x_1, y_1, z_1)$ be any point different from the origin O. Prove that the coördinates (x, y, z) of any point P on the line OP_1 have a unique representation in the form
$$x = t_1 x_1, \quad y = t_1 y_1, \quad z = t_1 z_1.$$
(See Ex. **15**.)

17. Prove that the parametric representation (1), § **42**, for the coördinates of a point in the plane through three given non-collinear points is unique. (See Exs. **15** and **16**.)

18. Let $P_1(x_1, y_1, z_1)$ and $P_2(x_2, y_2, z_2)$ be any two distinct points that are not collinear with the origin O. Prove that the coördinates (x, y, z) of **any** point P in the plane OP_1P_2 have a unique representation in the form
$$x = t_1 x_1 + t_2 x_2, \quad y = t_1 y_1 + t_2 y_2, \quad z = t_1 z_1 + t_2 z_2.$$
(See Exs. **15-17**.)

19. Let $P_1(x_1, y_1, z_1)$, $P_2(x_2, y_2, z_2)$, $P_3(x_3, y_3, z_3)$, and $P_4(x_4, y_4, z_4)$ be any four non-coplanar points. Prove that the coördinates (x, y, z) of any point P have a unique representation in the form

$$x = t_1 x_1 + t_2 x_2 + t_3 x_3 + t_4 x_4,$$
$$y = t_1 y_1 + t_2 y_2 + t_3 y_3 + t_4 y_4,$$
$$z = t_1 z_1 + t_2 z_2 + t_3 z_3 + t_4 z_4,$$

where $t_1 + t_2 + t_3 + t_4 = 1$. (See Exs. 15-18.)

20. Using the notation of Exercise **19**, prove that the points in the interior of the tetrahedron $P_1 P_2 P_3 P_4$, and only those points, correspond to positive values of t_1, t_2, t_3, t_4. To what values does the centroid correspond? (See Ex. **35**, § 13 and Exs. 15-19 of this section. Cf. Exs. **2** and **6**, § **43**.)

21. Let $P_1(x_1, y_1, z_1)$, $P_2(x_2, y_2, z_2)$, and $P_3(x_3, y_3, z_3)$ be any three points that are not coplanar with the origin O. Prove that the coördinates (x, y, z) of any point P have a unique representation in the form

$$x = t_1 x_1 + t_2 x_2 + t_3 x_3, \quad y = t_1 y_1 + t_2 y_2 + t_3 y_3, \quad z = t_1 z_1 + t_2 z_2 + t_3 z_3.$$

(See Exs. **4** and **15-19**.)

22. Assuming that the three planes $\Pi_1 = 0$, $\Pi_2 = 0$, and $\Pi_3 = 0$ have just one point P in common, prove that the family of planes

$$k_1 \Pi_1 + k_2 \Pi_2 + k_3 \Pi_3 = 0,$$

where k_1, k_2, and k_3 are parameters, consists of all planes through the point P. The family of all planes through a point is called a **bundle of planes.**

23. Assuming that the four planes $\Pi_1 = 0$, $\Pi_2 = 0$, $\Pi_3 = 0$, and $\Pi_4 = 0$ have no point in common and that three of these planes have just one point in common, prove that the family of planes

$$k_1 \Pi_1 + k_2 \Pi_2 + k_3 \Pi_3 + k_4 \Pi_4 = 0,$$

where k_1, k_2, k_3, and k_4 are parameters, consists of all planes in space.

24. Use the principle of linear dependence to *derive* the parametric equations of a plane, (1), § **42**.

* 62. Space of four or more dimensions. We have used the word *space* to refer to what is more precisely called *Euclidean three-dimensional space*. This space is based on a system of axioms given by Euclid and improved by Hilbert.† The principal fault in the axioms of Euclid is the omission from the list of postulates of certain assumptions which were necessary for many of the proofs, and which were made only tacitly.‡ Although the fact that we seem to be living in something like Euclidean three-dimensional space is a great advantage

---

† See David Hilbert, *The Foundations of Geometry* (The Open Court Publishing Co., 1938), or L. P. Eisenhart, *Coördinate Geometry* (Ginn and Co., 1939), p. 279.

‡ "From those same postulates it is easy to deduce, by irrefragable logic, spectacularly paradoxical consequences, such as 'all triangles are equilateral.'" Reprinted by permission from *The Development of Mathematics* by E. T. Bell, Copyrighted 1940, by the McGraw-Hill Book Company, Inc., p. 9.

in visualizing and studying problems in the geometry of space, on the other hand it can be a definite handicap by encouraging too prominent a place in mathematical proof for intuitive ideas. This is not to say that intuition should be completely submerged, for it is a powerful force for mathematical discovery. However, in any attempted proof we should know what is deductive and what is only heuristic. A good example of a non-mathematical concept which has been introduced in this book is that of right-handed and left-handed coördinate systems. It is meaningless to speak of a right-handed system without using spatial intuition. What *is* mathematically definable is whether two systems are similar or opposite. Since it is *convenient* to speak of right-handed and left-handed systems, and since proofs depend only on relative and not intrinsic right-handedness or left-handedness, we shall continue to use this language.

Because of the one-to-one correspondence existing between points in space and ordered triads of real numbers, a point could be *defined* to be an ordered triad. This is precisely the way higher dimensional spaces are defined, the concepts of *point, plane, line, angle, distance,* etc., resting ultimately on the fundamental notion of *number*. A point in n-dimensional Euclidean space, for example, is defined to be an ordered "n-tuple" of real numbers, $(x_1, x_2, \cdots, x_n)$. Much has been written regarding higher-dimensional spaces, and the student who is interested is referred to the literature.† We shall introduce at this point only a few of the basic ideas for Euclidean four-dimensional space, which we shall designate by the symbol E_4.

DEFINITIONS.

A **point** in E_4 is an ordered quadruple of real numbers, (x, y, z, w).

The **distance** d between the points (x_1, y_1, z_1, w_1) and (x_2, y_2, z_2, w_2) in E_4 is defined by the equation

$$d^2 = (x_2 - x_1)^2 + (y_2 - y_1)^2 + (z_2 - z_1)^2 + (w_2 - w_1)^2.$$

The **graph** of an equation (system of equations) in the variables x, y, z, w is the set of all points (x, y, z, w) whose coördinates satisfy the equation (system of equations simultaneously).

† A short bibliography:
E. Jouffret, *Traité Elémentaire de Géométrie à Quatre Dimensions* (Gauthier-Villars, 1903).
H. P. Manning, *Geometry of Four Dimensions* (The Macmillan Company, 1914).
P. R. Halmos, *Finite Dimensional Vector Spaces* (Princeton University Press, 1942).

A **hyperplane** (or 3-plane or plane) in E_4 is the graph of a linear equation, $ax + by + cz + dw + e = 0$ (where a, b, c, d are not all zero).

A **line** in E_4 is the graph of a system of equations

$$\frac{x - x_0}{a} = \frac{y - y_0}{b} = \frac{z - z_0}{c} = \frac{w - w_0}{d},$$

where the denominators are not all zero. The numbers a, b, c, d are called a set of **direction numbers** of the line. A set of direction numbers (λ, μ, ν, ξ) the sum of whose squares is 1 is called a set of **direction cosines** of the line, and is also called a **direction**.

The **angle** θ between two directions in E_4, $(\lambda_1, \mu_1, \nu_1, \xi_1)$ and $(\lambda_2, \mu_2, \nu_2, \xi_2)$, is the angle satisfying the inequalities $0° \leq \theta \leq 180°$ and the relation

$$\cos \theta = \lambda_1\lambda_2 + \mu_1\mu_2 + \nu_1\nu_2 + \xi_1\xi_2.$$

There is almost nothing more helpful in freeing one from the influence of spatial concepts than contemplation of spaces different from that of experience, and it is hoped that the student will spend a little time thinking about the Exercises of § **64,**

* **63. Complex space.** Later in this book it will be convenient to speak of such things as imaginary points and planes. These quantities arise in a fairly natural way algebraically. To provide a background for future discussion as well as another example of a space that is different from those previously mentioned, we present what might be called *complex Euclidean three-dimensional space*, denoted C_3:

DEFINITIONS.

A **point** in C_3 is an ordered triad of complex numbers,

$$(x, y, z) = (x' + ix'', y' + iy'', z' + iz'').$$

The **distance** d between the points (x_1, y_1, z_1) and (x_2, y_2, z_2) in C_3 is defined by the equation

$$\begin{aligned}d^2 &= |x_2 - x_1|^2 + |y_2 - y_1|^2 + |z_2 - z_1|^2 \\ &= (x_2' - x_1')^2 + (x_2'' - x_1'')^2 + (y_2' - y_1')^2 + (y_2'' - y_1'')^2 \\ &\quad + (z_2' - z_1')^2 + (z_2'' - z_1'')^2.\end{aligned}$$

The **graph** of an equation (system of equations) in the variables x, y, z is the set of all points (x, y, z) whose coördinates satisfy the equation (system of equations simultaneously).

A **plane** in C_3 is the graph of a linear equation $ax + by + cz + d = 0$, where a, b, and c are complex numbers (not all zero).

A **line** in C_3 is the graph of a system of equations

$$\frac{x - x_0}{l} = \frac{y - y_0}{m} = \frac{z - z_0}{n},$$

where the denominators are not all zero. The numbers l, m, n are called a set of **direction numbers** of the line. A set of direction numbers (λ, μ, ν) the sum of the squares of whose absolute values is 1 is called a set of **direction cosines** of the line, and is also called a **direction**.

Definition of *angle* between two directions in C_3 cannot be given without too extended a treatment of complex numbers. However, we can easily define **perpendicularity** of two lines with direction numbers l_1, m_1, n_1 and l_2, m_2, n_2 by the condition

$$\overline{l_1}l_2 + \overline{m_1}m_2 + \overline{n_1}n_2 = 0.\dagger$$

EXAMPLE. Find the distance between the points in C_3: $(i, 3 + 4i, 6 + i)$ and $(-i, 1 + i, -2 + i)$.

Solution. The square of the distance is

$|2i|^2 + |2 + 3i|^2 + |8|^2 = 4 + 4 + 9 + 64 = 81$, and therefore $d = 9$.

* **64. Exercises.**

Exercises **1-13** concern Euclidean four-dimensional space, E_4.

1. Define parallelism for hyperplanes and prove that two distinct hyperplanes have no points in common if and only if they are parallel.

2. Define perpendicularity between a line and a hyperplane. Show that the equation of the hyperplane through the point (x_1, y_1, z_1, w_1) perpendicular to the line

$$\frac{x - x_2}{a} = \frac{y - y_2}{b} = \frac{z - z_2}{c} = \frac{w - w_2}{d}$$

can be written

$$a(x - x_1) + b(y - y_1) + c(z - z_1) + d(w - w_1) = 0.$$

3. Prove that through any two distinct points (x_1, y_1, z_1, w_1) and (x_2, y_2, z_2, w_2) passes one and only one line and that

$$x_2 - x_1, y_2 - y_1, z_2 - z_1, w_2 - w_1$$

is a set of direction numbers of the line.

4. Prove that a line perpendicular to a hyperplane is perpendicular to any line (or line segment) that lies in the hyperplane. (See Exs. **2** and **3**.)

† The *conjugate* of the complex number $z = x + iy$ is denoted $\bar{z} = x - iy$. It is readily shown that the conjugate of the sum (difference, product, quotient) of two complex numbers is equal to the sum (difference, product, quotient) of their conjugates, and that the product of a complex number and its conjugate is equal to the square of its absolute value.

5. Prove the Pythagorean relation; that is if P_1P_2 is perpendicular to P_2P_3, show that
$$d_{12}^2 + d_{23}^2 = d_{13}^2,$$
where d_{ij} is the distance between P_i and P_j.

6. Write equations of a line in parametric form.

7. Prove that if two distinct points of a line lie in a hyperplane, the entire line lies in the hyperplane. (See Exs. 3 and 6.)

* 8. Discuss the normal form for the equation of a hyperplane and obtain a formula for the distance between a hyperplane and a point. Show that this distance is actually the distance between the given point and some point of the hyperplane, and is less than the distance between the given point and any other point of the hyperplane. (See Exs. 2-7.) *Suggestion:* Let the given point be P_1, let its projection on the hyperplane be P_2, and find d_{12}. If P_3 is any other point of the hyperplane, $P_1P_2 \perp P_2P_3$ and therefore
$$d_{13} = \sqrt{d_{12}^2 + d_{23}^2} > d_{12}.$$

9. Show that the formula for $\cos \theta$, used in defining the angle between two directions, always gives a number whose absolute value $\leqq 1$. *Suggestion:* Expand the inequality
$$(\lambda_1 \pm \lambda_2)^2 + (\mu_1 \pm \mu_2)^2 + (\nu_1 \pm \nu_2)^2 + (\xi_1 \pm \xi_2)^2 \geqq 0.$$

* 10. In what sense is a hyperplane three-dimensional? In what sense is the intersection of two non-parallel hyperplanes two-dimensional?

11. Prove that if three hyperplanes have equations
$$a_ix + b_iy + c_iz + d_iw + e_i = 0,$$
where $i = 1, 2, 3$, with a coefficient matrix of rank 3, then there is defined a set of direction numbers of a line parallel to each of the three hyperplanes, and that these direction numbers must be proportional to
$$\begin{vmatrix} b_1 & c_1 & d_1 \\ b_2 & c_2 & d_2 \\ b_3 & c_3 & d_3 \end{vmatrix}, \quad -\begin{vmatrix} a_1 & c_1 & d_1 \\ a_2 & c_2 & d_2 \\ a_3 & c_3 & d_3 \end{vmatrix}, \quad \begin{vmatrix} a_1 & b_1 & d_1 \\ a_2 & b_2 & d_2 \\ a_3 & b_3 & d_3 \end{vmatrix}, \quad -\begin{vmatrix} a_1 & b_1 & c_1 \\ a_2 & b_2 & c_2 \\ a_3 & b_3 & c_3 \end{vmatrix}.$$

12. Prove that the five points (x_i, y_i, z_i, w_i), $i = 1, \cdots, 5$, lie in a hyperplane if and only if
$$\begin{vmatrix} x_1 & y_1 & z_1 & w_1 & 1 \\ x_2 & y_2 & z_2 & w_2 & 1 \\ x_3 & y_3 & z_3 & w_3 & 1 \\ x_4 & y_4 & z_4 & w_4 & 1 \\ x_5 & y_5 & z_5 & w_5 & 1 \end{vmatrix} = 0.$$

13. Find a condition on the rank of a matrix involving the coördinates of four points that is necessary and sufficient for them to lie in just one hyperplane. Write an equation of this hyperplane in determinant form. (Cf. Ex. **12.**)

Exercises 14-23 concern complex Euclidean space C_3.

14. Define parallelism for planes and prove that two distinct planes have no points in common if and only if they are parallel.

15. A line is defined to be perpendicular to a plane if and only if the conjugates of the coefficients of x, y, and z in the equation of the plane are direction numbers of the line. Show that the equation of the plane through the point (x_1, y_1, z_1) perpendicular to the line

$$\frac{x - x_2}{a} = \frac{y - y_2}{b} = \frac{z - z_2}{c}$$

can be written

$$\bar{a}(x - x_1) + \bar{b}(y - y_1) + \bar{c}(z - z_1) = 0.$$

16-20. State and prove theorems corresponding to those given in Exercises **3-7**.

* 21. Discuss the normal form for the equation of a plane and show that the distance between the plane

$$ax + by + cz + d = 0$$

and the point (x_1, y_1, z_1) is

$$\frac{|\, ax_1 + by_1 + cz_1 + d\, |}{\sqrt{|\, a\, |^2 + |\, b\, |^2 + |\, c\, |^2}}.$$

Complete the discussion as suggested in Exercise **8**.

22. Prove that two non-parallel planes

$$a_1 x + b_1 y + c_1 z + d_1 = 0 \quad \text{and} \quad a_2 x + b_2 y + c_2 z + d_2 = 0$$

have a line of points in common, and that a set of direction numbers of this line is

$$\begin{vmatrix} b_1 & c_1 \\ b_2 & c_2 \end{vmatrix}, \quad \begin{vmatrix} c_1 & a_1 \\ c_2 & a_2 \end{vmatrix}, \quad \begin{vmatrix} a_1 & b_1 \\ a_2 & b_2 \end{vmatrix}.$$

23. Explain how complex space, C_3, might be thought of as six-dimensional.

CHAPTER IV

SURFACES AND CURVES

65. Surfaces.† It is not easy to give a definition of *surface* that is completely satisfactory from all points of view. Since the surfaces with which we shall be primarily concerned in this book are what are called *algebraic surfaces*, which are defined in terms of *algebraic equations*, we shall limit our treatment accordingly.

Before giving any definitions, let us look at the route that must be followed. As might be expected, algebraic surfaces are studied principally by algebraic methods. We shall confine ourselves in this book almost entirely to the algebraic type of proof. Let us illustrate what is meant by an algebraic method, as distinguished from that of the calculus, by considering a line tangent to a conic in plane geometry. The method of the calculus is to define a tangent line as the limit of a variable secant line. The method of algebra is to define a tangent as a line that meets the conic in "two coincident points."

It might be worth mentioning that there is an extensive branch of mathematics called *algebraic geometry*, where the values of the variables need not be ordinary numbers, but may be arbitrary elements of a "field"‡ where the notion of *limit* may be completely meaningless. In this case the algebraic method is not only preferable, but essential.

Since we have adopted a program with an algebraic emphasis, we shall find it highly important to distinguish between certain surfaces which consist of the same points. For example, although the graphs of the equations $xy = 0$, $x^2y = 0$, and $xy^2 = 0$ are the same, according to the definition of the graph of an equation given in Chapter II (being the points in the xz and yz planes), we do not wish to consider the *surfaces* $xy = 0$, $x^2y = 0$, and $xy^2 = 0$ as identical.

---

† For a first course, §§ **65–67** may be regarded rather as "reading sections" than as "study sections."

‡ For a discussion of fields see Garrett Birkhoff and Saunders MacLane, *A Survey of Modern Algebra* (The Macmillan Company, 1944).

DEFINITION I. *An **algebraic equation**, or **integral rational equation**, in the variables x, y, and z is an equation of the form*

(1) $$g(x, y, z) = h(x, y, z),$$

where $g(x, y, z)$ and $h(x, y, z)$ are polynomials in x, y, and z. The equation (1) *is **real** if and only if the coefficients in the polynomials g and h are real. The **degree** of the equation* (1) *is the degree in x, y, and z of the polynomial $g - h$.*

THEOREM I. *Equivalent algebraic equations (see § 27) have the same degree. Any [real] algebraic equation is equivalent to an algebraic equation of the form*

(2) $$f(x, y, z) = 0,$$

where f is a polynomial in x, y, and z [with real coefficients].

For brevity we shall call a polynomial of at least the first degree, that is, a polynomial that is not a constant, a **variable** polynomial.

DEFINITION II. *A polynomial $f(x, y, z)$ is **reducible** if and only if it is the product of two variable polynomials; otherwise it is **irreducible**. If the kth power of a polynomial g is a factor of f and if the $(k + 1)$th power is not, g is called a factor of **multiplicity** k.*

For example, the polynomial $(2x^3 - z + 9)(2y + z^2)^3 y^4 z^8$ is reducible; the factors $2x^3 - z + 9$, $2y + z^2$, y, and z are irreducible and have multiplicities 1, 3, 4, and 8, respectively.

We shall define an algebraic surface in two stages.

DEFINITION III. *An **irreducible algebraic surface** is the graph of an equation of the form* (2), *where $f(x, y, z)$ is an irreducible variable polynomial.*

DEFINITION IV. *Let $f(x, y, z)$ be a polynomial that can be represented in the form*

$$f \equiv f_1^{k_1} f_2^{k_2} \cdots f_m^{k_m},$$

*where $f_1, f_2, \cdots, f_m$ are irreducible variable polynomials no two of which are proportional, and $k_1, k_2, \cdots, k_m$ are positive integral exponents. The **algebraic surface** $f = 0$ consists of the m irreducible algebraic surfaces*

$$f_1 = 0, \quad f_2 = 0, \quad \cdots, \quad f_m = 0,$$

*called **factors** of the surface $f = 0$, each taken with the corresponding multiplicity k_i. Two algebraic surfaces are **identical** if and only if*

§ 65] SURFACES 85

they consist of the same factors, each factor having the same multiplicity in both cases. The algebraic surface $fg = 0$ is called the **product** of the algebraic surfaces $f = 0$ and $g = 0$.

Of basic importance for the definition just stated is the fact that a variable polynomial has essentially only one decomposition into irreducible factors; more precisely, the factors of any two such decompositions differ only by constant factors. Proofs of this and the following theorem are given in the reference cited in the footnote†. For the case in which we shall be primarily interested, a polynomial of the second degree, this *unique factorization* will be readily established with the aid of Chapter VIII. (See Ex. **60**, **§ 156**.)

THEOREM II. *If $f(x, y, z)$ and $g(x, y, z)$ are variable polynomials, the surfaces $f = 0$ and $g = 0$ are identical if and only if f and g differ only by constant factors. In other words, two algebraic surfaces are identical if and only if their defining equations are equivalent.*

Although we shall not prove this theorem in general, we have already established a special case of it in Chapter II, where f and g are first degree polynomials. In Chapter VIII, we shall give a proof for the case where f and g are both second degree polynomials (and, incidentally, of the more trivial case where one of the polynomials is of the first degree and the other is of the second degree). Consequently, we shall have established the theorem for surfaces defined by equations whose degrees do not exceed 2.

As a consequence of Theorem II, it is possible to speak of *the* equation of an algebraic surface in the same way that it is possible to speak of *the* equation of a plane. Furthermore, we can define the **degree** of an algebraic surface to be the degree of its equation.

It is important to understand that in order that Theorem II hold it is necessary that imaginary numbers be admitted; that is that the graphs be taken in complex space. For example, although the equations

$$x^2 + y^2 + z^2 + 1 = 0 \quad \text{and} \quad x^2 + y^2 + z^2 + 4 = 0$$

are not equivalent, their graphs in real three-dimensional space, being vacuous, are identical. However, their graphs in complex space are distinct. They are called *imaginary spheres.* (See § **81**.)

---

† Maxime Bôcher, *Introduction to Higher Algebra* (The Macmillan Company, 1935), § 76, p. 212.

DEFINITION V. *An algebraic surface of the second degree is called a* **quadric surface,** *or* **quadric,** *or* **conicoid.** *That is, a quadric surface is a surface whose equation has the form*

$$ax^2 + by^2 + cz^2 + 2fyz + 2gxz + 2hxy + 2px + 2qy + 2rz + d = 0,$$

where a, b, c, f, g, and h are not all zero.

We shall henceforth assume that *all quadric surfaces under discussion are defined by real equations.*

EXAMPLE 1. The equation

$$x^2 - 3x = xy - 5xz$$

is equivalent to $x(x - y + 5z - 3) = 0$. Therefore its graph is a reducible algebraic surface, whose two factors are the planes $x = 0$ and $x - y + 5z - 3 = 0$.

EXAMPLE 2. The surface

$$x^2 + y^2 + z^2 = 1$$

consists of all points a unit distance from the origin, and is therefore a sphere with center at the origin and radius 1.

EXAMPLE 3. The only point of the surface

$$x^2 + y^2 + z^2 = 0$$

with real coördinates is the origin. This type of surface, called a *point sphere* or *imaginary cone*, is discussed more thoroughly in § **81** and in Chapter V.

EXAMPLE 4. The real points of the surface

$$x^2 + y^2 = 0$$

have coördinates of the form $(0, 0, z)$, where z is any real number. The real graph is therefore the z axis, a straight line, although the surface, as treated in Chapter V, is more properly called a *pair of imaginary planes*. These planes are

$$x + iy = 0 \quad \text{and} \quad x - iy = 0,$$

which intersect in the z axis.

* 66. Parametric equations of a surface. The parametric method of representing a surface is commonly used in certain types of investigation, such as differential geometry, and, in fact, is a standard means of defining a surface. According to this definition, a surface is represented by equations giving the coördinates of a point in terms of three suitably well-behaved functions of two parameters, u and v:

$$x = f(u, v),$$
$$y = g(u, v),$$
$$z = h(u, v).$$

If the surface is the graph of an equation of the form

$$z = h(x, y),$$

the parameters can be chosen to be x and y, with $f(x, y) \equiv x$ and $g(x, y) \equiv y$.

The parametric equations of a plane through three points (§ **42**) assume the form given above if one of the parameters is eliminated by means of the relation $t_1 + t_2 + t_3 = 1$. For example, one representation is

$$x = (x_1 - x_3)t_1 + (x_2 - x_3)t_2 + x_3,$$
$$y = (y_1 - y_3)t_1 + (y_2 - y_3)t_2 + y_3,$$
$$z = (z_1 - z_3)t_1 + (z_2 - z_3)t_2 + z_3,$$

the parameters being t_1 and t_2. In this case the functions f, g, and h are *linear*.

67. Curves. A *curve* is usually defined as the set of points whose coördinates are given parametrically by a system of equations

$$x = f(t),$$
$$y = g(t),$$
$$z = h(t),$$

where f, g, and h are suitably well-behaved functions of the parameter t. As we have seen (§ **39**), any straight line can be represented in this manner, where the functions f, g, and h are *linear*. If these functions are not sufficiently restricted, they may define something completely at variance with our ideas of what a curve should be. This is true even with continuous functions, as was shown by the celebrated space-filling curve of Peano.[†]

In this book a curve will usually be represented as a set of points common to two surfaces. Specifically, we shall define an *algebraic curve*, the type of curve that will receive our greatest attention:

DEFINITION. *If two algebraic surfaces have points in common but no factors in common, their intersection is called an* **algebraic curve.**

As in the case of a straight line, which is the intersection of infinitely many pairs of planes, a curve may be obtained as the intersection of a pair of surfaces in infinitely many ways. For example, a circle may be the curve of intersection of two spheres, a sphere and a plane, or a right circular cone and a plane, each in infinitely many ways.

Like a surface, a curve may contain only one real point, or no real points.

[†] For an example of a curve that passes through every point of a plane square, see Philip Franklin, *A Treatise on Advanced Calculus* (John Wiley and Sons, 1940), p. 56.

EXAMPLE 1. The curve
$$x^2 + y^2 + z^2 = 1,$$
$$2x = 1$$
consists of all points that lie on the sphere $x^2 + y^2 + z^2 = 1$ and the plane $2x = 1$, and is therefore a circle.

EXAMPLE 2. If the coördinates of a point satisfy the two equations
$$x^2 + y^2 + z^2 = 1$$
and
$$x = 2,$$
they must also satisfy the equation
$$y^2 + z^2 = -3.$$
Therefore the curve defined by the first two equations does not contain any real points, although it does contain points with imaginary coördinates. This curve is called an *imaginary circle*. (See § **83**.)

EXAMPLE 3. The graph of the system
$$x + 2y + z = 1,$$
$$2x + 4y + 2z = 2$$
is a plane, since the two equations are equivalent. Since the planes are identical they have a factor (the plane) in common, and the graph of the system is not a curve.

EXAMPLE 4. Any solution of the system
$$x + 2y + z = 1,$$
$$x + 2y + z = 2$$
must be a solution of the equation $0 = 1$. Therefore the **graph contains no** points, real or imaginary, and is not a curve.

EXAMPLE 5. The graph of the system
$$x^2 + y^2 + z^2 + 1 = 0,$$
$$x^2 + y^2 + z^2 + 4 = 0$$
also contains no points, real or imaginary.

The examples just given emphasize the fact that the words *imaginary* and *non-existent* mean different things. Another word that is sometimes confused with these two is *infinite*. The concept of *point at infinity*, although of great importance in projective geometry and of considerable general interest and usefulness, is consistently avoided in this book, since an adequate discussion would require an extensive treatment of homogeneous coördinates. Let us at least remark that it is possible to study "real points at infinity" and "imaginary points at infinity," that the two planes of Example 4 can be thought of as intersecting in a "real line at infinity," and that the system of Example 5 defines an "imaginary circle at infinity."

Although we shall consider later a few examples of surfaces and curves that are not algebraic, *we shall henceforth use the terms surface and curve to mean, in general, algebraic surface and algebraic curve.*

68. Sections, traces, and intercepts. The curve of intersection of a surface and a plane is called a **section** of that surface by the plane. The section of a surface by a coördinate plane is called the **trace** of the surface in that plane. An equation of a trace of a surface in the two variables of the coördinate plane of the trace is easily obtained by setting the third variable equal to zero in the given equation of the surface.

If a coördinate axis intersects a surface, such a point of intersection (or the appropriate coördinate) is called an **intercept**. The intercepts of a surface on any axis can be found by setting equal to zero the variables corresponding to the other axes.

EXAMPLE. Find the traces and intercepts of the surface
$$x^2 + y^2 - z^2 - 2x = 0.$$

Solution. By equating to zero in turn x, y, and z, we obtain the equations of the traces in the yz, xz, and xy planes:
$$y^2 - z^2 = 0, \ x^2 - z^2 - 2x = 0, \ x^2 + y^2 - 2x = 0.$$
These traces are two straight lines, a hyperbola, and a circle, respectively. The x intercepts are 0 and 2, and the y and z intercepts are both zero.

Sections of a surface are very useful in determining its general shape or in drawing a sketch of it. They are discussed again later in this chapter, and used repeatedly in Chapter V.

69. Cylinders. A *cylinder* is any surface generated by a line moving parallel to a fixed line. A more precise definition is the following:

DEFINITION. *Let C be a plane curve and let F denote the family of all lines through points of C perpendicular to the given plane. The surface consisting of all points of these lines is called a **cylinder**. Each line of the family F is called a **ruling** or **generator**, and the curve C is called a **directrix**.*† *A cylinder is said to be **parallel** to its rulings and **perpendicular** to any plane perpendicular to them.*

Since the normal sections of a cylinder cut by planes perpendicular to the rulings are congruent, there is no ambiguity in speaking of *the* directrix. (See Fig. 23.)

† A directrix is often defined to be *any* curve that intersects every line of the family F. (See Ex. **36**, § **70**.) For simplicity, however, in this book we shall limit our use of the term to *normal sections*.

If the directrix is one or more straight lines, the cylinder is one or more planes. Otherwise the cylinder is a curved surface, usually

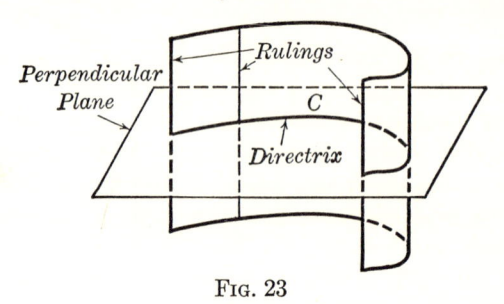

Fig. 23

named after the directrix. For example, if the directrix is a circle, the cylinder is a right circular cylinder. Similarly, we speak of elliptic, hyperbolic, and parabolic cylinders.

Although a cylinder need not be parallel to a coördinate axis, we shall consider in this chapter only those cylinders that are, and shall assume also that the directrix is in a coördinate plane. The principal theorem for such cylinders is the following (cf. Theorems V and VI, § 28):

Theorem. *A surface whose equation has a missing variable is a cylinder parallel to the axis of the missing variable. The curve corresponding to this equation, as an equation in the other two variables, is the directrix located in their coördinate plane.*

Proof. For definiteness, let

(1) $$f(x, y, z) = f(x, y) = 0$$

be an equation in which z is missing. This equation, regarded either as an equation in three variables or as an equation in two variables, will in general impose restrictions on x and y, but none on z. If (x_1, y_1, z_1) is any point on the graph of (1), then (x_1, y_1, z_2), where z_2 has any value whatsoever, is also on the graph. Therefore, corresponding to any point (x_1, y_1) on the curve

(2) $$f(x, y) = 0, \quad z = 0,$$

the line $x = x_1$, $y = y_1$, parallel to the z axis, lies in the surface (1), which must therefore be a cylinder parallel to the z axis. The curve (2) is a directrix of the surface (1).

If two variables are missing, the cylinder is parallel to the axes of both missing variables and consists of a collection of planes.

EXAMPLE 1. Discuss and sketch the surface
$$x^2 + 4y^2 = 16.$$

Solution. The surface is an *elliptic cylinder* parallel to the z axis. The normal sections are ellipses with semi-axes 4 and 2 with centers on the z axis. The z axis is called the *axis* of the cylinder. (See Fig. 24.)

EXAMPLE 2. Discuss the surface $z^4 = 1$.

Solution. The surface consists of two real horizontal planes, $z = 1$ and $z = -1$. In complex space the graph consists of four planes, $z = 1$, $z = -1$, $z = i$, and $z = -i$.

EXAMPLE 3. Discuss the graph of the equation $z = \sin x$.

Solution. The surface is a "sinusoidal" cylinder parallel to the y axis, lying half above and half below the xy plane. It is *not* an algebraic surface.

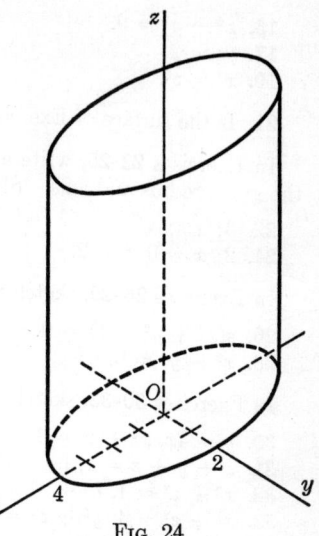

FIG. 24

70. Exercises.

In Exercises **1-6,** find an equation of the trace of the given surface in the specified coördinate plane. Identify the trace.

1. $x^2 + y^2 + z^2 - 2xz + 5z - 4 = 0$; xy plane.
2. $xy + xz + yz = 1$; xz plane.
3. $x^2 + y + z + 3 = 0$; yz plane.
4. $x^2 + z^2 - yz + xy + 1 = 0$; xz plane.
5. $x^2 + 4y^2 + z^2 + 4xy - 2xz - 2x - 4y + z + 1 = 0$; xy plane.
6. $x^2 + xy - 3xz - 2 = 0$; yz plane.

In Exercises **7-10,** find the intercepts of the given surface on the specified coördinate axis.

7. $x^2 + 3y^2 + 5xz - 2x + y - 3 = 0$; x axis.
8. $3x^2 - z^2 + xy - 8yz + y - 3z - 2 = 0$; y axis.
9. $2x^2 - y^2 + z^2 - xy + x + 6y + 1 = 0$; z axis.
10. $y^2 + 2z^2 + 3xy - 2yz + y - 2z + 1 = 0$; x axis.

In Exercises **11-14,** show that the surface is reducible, and find each factor and its multiplicity.

11. $x^2 + 3xy - 5xz + 2x = 0$.
12. $x^2 - xy + xz - yz = 0$.
13. $x^2 + 4y^2 + xz + 2yz + 4xy + x + 2y + z = 0$.
14. $x^2 + 9y^2 + 4z^2 - 12yz - 4xz + 6xy + 2x + 6y - 4z + 1 = 0$.

In Exercises **15-20,** show that the surface is a cylinder parallel to a coördinate axis. Name this coördinate axis and sketch the surface.

15. $x^2 + y^2 = 9$.
17. $yz = 2$.
19. $x^2 = z^2$.
16. $x^2 + 4z = 16$.
18. $x^2 + y^2 = 4x$.
20. $|y| + |z| = 1$.

21. Is the surface of Exercise **20** algebraic?

In Exercises **22-25**, write an equation of the right circular cylinder with the given radius and axis. Sketch the surface.

22. 3; x axis.
24. 2; $x = 0, z = 2$.
23. 5; z axis.
25. 5; $x = 3, y = 4$.

In Exercises **26-29**, sketch the surface.

26. $z(x^2 + y^2 - 1) = 0$.
28. $x^2 - y^2 = 0$.
27. $z(x^2 + y^2) = 0$.
29. $x^3 - y^3 = 0$.

In Exercises **30-33**, sketch the curve.

30. $y = 4x, x^2 + z^2 = 16$.
31. $x + y + z = 1, (x - 1)^2 + (y - 1)^2 = 1$.
32. $x^2 + z^2 = 1, y^2 + z^2 = 1$.
33. $x^2 + z^2 = 4, y^2 + z^2 = 9$.

In Exercises **34** and **35**, find the points where the curve meets the specified coördinate plane.

34. $4y^2 + z^2 + 2xz - xy - 5z - 3 = 0$,
$3xz - 4yz + x - 3y + z + 2 = 0$; xy plane.

35. $x^2 + y^2 + 2z^2 + 3x - 6 = 0$,
$xy + yz + xz + 2 = 0$; yz plane.

★ **36.** Give a definition of a *cylinder* in terms of a directrix that is not necessarily a plane curve.

★ **71. Projections of a curve. Elimination.** It has been remarked that any curve is the intersection of any number of pairs of surfaces. Among the surfaces containing a given curve, there are three of particular significance. They are the cylinders which are parallel to the coördinate axes and whose directrices are consequently the projections of the curve on the coördinate planes. In order to study these **projecting cylinders,** we must examine the process of eliminating a variable from a pair of equations. †

DEFINITION. *An equation*

(1) $$h(x, y) = 0$$

---

† In this discussion we shall regard any projecting cylinder merely as a set of points, without introducing the more complicated question of multiplicities. For a more complete treatment of elimination see Maxime Bôcher, *Introduction to Higher Algebra* (The Macmillan Company, 1935), § 70, p. 198.

is said to result from the elimination of z from the system
$$f(x, y, z) = 0, \qquad g(x, y, z) = 0 \tag{2}$$
if and only if the following two conditions are satisfied: (i) if (x_1, y_1, z_1) is a solution of the system (2), then (x_1, y_1) is a solution of the equation (1); (ii) if (x_1, y_1) is a solution of the equation (1), then there is a number z_1 such that (x_1, y_1, z_1) is a solution of the system (2).

A similar definition applies to elimination of x or y from the system (2).

It is not difficult to see that the graph of (1) is the projecting cylinder of the curve (2) parallel to the z axis. The projection of the curve (2) on the xy plane has equations $h(x, y) = 0$, $z = 0$. The projections on the other coördinate planes are found in a similar manner.

In connection with the question of elimination it is well to consider the related notions of *linear combination* and *pencil* of surfaces. (Cf. § **36**.) Accordingly, we give a definition and state two theorems, whose proofs are not difficult and should be supplied by the student (Exs. **11** and **12**, § **72**).

DEFINITION. *If S_1 and S_2 are two surfaces,*
$$S_1 : f_1(x, y, z) = 0, \qquad S_2 : f_2(x, y, z) = 0,$$
that intersect in a curve, the family of surfaces given by the equations
$$k_1 f_1(x, y, z) + k_2 f_2(x, y, z) = 0, \tag{3}$$
*where k_1 and k_2 are parameters, is called the **pencil of surfaces** defined by the surfaces S_1 and S_2. For any values of k_1 and k_2, the left member of (3) is called a **linear combination** of $f_1(x, y, z)$ and $f_2(x, y, z)$, and any surface of the family (3) is called a linear combination of the surfaces S_1 and S_2.*

THEOREM I. *The curve of intersection of two surfaces lies in every surface of the pencil of surfaces defined by them.*

THEOREM II. *The systems*
$$f(x, y, z) = 0, \qquad g(x, y, z) = 0 \tag{4}$$
and
$$\begin{aligned} k_1 f(x, y, z) + k_2 g(x, y, z) &= 0 \\ k_3 f(x, y, z) + k_4 g(x, y, z) &= 0 \end{aligned} \tag{5}$$
have the same solutions, and therefore the same graph, if
$$\begin{vmatrix} k_1 & k_2 \\ k_3 & k_4 \end{vmatrix} \neq 0.$$

A simple but useful special case of Theorem II is provided when $k_1 = 1$, $k_2 = 0$, $k_4 \neq 0$.

Elimination of a variable can often be obtained by means of linear combinations, as we shall see in Example 1, below. Another standard method is that of **elimination by substitution,** also used in Example 1. The basic theorem for this method, whose proof the student is invited to provide (Ex. **13, § 72**), is the following:

THEOREM III. *The systems*

(6) $$z = \phi(x, y), \ f(x, y, z) = 0$$
and
(7) $$z = \phi(x, y), \ f(x, y, \phi(x, y)) = 0$$

have the same solutions and therefore the same graph. The equation $f(x, y, \phi(x, y)) = 0$ *results from elimination of z from the system* (6).

A similar theorem applies to elimination by substitution of the variable x or the variable y.

EXAMPLE 1. Find the projections on the coördinate planes of the curve
$$x^2 + 2y^2 + 3z^2 = 9, \ x^2 - y^2 + z = 4.$$

Solution. Elimination of x and elimination of y are achieved by forming the linear combinations
$$(x^2 + 2y^2 + 3z^2 - 9) - (x^2 - y^2 + z - 4)$$
and
$$(x^2 + 2y^2 + 3z^2 - 9) + 2(x^2 - y^2 + z - 4),$$
respectively. The projections on the yz and xz planes are the circles
$$3y^2 + 3z^2 - z = 5, \ x = 0$$
and
$$3x^2 + 3z^2 + 2z = 17, \ y = 0,$$
respectively. Solving for z in the second equation, and substituting this expression for z in the first equation, we find that the projection of the given curve on the xy plane is the fourth degree curve
$$3x^4 - 6x^2y^2 + 3y^4 - 23x^2 + 26y^2 + 39 = 0, \ z = 0.$$

EXAMPLE 2. Show that the curve
$$x^2 + 2y^2 - z^2 - x = 0, \ 2x^2 + 4y^2 - 2z^2 + 3y = 0$$
is a plane curve.

Solution. A linear combination of the left members of the given equations is $2x + 3y$. Therefore the curve lies in the plane $2x + 3y = 0$.

EXAMPLE 3. Prove that the helix
$$x = \cos t,\ y = \sin t,\ z = t,$$
where t is a parameter, is the curve of intersection of **two sinusoidal cylinders**. (See Example 3, § 69.)

Solution. Elimination of t gives the two equations
$$x = \cos z,\ y = \sin z,$$
each of which defines a sinusoidal cylinder.

* 72. Exercises.

In Exercises **1-4,** find equations of the projection of the given curve on the specified coördinate plane. If the projection is a conic, name it.

1. $y^2 + z^2 = 9,\ x^2 + z^2 = 4$; xy plane.
2. $5y^2 - 2z^2 + 3x + y - 5z = 0,\ y^2 + 2z^2 + x - 2y + 3z = 0$; yz plane.
3. $x^2 + z^2 = 4,\ z = x^2 + y^2$; xy plane.
4. $xy + xz + yz = 2,\ xy - 2xz + 3yz = 6$; xz plane.

In Exercises **5** and **6,** find equations of the projection of the section of the given surface by the given plane on the specified coördinate plane. Draw a figure.

5. $y = z^2;\ 2x = y$; xz plane.
6. $x^2 + z^2 = 16;\ x + y + z = 0$; xy plane.

In Exercises **7** and **8,** show that the curve is a plane curve.

7. $xy + z = 0,\ xy - x - y = 0.$
8. $x^2 + y^2 + z^2 = 4,\ x^2 + y^2 + z^2 + 2x - 4y - 5 = 0.$

9. Write in standard form the equation of the plane defined by the equations
$$x = u + 2v + 2,\quad y = 2u - v - 1,\quad z = u + 3v,$$
where u and v are parameters.

10. Write equations of the three projecting cylinders for the curve
$$x = u + 1,\quad y = u - 3,\quad z = u^2 + 2.$$

11. Prove Theorem I, § 71.
12. Prove Theorem II, § 71.
13. Prove Theorem III, § 71.

* **14.** Prove that the degree of the projection on any coördinate plane of the curve of intersection of two quadric surfaces is at most 4.

73. Symmetry. In the first chapter we defined what we meant by saying that two points are located symmetrically with respect to a plane, a line, or a point. We can easily extend the notion of symmetry to any set of points.

DEFINITION. *A set of points is* **symmetrical** *with respect to a plane (or line or point) if and only if corresponding to any point of the set there is a point of the set symmetrical to it with respect to the plane (or line or point).* Under the appropriate circumstances we say that the set has a **plane of symmetry**, a **line of symmetry** or **axis**, or a **point of symmetry** or **center**.

A sphere is symmetrical with respect to any plane or line through its center, and with respect to its center. For a (toy) top any plane through the axis is a plane of symmetry, but there is no point of symmetry. The surface of a baseball or tennis ball, counting the seams as part of the pattern, has two perpendicular planes of symmetry intersecting in one of three mutually perpendicular axes, but it has no center.

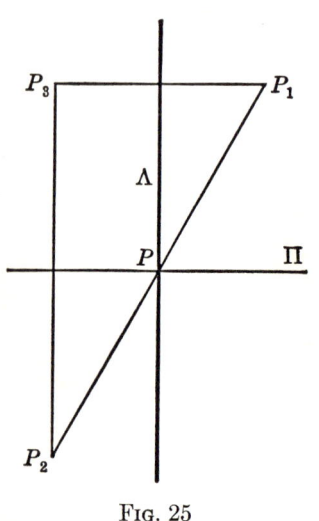

EXAMPLE. Prove that if a set of points S has a center P located on a plane of symmetry Π, the line Λ through P perpendicular to Π is an axis of the set S.

Solution. Let P_1 be any point of the set S that is not on Λ, and let P_2 and P_3 be the points symmetrical to P_1 with respect to P and Λ respectively. Fig. 25 shows the plane of Λ and P_1, which contains the points P_2 and P_3. The plane Π is represented by the line that it has in common with the plane of the paper. Since P is a center P_2 must belong to the set S, and therefore, since Π is a plane of symmetry perpendicular to the plane of the paper, P_3 must belong to S. This means that Λ is an axis of the set S.

FIG. 25

74. Exercises.

1. Give an example of an object or set of points having the following symmetries:

(a) a plane of symmetry but no axis or center;
(b) an axis but no plane of symmetry or center;
(c) a center but no plane of symmetry or axis.

2. Prove that if a set of points has two perpendicular planes of symmetry, their line of intersection is an axis.

3. Prove that if a set of points has two perpendicular intersecting axes, the line through their point of intersection perpendicular to their plane is also an axis.

4. Prove that if a set of points has a center on an axis, the plane through the center perpendicular to the axis is a plane of symmetry.

75. Symmetry for graphs.

From the definition of symmetry for a set of points, it follows immediately that the graph of the equation $f(x, y, z) = 0$ is symmetrical with respect to a plane (or line or point) if and only if corresponding to any point P whose coördinates satisfy the equation there is a point Q whose coördinates satisfy the equation, such that P and Q are symmetrically located with respect to the plane (or line or point).

Since the points (x, y, z) and $(-x, y, z)$ are located symmetrically with respect to the yz plane, we have the theorem:

Theorem I. *The graph of the equation $f(x, y, z) = 0$ is symmetrical with respect to the yz plane if and only if it is identical with the graph of the equation $f(-x, y, z) = 0$.*

A similar theorem holds for symmetry with respect to either of the other coördinate planes.

In an analogous manner we can set up criteria for symmetry with respect to the coördinate axes or the origin, typified as follows:

Theorem II. *The graph of the equation $f(x, y, z) = 0$ is symmetrical with respect to the x axis if and only if it is identical with the graph of the equation $f(x, -y, -z) = 0$.*

Theorem III. *The graph of the equation $f(x, y, z)$ is symmetrical with respect to the origin if and only if it is identical with the graph of the equation $f(-x, -y, -z) = 0$.*

These tests for symmetry are adequate for most of the graphs that one normally encounters. In particular, they provide a workable method for finding symmetry properties of quadric surfaces. For algebraic surfaces in general, and particularly for algebraic surfaces with multiple factors, the problem is more complicated. To compensate for the added difficulties, however, the tests (when they are finally established) contain refinements which greatly simplify the determination of symmetry properties. This problem of symmetry for algebraic surfaces is discussed in §77.

Example. State which coördinate planes are planes of symmetry, which coördinate axes are lines of symmetry, and whether the origin is a center:

(a) $2x^2 + y^2 - 5z^2 + 3xz + 12 = 0$;
(b) $xz + 8yz - 3z = 0$.

Solution. Let $f(x, y, z)$ denote the left member of the given equation in each case. (a) In this case, since $f(x, -y, z) \equiv f(x, y, z)$, the graph of $f(x, -y, z) = 0$

is identical with the graph of $f(x, y, z) = 0$, and the xz plane is therefore a plane of symmetry. It is the only coördinate plane that is a plane of symmetry. (For example, if the yz plane were a plane of symmetry, the equations

$$2x^2 + y^2 - 5z^2 + 3xz + 12 = 0 \quad \text{and} \quad 2x^2 + y^2 - 5z^2 - 3xz + 12 = 0$$

would have the same solutions, whereas $x = 1$, $y = 0$, $z = 2$ is a solution of the first equation but not of the second.) Similarly, since $f(-x, y, -z) \equiv f(x, y, z)$, the y axis is a line of symmetry. No other coördinate axis is a line of symmetry. The origin is a center. (b) With this equation, $f(x, y, -z) \equiv -f(x, y, z)$, and therefore the xy plane is a plane of symmetry. No other coördinate plane is a plane of symmetry. No coördinate axis is a line of symmetry, and the origin is not a center.

76. Exercises.

In Exercises **1-6**, state which coördinate planes are planes of symmetry, which coördinate axes are lines of symmetry, and whether the origin is a center.

1. $x^2 - xy - 2y^2 + 5z^2 - 6 = 0$.
2. $x^2 - 2y^2 + 4z^2 - 3xz + y - 5 = 0$.
3. $xy + xz + yz - 1 = 0$.
4. $x^2 + 4y^2 - z^2 + 5 = 0$.
5. $4xy - 2xz + x = 0$.
6. $xyz + x + y + z = 0$.

★ **77. Symmetry for algebraic surfaces.** In order to be able to discuss questions of multiplicity for surfaces, we have made a distinction between graphs and algebraic surfaces. For an algebraic surface that has factors of multiplicity greater than 1, this dissimilarity again receives accent in the subject of symmetry. For example, the graph of the equation

$$(z - 1)^2(z + 1) = 0$$

is symmetrical with respect to the xy plane, since it consists of two planes, a unit distance on each side of it. However, one of these planes has the multiplicity 2, and therefore we do not wish to call the xy plane a plane of symmetry for the *algebraic surface* $(z - 1)^2(z + 1) = 0$.

DEFINITION. *An algebraic surface $f(x, y, z) = 0$, each factor of which has multiplicity 1, is symmetrical with respect to a plane (or line or point) if and only if the graph of $f(x, y, z) = 0$ is symmetrical with respect to the plane (or line or point). In general, an algebraic surface is symmetrical with respect to a plane (or line or point) if and only if it can be expressed as the product of algebraic surfaces, none of which* has repeated factors and each of which is symmetrical with respect to the plane (or line or point).

For example, since the equation
$$(z-1)^2(z+1)^2 = 0$$
can be written
$$(z^2-1)(z^2-1) = 0,$$
the surface defined by this equation is the product of the two surfaces $z^2 - 1 = 0$ and $z^2 - 1 = 0$, neither of which has repeated factors and each of which is symmetrical with respect to the xy plane. Therefore the original surface is also symmetrical with respect to the xy plane.

The following three theorems are concerned with algebraic surfaces symmetrical with respect to coördinate planes and axes and the origin (cf. § 75):

THEOREM I. *Each of the following conditions is necessary and sufficient for the algebraic surface*

(1) $$f(x, y, z) = 0$$

to be symmetrical with respect to the yz plane:

  (i) *The surface* (1) *is identical with the surface*

(2) $$f(-x, y, z) = 0.$$

  (ii) *The equations* (1) *and* (2) *are equivalent.*

  (iii) *The coefficients of the terms of* (1) *of even degree in x or of odd degree in x all vanish.*

A similar theorem holds for symmetry with respect to either of the other coördinate planes.

THEOREM II. *Each of the following conditions is necessary and sufficient for the algebraic surface* (1) *to be symmetrical with respect to the x axis:*

  (i) *The surface* (1) *is identical with the surface*

(3) $$f(x, -y, -z) = 0.$$

  (ii) *The equations* (1) *and* (3) *are equivalent.*

  (iii) *The coefficients of the terms of* (1) *of even degree in y and z or of odd degree in y and z all vanish.*

A similar theorem holds for symmetry with respect to either of the other coördinate axes.

Theorem III. *Each of the following conditions is necessary and sufficient for the algebraic surface* (1) *to be symmetrical with respect to the origin:*

(i) *The surface* (1) *is identical with the surface*
(4) $$f(-x, -y, -z) = 0.$$

(ii) *The equations* (1) *and* (4) *are equivalent.*

(iii) *The coefficients of the terms of* (1) *of even degree in x, y, and z or of odd degree in x, y, and z all vanish.*

Proof of Theorem I. By Theorem II, § 65, conditions (i) and (ii) are equivalent. (That is, each condition implies the other.) To show that (i) is a necessary condition for symmetry, assume that the surface (1) is symmetrical with respect to the yz plane, and let it be represented as the product of surfaces none of which has repeated factors and each of which is symmetrical with respect to the yz plane:

$$f(x, y, z) \equiv f_1(x, y, z) f_2(x, y, z) \cdots f_p(x, y, z) = 0.$$

Then
$$f(-x, y, z) \equiv f_1(-x, y, z) f_2(-x, y, z) \cdots f_p(-x, y, z).$$

Our task is to show that the surfaces

$$f_i(x, y, z) = 0 \text{ and } f_i(-x, y, z) = 0, \qquad i = 1, 2, \cdots, p,$$

are identical. This follows immediately from the fact that these two equations, for any i, have identical graphs. (See Theorem I, § 75.)

Let us now show that (ii) is a sufficient condition. If equations (1) and (2) are equivalent, any factor of either surface must be a factor of the other, with the same multiplicity. Therefore, if $g(x, y, z) = 0$ is a factor of the surface (1), then $g(-x, y, z) = 0$ must also be a factor, with the same multiplicity. But these two factors are symmetrical to each other with respect to the yz plane, and therefore the surface (1) is itself symmetrical with respect to the yz plane.

It remains to be shown that the conditions (ii) and (iii) are equivalent. In the first place, if equations (1) and (2) are equivalent either $f(x, y, z) \equiv f(-x, y, z)$ or $f(x, y, z) \equiv -f(-x, y, z)$, and in either case the third condition holds. Conversely, if the third condition holds, either $f(x, y, z) \equiv f(-x, y, z)$ or $f(x, y, z) \equiv -f(-x, y, z)$, and equations (1) and (2) are equivalent. This completes the proof of Theorem I. Proofs of Theorems II and III are similar.

Theorems I, II, and III of this section depend for their proof on Theorem II, § 65, whose general proof is not given in this book. However, the algebraic surfaces in which we are primarily interested are planes and quadric surfaces.

As was remarked at the end of § **65,** proof of Theorem II of that section has been given for planes and will be given for quadric surfaces as well in Chapter VIII. Our treatment will therefore be logically complete for algebraic surfaces of the first and second degree.

★ 78. Exercises.

1. Apply the methods of § **77** to the Exercises of § **76.**

In Exercises **2-5,** formulate tests, similar to parts (*ii*) of Theorems I, II, and III, § **77,** for determining whether the surface $f(x, y, z) = 0$ is symmetrical with respect to the given plane, line, or point.

2. Plane $x = k$.
3. Plane $x = y$.
4. Line $x = y = z$.
5. Point (x_0, y_0, z_0).

79. Surfaces of revolution.

A surface which is symmetrical with respect to a line and which is cut only in circular sections by planes perpendicular to the line of symmetry is called a **surface of revolution.** Such a surface can be thought of as being generated by a plane curve, called a **generatrix,** revolved about a fixed line, called the **axis of revolution.** Particularly well adapted to algebraic treatment is a surface of revolution having a coördinate axis as axis of revolution.

Fig. 26 illustrates the surface obtained by revolving about the x axis a portion of the graph of a function $f(x)$.† Fig. 27 indicates the graphs of the equation $y = f(x)$ and those of the related equations $y = -f(x)$, $y = |f(x)|$, and $y^2 = (f(x))^2$.

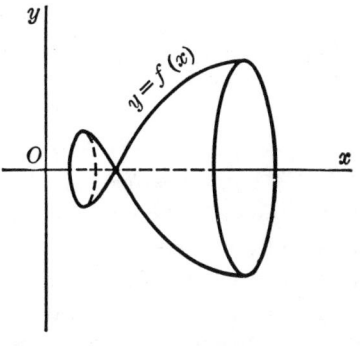

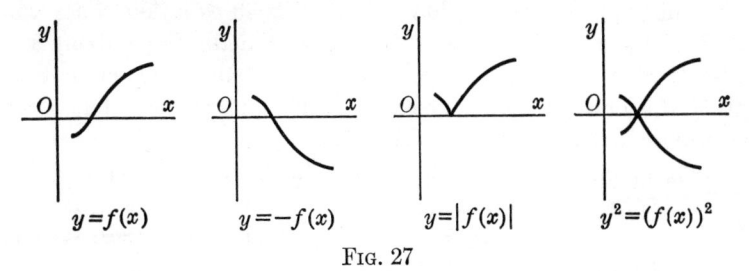

FIG. 26

$y = f(x)$ $y = -f(x)$ $y = |f(x)|$ $y^2 = (f(x))^2$

FIG. 27

† For this figure the observer is thought of as being a point in the *horizontal xz* plane, with positive x and z coördinates. The z axis would therefore appear as a horizontal line directed to the left. For simplicity, the z axis has not been labeled.

It is readily seen that the surfaces obtained by revolving about the x axis the curves shown in Fig. 27 are all the same, and that therefore it is not quite reasonable to speak of *the* generatrix, although it is sometimes convenient to do so.

We shall obtain the equation of the surface of revolution by considering the last curve of Fig. 27, which can be described as the locus of a point in the xy plane moving so that the square of its distance from the x axis is equal to the square of $f(x)$, where x is the directed distance of the point from the y axis. As the curve is revolved about the x axis, this point remains at a constant distance from the x axis and also at a constant distance from the yz plane. The surface of revolution can therefore be described as the locus of a point in space moving so that the square of its distance from the x axis is equal to the square of $f(x)$, where x is now the directed distance of the point from the yz plane. Since the square of the distance between the point (x, y, z) and the x axis is $y^2 + z^2$, the equation of the surface of revolution is

$$y^2 + z^2 = (f(x))^2.$$

In other words, to find the equation of the surface obtained by revolving the curve $y = f(x)$ about the x axis, we square each side of the given equation and replace y^2 by $y^2 + z^2$. The student will undoubtedly observe with pleasure that this is one occasion when he can square both sides of an equation with impunity—there is no danger that anything extraneous will be introduced in the process.

It is obvious that the graph of any equation in which y and z appear only in the form $y^2 + z^2$ is a surface of revolution about the x axis. We can now say, conversely, that any such surface is the graph of such an equation.

A similar procedure is followed in finding an equation of a surface obtained by revolving a curve in any coördinate plane about a coordinate axis lying in it. The results are tabulated below, not as a substitute for separate analyses of individual problems, but to emphasize the naturalness of the formulas.

Curve in the	revolved about	replace	by
xy plane	x axis	y^2	$y^2 + z^2$
	y axis	x^2	$x^2 + z^2$
xz plane	x axis	z^2	$y^2 + z^2$
	z axis	x^2	$x^2 + y^2$
yz plane	y axis	z^2	$x^2 + z^2$
	z axis	y^2	$x^2 + y^2$

A check on one's work is given by comparing the generatrix with the trace of the surface of revolution in the plane of the generatrix.

EXAMPLE 1. Write the equation of the surface of revolution obtained by revolving the curve
$$x = z^2, \; y = 0$$
(a) about the x axis; (b) about the z axis. Sketch the surface in each case.

Solution. (a) Replace z^2 by $y^2 + z^2$. The equation is $x = y^2 + z^2$. (b) Square both members of the equation $x = z^2$ and replace x^2 by $x^2 + y^2$. The equation is $x^2 + y^2 = z^4$. (See Fig. 28. For part (b) of Fig. 28, the observer is assumed to be in the xz plane, so that the x axis would be represented by a vertical line directed downward. It is not labeled on the figure.)

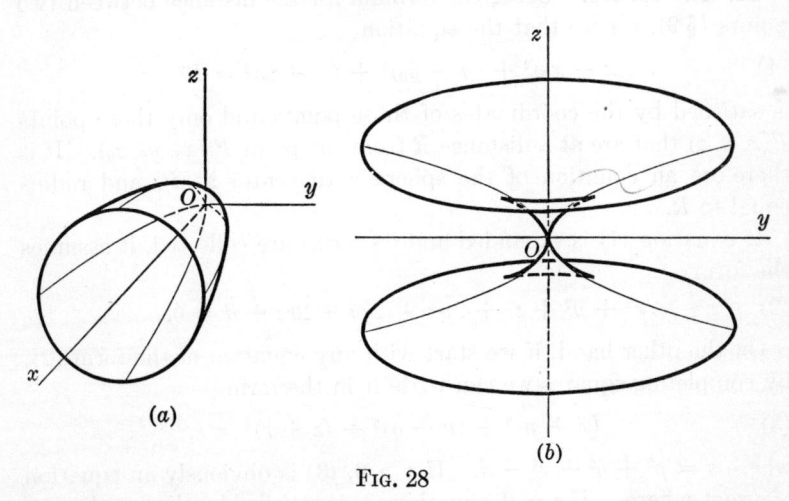

FIG. 28

EXAMPLE 2. Show that $x^2 + y^2 = z^2$ is the equation of a right circular cone.

Solution. The surface is obtained by revolving about the z axis (i) the curve $x^2 = z^2$ or $x = z$ or $x = -z$ in the xz plane, or (ii) the curve $y^2 = z^2$ or $y = z$ or $y = -z$ in the yz plane. The surface obtained by revolving one of two intersecting lines about the other is a right circular cone.

80. Exercises.

In Exercises 1-6, write an equation of the surface obtained by revolving the given curve about the specified axis. Draw a figure in each case.

1. $x^2 + 2y^2 = 8, \; z = 0$; (a) x axis; (b) y axis.
2. $4x^2 - 9z^2 = 5, \; y = 0$; (a) x axis; (b) z axis.
3. $6y^2 + 6z^2 = 7, \; x = 0$; (a) y axis; (b) z axis.
4. $2x + 3y = 6, \; z = 0$; (a) x axis; (b) y axis.
5. $y = 2, \; x = 0$, (a) y axis; (b) z axis.
6. $x^{\frac{2}{3}} + z^{\frac{2}{3}} = 1, \; y = 0$; (a) x axis, (b) z axis.

In Exercises **7-10**, state which coördinate axis is the axis of revolution for the surface, and write equations of a generatrix in the specified coördinate plane. Draw a figure.

7. $x^2 + y^2 + z = 2$; xz plane.   **8.** $x^2 - 4y^2 - 4z^2 = 8$; xy plane.
9. $x^2 - 4y^2 - 4z^2 = 0$; xz plane.
10. $x^2 + z^2 = 4y^4 - 4y^3 + 5y^2 - 2y + 1$; yz plane.

11. Find an equation of the torus obtained by revolving about the y axis the circle in the xy plane with center $(a, 0, 0)$ and radius b, where $b < a$.

* **12.** Formulate a rule for finding an equation of the surface obtained by revolving a curve in the xy plane about the line $x = a$, $z = 0$.

81. The sphere. Using the formula for the distance between two points (§ **9**), we see that the equation

$$(1) \qquad (x - x_0)^2 + (y - y_0)^2 + (z - z_0)^2 = R^2$$

is satisfied by the coördinates of those points and only those points $P(x, y, z)$ that are at a distance R from the point $P_0(x_0, y_0, z_0)$. It is therefore an equation of the sphere with center at P_0 and radius equal to R.

If equation (1) is expanded and its terms are collected, it assumes the form

$$(2) \qquad x^2 + y^2 + z^2 + 2px + 2qy + 2rz + d = 0.$$

On the other hand, if we start with any equation of the form (2), by completing squares we can write it in the form

$$(3) \qquad (x + p)^2 + (y + q)^2 + (z + r)^2 = s,$$

where $s = p^2 + q^2 + r^2 - d$. If $s > 0$, (3) is obviously an equation of a **real sphere**. If $s = 0$, equation (3) is satisfied by the coördinates of only one real point, $(-p, -q, -r)$. In this case the real graph consists of this one point and is called a **point sphere**, although, as we shall see in Chapter V, there is perhaps more reason for it to be called an *imaginary cone*. If $s < 0$, equation (3) has only imaginary solutions and is called an **imaginary sphere**. In any case, we call the surface (2) a **sphere** and the real point $(-p, -q, -r)$ its **center**.

Except for minor details, which the student can supply, we have proved the theorem:

Theorem. *A necessary and sufficient condition for the quadratic equation*
$$ax^2 + by^2 + cz^2 + 2fyz + 2gxz + 2hxy + 2px + 2qy + 2rz + d = 0$$
to be an equation of a sphere is that
$$a = b = c \neq 0 \quad and \quad f = g = h = 0.$$

It is rather easily seen that two equations of the type (2) represent the same sphere (that is, have the same solutions, real or imaginary) if and only if they are identical. From this it follows that if a second degree equation, $f(x, y, z) = 0$, represents a given sphere, its coefficients are uniquely determined except for a constant factor. We can therefore speak of *the* equation of a sphere in the same way that we speak of *the* equation of a plane.

EXAMPLE. Describe the graph of the equation
$$4x^2 + 4y^2 + 4z^2 - 16x + 24y + 8z + d = 0,$$
where d is in turn equal to 47, 56, and 65.

Solution. We divide each member of the equation by 4 and complete squares to obtain the equivalent equation
$$(x - 2)^2 + (y + 3)^2 + (z + 1)^2 = \frac{56 - d}{4}.$$

Regardless of the value of d the graph is a sphere with center $(2, -3, -1)$. If $d = 47$, the sphere is real with radius equal to $\frac{3}{2}$. If $d = 56$, the sphere is a point sphere whose only real point is its center. If $d = 65$, the sphere is imaginary.

82. Sphere through four points.

THEOREM. *The equation of the sphere through four non-coplanar points $P_1(x_1, y_1, z_1)$, $P_2(x_2, y_2, z_2)$, $P_3(x_3, y_3, z_3)$, and $P_4(x_4, y_4, z_4)$ can be written*

(1)
$$\begin{vmatrix} x^2 + y^2 + z^2 & x & y & z & 1 \\ x_1^2 + y_1^2 + z_1^2 & x_1 & y_1 & z_1 & 1 \\ x_2^2 + y_2^2 + z_2^2 & x_2 & y_2 & z_2 & 1 \\ x_3^2 + y_3^2 + z_3^2 & x_3 & y_3 & z_3 & 1 \\ x_4^2 + y_4^2 + z_4^2 & x_4 & y_4 & z_4 & 1 \end{vmatrix} = 0.$$

Instead of proving this theorem, we shall ask three questions whose correct answers will provide a proof:

(i) What degree does equation (1) have when the determinant is expanded with respect to the elements of the first row?

(ii) How do you know that the coefficients of x^2, y^2, and z^2 are equal to each other, but not equal to zero?

(iii) What is the value of the determinant if the coördinates of any one of the four given points are substituted for x, y, and z?

This equation can also be obtained by substituting the coördinates of each of the four points in the equation (2), § **81,** and solving for the four unknowns p, q, r, and d, in the resulting system of four linear equations. This system has a unique solution, since the coefficient matrix is non-singular. (Why?)

83. The circle. As in the case of any curve, a circle is determined by any two surfaces that have just the points of the curve in common. Two of the most easily handled pairs of surfaces are (*i*) two spheres and (*ii*) a sphere and a plane. In either case, the curve of intersection is called a **real circle,** a **point circle,** or an **imaginary circle** according as the two surfaces have in common more than one real point, just one real point, or no real points.

If two given spheres or a given sphere and a given plane intersect in a real circle, it is possible to determine the center and radius of this circle. This is illustrated by the following example:

EXAMPLE. Find the center and radius of the circle
$$x^2 + y^2 + z^2 + 2x - 2y - 6z - 14 = 0,$$
$$2x + 2y - z - 6 = 0.$$

Solution. We complete squares and find the sphere has center $P_0(-1, 1, 3)$ and radius 5. The distance between P_0 and the given plane is $\frac{1}{3}|-2 + 2 - 3 - 6|$ = 3. Therefore the radius of the circle is $\sqrt{25 - 9} = 4$. The center of the circle is the intersection of the given plane and the line through $(-1, 1, 3)$ with direction numbers 2, 2, −1. The parametric equations of this line are
$$x = -1 + 2t, y = 1 + 2t, z = 3 - t.$$

Substituting these expressions for x, y, and z in the equation of the plane, we find that $t = 1$, and the center of the circle is $(1, 3, 2)$. As a check, we see that the distance between the centers of the sphere and the circle is 3. The student should draw a figure to illustrate this example.

*84. Pencil of spheres. If
$$S_1 \equiv x^2 + y^2 + z^2 + 2p_1x + 2q_1y + 2r_1z + d_1 = 0$$
and
$$S_2 \equiv x^2 + y^2 + z^2 + 2p_2x + 2q_2y + 2r_2z + d_2 = 0$$
are equations of two non-concentric spheres, the family of spheres
(1) $$k_1S_1 + k_2S_2 = 0,$$
where k_1 and k_2 are parameters, not both zero, is called the **pencil of spheres** defined by the spheres $S_1 = 0$ and $S_2 = 0$. If the two given spheres intersect in a real circle, the pencil (1) consists of all spheres through this circle of intersection. (Prove this.) As a limiting case, the plane through this circle is given, when $k_1 + k_2 = 0$, by the equation
(2) $$S_1 - S_2 = 0.$$

Whether the given spheres intersect in a real circle or not, if they are not concentric the graph of the equation (2) is a plane. (Prove this.) This plane (2) is called the **radical plane** of the two spheres.

EXAMPLE. Find the radical plane of the two spheres

$$x^2 + y^2 + z^2 + 3x - 2y - z + 5 = 0, \quad x^2 + y^2 + z^2 + x + 3y - 4z - 1 = 0.$$

Solution. We equate to zero the difference between the left members of the given equations:
$$2x - 5y + 3z + 6 = 0.$$

85. Exercises.

In Exercises **1** and **2**, write the equation of the sphere with the given center and radius.

1. $(2, -5, 1); 6$.   **2.** $(3, 0, -4); 2$.

In Exercises **3** and **4**, write the equation of the sphere having the first point as center and passing through the second point.

3. $(4, 1, 3), (2, -1, 2)$.   **4.** $(-2, -1, 5), (8, -7, 3)$.

In Exercises **5-8**, find the center, and state whether the sphere is real, a point sphere, or imaginary. If it is real find the radius.

5. $x^2 + y^2 + z^2 - 6x + 2y + 4z - 11 = 0.$
6. $x^2 + y^2 + z^2 + 2x + 14y - 10z + 94 = 0.$
7. $2x^2 + 2y^2 + 2z^2 + 6x - 8y - 2z + 13 = 0.$
8. $9x^2 + 9y^2 + 9z^2 - 6x + 30y - 18z + 31 = 0.$

In Exercises **9** and **10**, find the equation of the sphere through the four given points.

9. $(0, 0, 0), (1, 0, 0), (0, 1, 0), (0, 0, 1).$
10. $(3, -2, 1), (5, -2, 2), (3, -1, 4), (4, -1, 5).$

11. Prove that the plane tangent to the real sphere

(1) $$x^2 + y^2 + z^2 + 2px + 2qy + 2rz + d = 0$$

at the point (x_1, y_1, z_1) has the equation

$$x_1 x + y_1 y + z_1 z + p(x_1 + x) + q(y_1 + y) + r(z_1 + z) + d = 0.$$

In Exercises **12** and **13**, find the equation of the plane tangent to the given sphere at the given point. (See Ex. **11**.)

12. $x^2 + y^2 + z^2 - 6x - 2y - 4z - 7 = 0; (5, 0, 6).$
13. $x^2 + y^2 + z^2 - 10x - 12z + 40 = 0; (3, 1, 2).$

In Exercises **14-16**, describe the graph of the system of equations.

14. $x^2 + y^2 + z^2 - 8x - 2y + 4z - 79 = 0, 7x + 4y - 4z + 41 = 0.$
15. $x^2 + y^2 + z^2 - 2x - 10y - 12z + 26 = 0,$
$x^2 + y^2 + z^2 + 4x + 8y - 6z - 40 = 0.$
16. $x^2 + y^2 + z^2 + 8x - 2y - 12z + 37 = 0,$
$x^2 + y^2 + z^2 - 10x + 2y - 23 = 0.$

*Supplementary Exercises

In Exercises **17** and **18**, find the equation of the sphere through the circle of intersection of the two given spheres and through the specified point.

17. $x^2 + y^2 + z^2 + 2x + 8y - 4z - 20 = 0$,
$x^2 + y^2 + z^2 - 10x - 8y - 2z = 0$; $(1, 3, -1)$.

18. $x^2 + y^2 + z^2 - 6x + 4y + z + 1 = 0$,
$x^2 + y^2 + z^2 - 2x + 3y + 6z + 1 = 0$; $(3, -2, -1)$.

In Exercises **19** and **20**, find the equation of the radical plane of the two given spheres.

19. $x^2 + y^2 + z^2 - 5x + 6y + z - 13 = 0$,
$x^2 + y^2 + z^2 + 8x - y + 2z + 5 = 0$.

20. $2x^2 + 2y^2 + 2z^2 + 4x + 10y - 2z - 11 = 0$,
$3x^2 + 3y^2 + 3z^2 + 18x - 13y - z + 1 = 0$.

21. Find the equation of the sphere whose center is $(2, -3, 5)$ and which belongs to the pencil determined by the two spheres
$$x^2 + y^2 + z^2 + 6y - 4z + 6 = 0,$$
$$x^2 + y^2 + z^2 + 2x + 6y - z - 5 = 0.$$

22. Find the equations of the spheres whose radii are 5 and which belong to the pencil determined by the spheres
$$x^2 + y^2 + z^2 + 2x + 6y - 6z = 17,$$
$$x^2 + y^2 + z^2 - 2x + 2y + z = 8.$$

23. Find the equation of the sphere with center $(3, -1, 6)$ tangent to the plane $4x - 8y - z + 22 = 0$.

24. Find the equations of the spheres with center $(3, -1, 6)$ tangent to the sphere $x^2 + y^2 + z^2 - 4x + 8y - 16z - 42 = 0$.

25. Prove that the left member of equation (1) of Exercise **11** is negative for the points inside the real sphere (1) and positive for the points outside the real sphere (1). Consider the cases where the sphere (1) is a point sphere or imaginary.

26. Prove that if $P(x, y, z)$ is a point outside the real sphere (1) of Exercise **11**, the left member of (1) is the square of the length of the tangent line segment from P to the sphere. *Suggestion:* If r is the radius of the sphere, t the length of the tangent line segment, and s the distance between P and the center of the sphere, use the fact that $t^2 = s^2 - r^2$.

27. Prove that the real spheres
$$x^2 + y^2 + z^2 + 2p_1x + 2q_1y + 2r_1z + d_1 = 0,$$
$$x^2 + y^2 + z^2 + 2p_2x + 2q_2y + 2r_2z + d_2 = 0$$
intersect orthogonally if and only if
$$2(p_1p_2 + q_1q_2 + r_1r_2) = d_1 + d_2.$$

Suggestion: Prove and make use of the fact that two spheres intersect orthogonally if and only if the square of the distance between their centers is equal to the sum of the squares of their radii.

28. Prove that the centers of the spheres of a pencil are collinear.

29. Prove that if S_1 and S_2 are distinct concentric spheres, the family (1), § 84, consists of all spheres concentric with S_1 and S_2. Is there a radical plane in this case?

30. Prove that the radical plane of two non-concentric spheres is perpendicular to the line through their centers.

31. Prove that the radical plane of two distinct point spheres is the perpendicular bisector of the line segment joining them. (Cf. Ex. **30.**)

32. Prove that the three radical planes obtained from different pairs of three spheres with non-collinear centers have exactly one line in common. *Suggestion:* Let the equations of the spheres be

$$x^2 + y^2 + z^2 + 2p_i x + 2q_i y + 2r_i z + d_i = 0, \quad i = 1, 2, 3,$$

and consider the matrix

$$\begin{pmatrix} p_1 - p_2 & q_1 - q_2 & r_1 - r_2 & d_1 - d_2 \\ p_1 - p_3 & q_1 - q_3 & r_1 - r_3 & d_1 - d_3 \\ p_2 - p_3 & q_2 - q_3 & r_2 - r_3 & d_2 - d_3 \end{pmatrix},$$

obtained by dividing by 2 the first three columns of the augmented matrix of the equations of the three radical planes. The three rows are linearly dependent, and the rank is less than 3. What is the rank of the coefficient matrix?

33. Prove that the six radical planes obtained from different pairs of four spheres with non-coplanar centers have exactly one point in common. What is this point if the four spheres are point spheres? (Cf. Ex. **8,** § **61,** and Exs. **31** and **32** of this section.)

34. Prove analytically that through four non-coplanar points passes *just one* sphere. (Cf. Ex. **33.**)

35. If $S_1 = 0$, $S_2 = 0$, and $S_3 = 0$ are equations of three spheres that do **not belong to the** same pencil, the spheres represented by the equations

(2) $$k_1 S_1 + k_2 S_2 + k_3 S_3 = 0,$$

where k_1, k_2 and k_3 are parameters, form a **bundle of spheres.** Under the assumption that the centers of the three given spheres are non-collinear, prove the following facts:

(*i*) The centers of the spheres of the bundle (2) form a plane, called the *plane of centers* of the bundle.

(*ii*) If the three given spheres have just two distinct real points in common, the bundle (2) consists of all spheres through these two points.

(*iii*) The planes defined by equation (2) form a pencil of planes through a line perpendicular to the plane of centers.

Suggestion for part (*iii*): Let $k_1 + k_2 + k_3 = 0$, and write

$$k_1 S_1 + k_2 S_2 + k_3 S_3 = k_1 (S_1 - S_3) + k_2 (S_2 - S_3).$$

36. If $S_1 = 0$, $S_2 = 0$, $S_3 = 0$, and $S_4 = 0$ are equations of four spheres that do not belong to the same bundle (see Ex. **35**), the spheres represented by the equations

(3) $$k_1 S_1 + k_2 S_2 + k_3 S_3 + k_4 S_4 = 0,$$

where k_1, k_2, k_3, and k_4 are parameters, form a **complex of spheres**. Under the assumption that the centers of the four given spheres are non-coplanar, prove the following facts:

(*i*) Any point in space is the center of a sphere of the complex (3).

(*ii*) The planes defined by equation (3) form a bundle of planes (Ex. **22**, § 61). (The point of intersection of the planes of this bundle is called the *radical center* of the four given spheres.)

Suggestion for part (*ii*): Let $k_1 + k_2 + k_3 + k_4 = 0$, and write $k_1 S_1 + k_2 S_2 + k_3 S_3 + k_4 S_4 = k_1(S_1 - S_4) + k_2(S_2 - S_4) + k_3(S_3 - S_4).$

86. Locus problems. Although locus problems of solid analytic geometry are in many respects similar to those of plane analytic geometry, there is one difference to be noted. Whereas a point in the plane case is considered as moving subject to one condition, thus describing a curve, a point in space is thought of as moving subject either to one condition or to two conditions, thus describing either a surface or a curve. That is, loci in space can be divided into two categories, surfaces and curves. A surface may be regarded as the locus of a point whose coördinates satisfy one equation, and a curve may be regarded as the locus of a point whose coördinates satisfy a system of two equations. Since a geometric condition on a point usually corresponds to an equation in the coördinates of the point, a surface can be considered as the locus of a point that moves subject to one geometric condition, and a curve as the locus of a point subject to two such conditions.

As in plane analytic geometry, locus problems in space are of two principal types: (*i*) to find a specific surface or curve defined in terms of certain fixed geometric objects, whose positions with respect to the coördinate system are specified (see Examples 1 and 3, below); (*ii*) to determine the *nature* of a surface or curve defined in terms of certain geometric objects whose location is not specified (see Examples 2 and 4, below). For the second type of problem it is expedient to consider the given geometric objects located with respect to the coördinate planes and axes in some especially advantageous manner. This process can be thought of in two ways: (*i*) placing the geometric objects in a fixed coördinate system, or (*ii*) choosing a coördinate system relative to the geometric objects considered now as fixed.

§ 86] LOCUS PROBLEMS

All locus problems in this book refer to real loci. The formula to be used for the distance between two points is therefore that for real three-dimensional space and not that for the complex space C defined in § 63.

EXAMPLE 1. Find the locus of a point equidistant from the points $(5, -1, 2)$ and $(1, 7, -4)$.

Solution. Let the coördinates of the point be (x, y, z). The required condition is represented by the equation

$$(x - 5)^2 + (y + 1)^2 + (z - 2)^2 = (x - 1)^2 + (y - 7)^2 + (z + 4)^2,$$

which is equivalent to the equation

$$2x - 4y + 3z + 9 = 0.$$

The locus sought is the plane that is the perpendicular bisector of the segment joining the two given points.

EXAMPLE 2. Find the locus of a point equidistant from two distinct points.

Solution. In this problem the location of the points is not specified. Let the points be $(a, 0, 0)$ and $(-a, 0, 0)$. (Why not $(1, 0, 0)$ and $(-1, 0, 0)$?) Proceeding as in Example 1, we obtain the equation $x = 0$, and the general result: The locus of a point equidistant from two distinct points is the plane that is the perpendicular bisector of the segment joining them.

EXAMPLE 3. A line segment extends from the point $(2, 4, 6)$ to a point on the sphere

$$x^2 + y^2 + z^2 + 12x - 8y - 16z + 80 = 0.$$

Find the locus of the midpoint of this segment.

Solution. Let $P(x, y, z)$ be the point sought and $P_1(x_1, y_1, z_1)$ be the variable point on the sphere. Then

$$x = \frac{x_1 + 2}{2}, \quad y = \frac{y_1 + 4}{2}, \quad z = \frac{z_1 + 6}{2}$$

or

$$x_1 = 2x - 2, \quad y_1 = 2y - 4, \quad z_1 = 2z - 6.$$

Substituting these expressions for x_1, y_1 and z_1 in the equation of the given sphere, we obtain

$$(2x - 2)^2 + (2y - 4)^2 + (2z - 6)^2 + 12(2x - 2) - 8(2y - 4) - 16(2z - 6) + 80 = 0,$$

or

$$x^2 + y^2 + z^2 + 4x - 8y - 14z + 60 = 0,$$

as the equation of the desired locus. The locus is a sphere whose radius is half that of the given sphere and whose center is the midpoint of the segment joining the given fixed point and the center of the given sphere.

EXAMPLE 4. Find the locus of a point whose distance from a fixed line is a given positive number a and whose distance from a fixed point on this line is $2a$.

Solution. Let the fixed line and point be the z axis and the origin, respectively, and the variable point be (x, y, z). Then the two given conditions are expressed by the equations
$$x^2 + y^2 = a^2, \quad x^2 + y^2 + z^2 = 4a^2.$$
Therefore the locus consists of the two circles
$$x^2 + y^2 = a^2, \quad z = \pm\sqrt{3}\, a.$$

87. Exercises.

In Exercises **1-10**, write an equation of the described locus of a point P. Name the locus.

1. The point P is equidistant from the planes
$$x - 2y + 2z + 1 = 0 \text{ and } 4x + y - 8z - 4 = 0.$$
2. The point P is equidistant from the planes
$$5x - 3y - z + 11 = 0 \text{ and } 5x - 3y - z - 3 = 0.$$
3. The point P is twice as far from the first plane of Ex. **1** as it is from the second.
4. The point P is twice as far from the first plane of Ex. **2** as it is from the second.
5. The point P is equidistant from the z axis and the origin.
6. The sum of the squares of the distances of P from the planes $x + y = 0$ and $x - y = 0$ is a constant a^2.
7. The sum of the squares of the distances of P from the points $(3, 0, 1)$ and $(5, 4, -3)$ is equal to 58.
8. The square of the distance of the point P from the point $(1, 0, 0)$ is equal to the distance of P from the plane $4x + 8y + 8z - 11 = 0$. (Cf. Ex. 24.)
9. The square of the distance of the point P from the point $(1, 0, 0)$ is equal to the distance of P from the plane $4x + 8y + 8z - 3 = 0$. (Cf. Ex. 24.)
10. The square of the distance of the point P from the point $(1, 0, 0)$ is equal to the distance of P from the plane $4x + 8y + 8z - 7 = 0$. (Cf. Ex. 24.)

11. Find the locus of a point equidistant from the three points $(6, 1, 1)$, $(8, 3, 3)$, and $(2, 1, -1)$.
12. Find the locus of a point equidistant from the origin and the xy plane.
13. Find the locus of a point whose distances from the yz, xz, and xy planes, in that order, are proportional to $1 : 2 : 3$.
* **14.** Find the locus of a point whose distances from the points $(1, 0, 0)$, $(0, 0, 0)$, and $(0, 1, 0)$, in that order, are proportional to $2 : 1 : 2$.
* **15.** Find the locus of a point whose distances from the points $(-5, 0, 0)$, $(0, 0, 0)$, and $(3, 0, 0)$, in that order, are proportional to $1 : 2 : 3$.
16. A line segment extends from the point $(3, 0, 0)$ to a point on the sphere $(x + 3)^2 + y^2 + z^2 = 4$. Find the locus of the midpoint of this segment.
17. A moving right circular cylinder contains the line $x = 3$, $y = 0$ and is tangent to the cylinder $(x + 3)^2 + y^2 = 4$. Find an equation of the surface generated by the axis of the moving cylinder.

In Exercises **18-21**, determine the nature of the described locus of a point P.

18. The distances of P from two distinct fixed points have a constant ratio.

19. The distances of P from three fixed non-collinear points are equal.

20. The sum of the squares of the distances of P from two distinct fixed points is constant.

21. The sum of the squares of the distances of P from two perpendicular planes is constant.

22. Prove that the locus of a point the sum of the squares of whose distances from the n points $P_i(x_i, y_i, z_i)$, $i = 1, \cdots, n$, is constant is a sphere with center
$$\left(\frac{x_1 + \cdots + x_n}{n}, \frac{y_1 + \cdots + y_n}{n}, \frac{z_1 + \cdots + z_n}{n}\right).$$

23. Prove that the locus of a point the sum of the squares of whose distances from two intersecting planes is constant is an elliptic cylinder whose axis is the line of intersection of the given planes. *Suggestion:* Let the planes be $x \pm ky = 0$.

24. The distance between a plane Π and a point P_0 is a. Prove that the locus of a point the square of whose distance from the point P_0 is equal to its distance from the plane Π is a sphere if $a > \frac{1}{4}$, a sphere and a point if $a = \frac{1}{4}$, and two spheres if $a < \frac{1}{4}$. (Cf. Exs. **8-10**.)

* 88. Cylindrical and spherical coördinates. Many problems in the geometry of space are more easily attacked with the aid of a coördinate system different from the rectangular system defined in Chapter I. We shall discuss two of the most common in this section, referring each to a basic rectangular system.

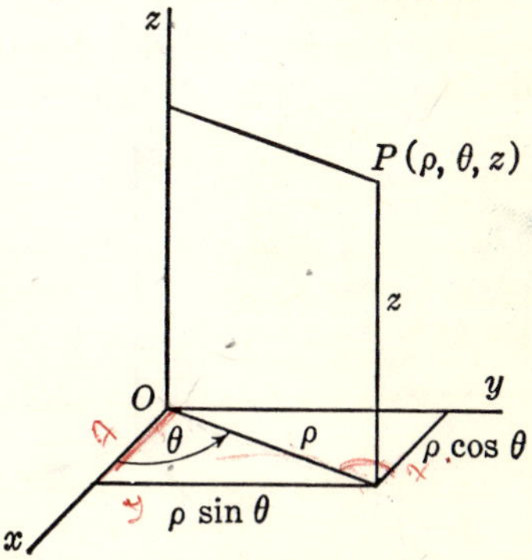

I. Cylindrical coördinates.

In this system a point is specified by the polar coördinates (ρ, θ) of its projection on the xy plane and its directed distance z from the xy plane. (See Fig. 29.) With the restrictions

$$\rho \geqq 0,$$
$$0° \leqq \theta < 360°,$$

any point not on the z axis has a unique representation (ρ, θ, z).

Fig. 29

If the cylindrical coördinates of a point are given, its rectangular coördinates can be obtained by means of the relations

(1) $$\begin{aligned}x &= \rho \cos \theta,\\ y &= \rho \sin \theta,\\ z &= z.\end{aligned}$$

Conversely, the cylindrical coördinates of a point are given in terms of the rectangular coördinates by the relations

(2) $$\begin{aligned}\rho^2 &= x^2 + y^2,\\ \tan \theta &= \frac{y}{x},\\ z &= z,\end{aligned}$$

the quadrant of θ being determined by the signs of x and y.

These relations can also be used to transform an equation of a surface in one coördinate system into an equation in the other coördinate system.

EXAMPLE. Write an equation in rectangular coördinates of the surface whose equation in cylindrical coördinates is $z = \rho^2$.

Solution. Using relations (2) we have the equation
$$z = x^2 + y^2.$$

II. Spherical coördinates.

The spherical coördinates (r, ϕ, θ) of a point P, as indicated in Fig. 30, are (*i*) its distance r from the origin O, (*ii*) the angle ϕ be-

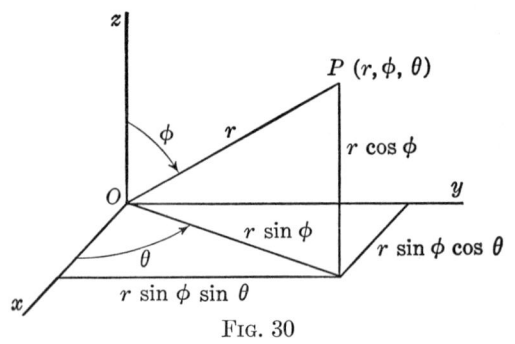

FIG. 30

tween OP and the z axis, and (*iii*) the angular polar coördinate θ of its projection on the xy plane, subject to the restrictions

$$\begin{aligned}r &\geq 0,\\ 0° \leq \phi &\leq 180°,\\ 0° \leq \theta &< 360°.\end{aligned}$$

Again the coördinates of any point not on the z axis are uniquely determined.

The equations of transformation are

(3)
$$x = r \sin \phi \cos \theta, \qquad r^2 = x^2 + y^2 + z^2,$$
$$y = r \sin \phi \sin \theta, \qquad \cos \phi = \frac{z}{\sqrt{x^2 + y^2 + z^2}},$$
$$z = r \cos \phi; \qquad \tan \theta = \frac{y}{x}.$$

EXAMPLE. Transform to an equation in spherical coördinates:
$$x^2 + y^2 + z^2 - 2z = 0.$$

Solution. Using relations (3) we have $r^2 - 2r \cos \phi = 0$, whose graph is identical with that of $r = 2 \cos \phi$.

★ 89. Exercises.

In Exercises **1-4**, write the rectangular coördinates of the point whose cylindrical coördinates are given. Draw a figure.

1. $(3, 0°, 2)$.
2. $(2, 270°, 3)$.
3. $(6, 60°, -1)$.
4. $(5, \text{arc tan } \frac{4}{3}, 0)$.

In Exercises **5-8**, write the cylindrical coördinates of the point whose rectangular coördinates are given. Draw a figure.

5. $(-4, 0, 1)$.
6. $(0, 5, -2)$.
7. $(6, 2\sqrt{3}, 0)$.
8. $(0, 0, 0)$.

In Exercises **9-12**, write the rectangular coördinates of the point whose spherical coördinates are given. Draw a figure.

9. $(7, 90°, 180°)$.
10. $(8, 45°, 90°)$.
11. $(12, 150°, 120°)$.
12. $(4, 45°, 45°)$.

In Exercises **13-16**, write the spherical coördinates of the point whose rectangular coördinates are given. Draw a figure.

13. $(0, -5, 0)$.
14. $(2, 2\sqrt{3}, 3)$.
15. $(0, 0, -3)$.
16. $(1, 1, 1)$.

In Exercises **17-20**, describe the graph of the equation in cylindrical coördinates. Draw a figure.

17. $\rho = $ constant.
18. $\theta = $ constant.
19. $\rho = \sin \theta$.
20. $z = \sin \theta$.

In Exercises **21-24**, describe the graph of the equation in spherical coördinates. Draw a figure.

21. $r = $ constant.
22. $\phi = $ constant.
23. $\theta = $ constant.
24. $f(r, \phi) = 0$.

In Exercises **25** and **26**, transform the equation from cylindrical to rectangular coördinates, and in Exercises **27** and **28**, transform the equation from rectangular to cylindrical coördinates. Draw a figure.

25. $z = \pm\rho$.
27. $x + 2y + 3z = 3$.
26. $\rho = \csc(\theta + 45°)$.
28. $x^2 + y^2 = x$.

In Exercises **29** and **30**, transform the equation from spherical to rectangular coördinates, and in Exercises **31** and **32**, transform the equation from rectangular to spherical coördinates. Draw a figure.

29. $r = 3$.
31. $z = 3$.
30. $r = \sin\phi$.
32. $x^2 + y^2 - 3z^2 = 0$.

In Exercises **33** and **34**, describe the curve defined by the equations in cylindrical coördinates. Draw a figure.

33. $\rho = $ constant, $z = $ constant.
34. $\rho = $ constant, $z = k\theta$.

In Exercises **35** and **36**, describe the curve defined by the equations in spherical coördinates. Draw a figure.

35. $\phi = $ constant, $\vartheta = $ constant.
36. $r = $ constant, $\phi = $ constant.

37. Write an equation in cylindrical coördinates of the locus of a point whose distance from the plane $z = 0$ is three times its distance from the z axis. Draw a figure and identify the surface.

38. Write an equation in spherical coördinates of the locus of a point the square of whose distance from the origin is equal to its distance from the plane $\phi = 90°$. Draw a figure and identify the surface.

CHAPTER V

THE SEVENTEEN QUADRIC SURFACES

90. Introduction. In this chapter we begin a systematic study of equations of the second degree in the variables x, y, and z. We shall study seventeen basic equations and determine the shapes of their graphs. It will be shown later that this list is complete; that is, that any quadric surface, with an appropriate choice of coördinate axes, corresponds to one of the fundamental seventeen equations. In other words, any equation of the second degree with real coefficients can be "transformed" into one of the fundamental seventeen, called the **canonical** or **standard** form of the equation.

In the listing below it is assumed that the letters a, b, c, and r that appear in the equations represent positive numbers. The surfaces and their equations are numbered consecutively from (1) to (17) through the chapter without regard to individual sections.

91. (1) Real ellipsoid.

(1) $$\frac{x^2}{a^2} + \frac{y^2}{b^2} + \frac{z^2}{c^2} = 1.$$

This surface is symmetrical with respect to each coördinate plane and axis and the origin. (See Fig. 31.) The section of the surface

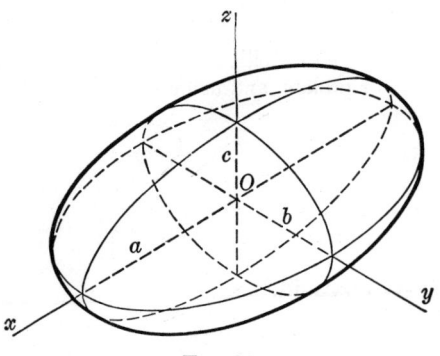

FIG. 31

by the plane $z = k$ is a real ellipse if $|k| < c$, a point (actually two imaginary lines intersecting in a real point) if $|k| = c$, and an imag-

inary ellipse if $|k| > c$. This can be seen if equation (1) is written in the form
$$\frac{x^2}{a^2} + \frac{y^2}{b^2} = \frac{c^2 - z^2}{c^2}.$$

Therefore the surface lies between the tangent planes $z = \pm c$. Similarly, it is bounded by the planes $x = \pm a$ and $y = \pm b$. The line segments on the axes with the intercepts as end points, or their lengths, $2a$, $2b$, and $2c$, are called the **axes** of the ellipsoid, and a, b, and c are called the **semi-axes**.

If the axes are all equal the ellipsoid is a sphere.

If two axes are equal and different from the third, the ellipsoid is called a **spheroid**, **oblate** if the third axis is shorter than the first two (like the earth or a pill), **prolate** if the third axis is longer than the first two (like a football or an egg).

92. Exercises.

1. Show that the sections of the ellipsoid (1) by planes parallel to a coördinate plane are similar ellipses (ellipses having the same eccentricity), decreasing in size as their distances from the origin increase.

2. Prove that a spheroid is a surface of revolution obtained by revolving an ellipse about an axis—about the minor axis for an oblate spheroid, and about the major axis for a prolate spheroid.

In Exercises **3** and **4**, determine the semi-axes of the ellipsoid. Draw a figure.

3. $4x^2 + y^2 + 9z^2 = 36$.
4. $x^2 + 2y^2 + 3z^2 = 6$.

In Exercises **5-8**, show that the ellipsoid is a spheroid. Determine whether the spheroid is oblate or prolate, and name the axis of revolution. Draw a figure.

5. $4x^2 + 9y^2 + 4z^2 = 36$.
6. $9x^2 + 9y^2 + 4z^2 = 36$.
7. $25x^2 + y^2 + z^2 = 25$.
8. $x^2 + 25y^2 + 25z^2 = 25$.

93. (2) Imaginary ellipsoid.

(2) $$\frac{x^2}{a^2} + \frac{y^2}{b^2} + \frac{z^2}{c^2} = -1.$$

Since this equation has only imaginary solutions and is similar in form to (1), its graph is called an imaginary ellipsoid.

94. (3) Hyperboloid of one sheet.

(3) $$\frac{x^2}{a^2} + \frac{y^2}{b^2} - \frac{z^2}{c^2} = 1.$$

This surface is symmetrical with respect to each coördinate plane and axis and the origin. (See Fig. 32.) Each section by a plane parallel to the xy plane is a real ellipse, and each section by a plane parallel to either of the other two coördinate planes is a hyperbola or a degenerate hyperbola consisting of two straight lines. (Which of these sections are degenerate? Cf. Ex. 2, § 95.) The surface is in one connected piece and accordingly is called a hyperboloid of *one sheet*.

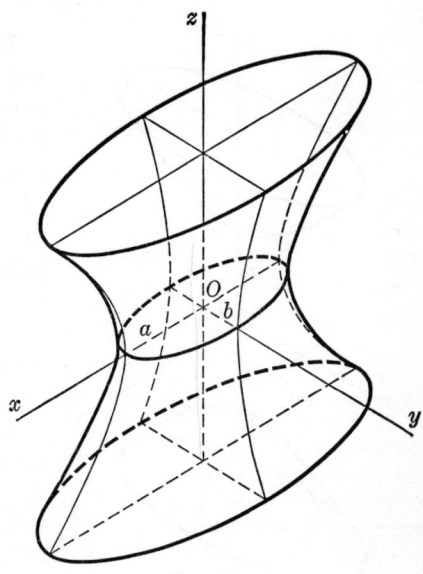

Fig. 32

Transverse and **conjugate axes** and **semi-axes** are defined by analogy with the terminology of plane analytic geometry. There are two transverse axes and one conjugate axis. The transverse semi-axes are a and b, and the conjugate semi-axis is c.

95. Exercises.

1. Show that the sections of (3) by planes parallel to the xy plane are similar ellipses, increasing in size as their distances from the origin increase. (Cf. Ex. 1, § 92.)

2. Show that the sections of (3) by planes parallel to, say, the xz plane are hyperbolas (or degenerate hyperbolas) whose projections on the xz plane have the same asymptotes. Are these projections similar? Which sections are degenerate, and what are their projections? (Cf. Ex. 1.)

3. Give further justification for the use of the terms "transverse axis" and "conjugate axis" by considering the traces of (3) in the xz and yz planes.

4. Prove that a hyperboloid of one sheet with equal transverse axes is the surface of revolution obtained by revolving a hyperbola about its conjugate axis.

96. (4) Hyperboloid of two sheets.

$$(4) \qquad \frac{x^2}{a^2} + \frac{y^2}{b^2} - \frac{z^2}{c^2} = -1.$$

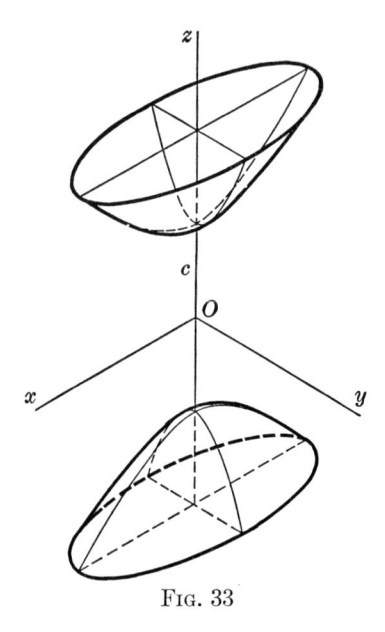

This surface is symmetrical with respect to each coördinate plane and axis and the origin. (See Fig. 33.) Each section by a plane $z = k$ is a real ellipse if $|k| > c$, a point if $|k| = c$, and an imaginary ellipse if $|k| < c$. Each section by a plane parallel to one of the other coördinate planes is a hyperbola. The surface consists of two separate pieces and accordingly is called a hyperboloid of *two sheets*.

Transverse and **conjugate axes** and **semi-axes** are defined as for the hyperboloid of one sheet. In this case there are one transverse axis and two conjugate axes. The transverse semi-axis is c and the conjugate semi-axes are a and b.

Fig. 33

Notice that the hyperboloids (3) and (4) might appropriately be called *conjugate* since a transverse axis of one is a conjugate axis of the other.

97. Exercises.

1. Formulate and prove statements for the hyperboloid (4), similar to those of Exercise 1 of § **92** and Exercises 1 and 2 of § **95**.

2. Prove that a hyperboloid of two sheets with equal conjugate axes is the surface of revolution obtained by revolving a hyperbola about its transverse axis.

In Exercises **3-6**, determine whether the hyperboloid is one of one sheet or one of two sheets and give the transverse and conjugate semi-axes in each case.

3. $4x^2 + y^2 - 9z^2 = 36.$
4. $x^2 - 4y^2 + 9z^2 + 36 = 0.$
5. $x^2 + 1 = y^2 + z^2.$
6. $z^2 - 1 = x^2 + y^2.$

98. (5) Real quadric cone.

$$\text{(5)} \qquad \frac{x^2}{a^2} + \frac{y^2}{b^2} - \frac{z^2}{c^2} = 0.$$

This surface is symmetrical with respect to each coördinate plane and axis and the origin. (See Fig. 34.) If $P_1(x_1, y_1, z_1)$ is any point on the surface, different from the origin O, then any point on the line OP_1 is also on the surface, since the coördinates of such a point have the form kx_1, ky_1, and kz_1, which satisfy (5) whenever x_1, y_1, and z_1 do. Therefore the surface (5) is a cone, according to the general definition of a cone as the surface generated by a moving line that passes through a fixed point, called the **vertex**. The vertex of (5) is the origin.

Each section of (5) by a plane parallel to the xz or yz plane is a hyperbola (degenerate if the section is the trace of the surface in either of these coördinate planes). Each section by a plane parallel to the xy plane is an ellipse (degenerating to a point, or two imaginary intersecting lines, if the section is the trace in the xy plane).

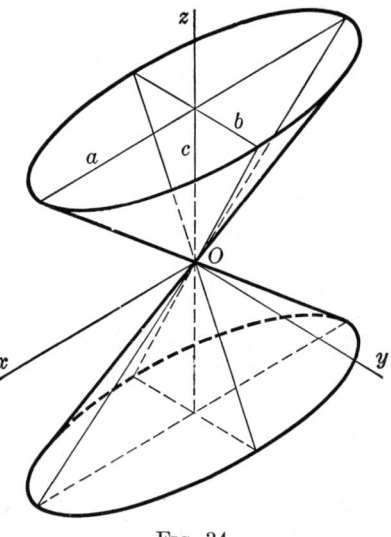

Fig. 34

The cone (5) is called the **asymptotic cone** of either of the hyperboloids (3) or (4). (See Ex. **5**, § 99.)

99. Exercises.

1. Formulate and prove statements for the quadric cone (5), similar to those of Exercise 1 of § 92 and Exercises 1 and 2 of § 95.

2. Show that if $a = b$, (5) is a right circular cone.

3. Show that the traces in the xz plane or in the yz plane of the surfaces (3), (4), and (5) are a pair of conjugate hyperbolas and their asymptotes.

4. Show that no two of the surfaces (3), (4), and (5) have a point in common.

*5. Study the relationship between the surfaces (3), (4), and (5) for points far from the origin, and show that (5) can be considered as the "asymptotic cone" for (3) and (4), and that a cone can be thought of as a "degenerate hyperboloid," or as the limiting form of a hyperboloid of either one or two sheets. *Suggestion:* Let k be a number whose absolute value is large, and consider the horizontal line through $(0, 0, k)$ whose equations are

$$x = \lambda t, \quad y = \mu t, \quad z = k,$$

the parameter t being the distance of the point (x, y, z) from the z axis. This line will meet the surfaces (3), (4), and (5) in points whose distances from the z axis are $\quad t_3 = s\sqrt{k^2 + c^2}, \quad t_4 = s\sqrt{k^2 - c^2}, \quad$ and $\quad t_5 = sk,$

where $$s = \left[c^2\left(\frac{\lambda^2}{a^2} + \frac{\mu^2}{b^2}\right)\right]^{-\frac{1}{2}}.$$

Show that as k increases numerically $t_3{}^2 - t_4{}^2$ remains constant, and consequently $t_3 - t_4$ can be made arbitrarily small. Notice finally that $0 < t_4 < t_5 < t_3$, and interpret these inequalities geometrically.

100. (6) Imaginary quadric cone.

$$\text{(6)} \qquad \frac{x^2}{a^2} + \frac{y^2}{b^2} + \frac{z^2}{c^2} = 0.$$

The real graph of this equation consists of just one point, the origin, and is sometimes called a *point ellipsoid*. However, in complex space the graph is similar to that of a real cone. For example, if (x_1, y_1, z_1) is any imaginary point on the surface and if k is any complex number, then (kx_1, ky_1, kz_1) is also a point on the surface. Therefore the surface is called an *imaginary cone*.

101. (7) Elliptic paraboloid.

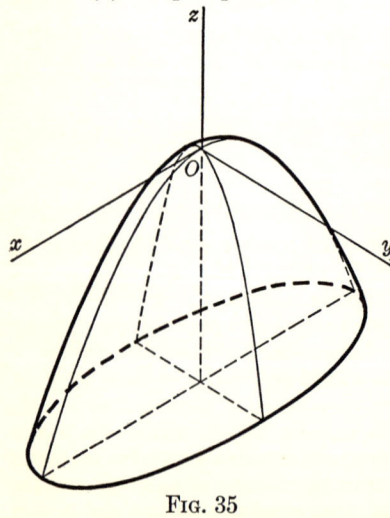

$$\text{(7)} \qquad \frac{x^2}{a^2} + \frac{y^2}{b^2} + 2z = 0.$$

This surface is symmetrical with respect to the xz plane, the yz plane, and the z axis. (See Fig. 35.) The origin is called the **vertex** of the paraboloid (7). The section of the surface by the plane $z = k$ is a real ellipse if $k < 0$, a point ellipse if $k = 0$, and an imaginary ellipse if $k > 0$. Sections by planes parallel to the other coördinate planes are parabolas with vertical axes, opening downward.

FIG. 35

102. Exercises.

1. Prove that the parabolic sections of (7) by planes parallel to the xz (or yz) plane are congruent.

2. Discuss the surface (7) if $a = b$.

103. (8) Hyperbolic paraboloid.

$$(8) \qquad \frac{x^2}{a^2} - \frac{y^2}{b^2} + 2z = 0.$$

This is a saddle-shaped surface symmetrical with respect to the xz plane, the yz plane, and the z axis. (See Fig. 36.) The origin is

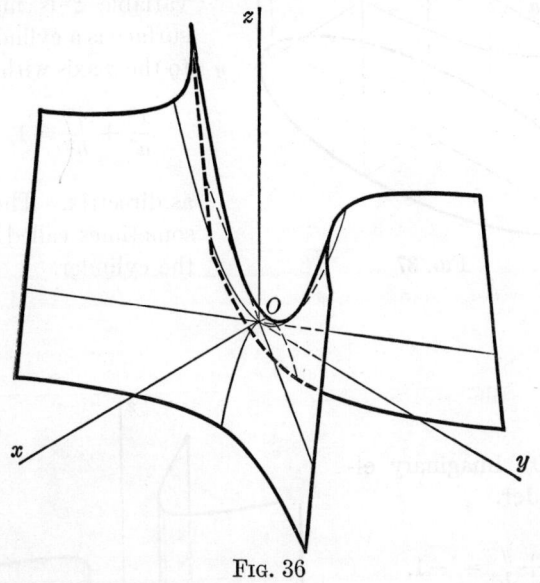

FIG. 36

called the **vertex** of the paraboloid (8). The section of the surface by the plane $z = k$ is a hyperbola, which is degenerate if $k = 0$. Sections by planes parallel to the other coördinate planes are parabolas with vertical axes, opening upward or downward.

104. Exercises.

1. Discuss similarity of hyperbolic sections of (8) by planes parallel to the xy plane.

2. Prove that the parabolic sections of (8) by planes parallel to the yz plane are congruent parabolas opening upward and that the parabolic sections by planes parallel to the xz plane are congruent parabolas opening downward.

105. (9) Real elliptic cylinder.

$$(9) \quad \frac{x^2}{a^2} + \frac{y^2}{b^2} = 1.$$

This surface is symmetrical with respect to the coördinate planes and axes and any point on the z axis. (See Fig. 37.) Since the variable z is missing, the surface is a cylinder parallel to the z axis with the ellipse

$$\frac{x^2}{a^2} + \frac{y^2}{b^2} = 1, \quad z = 0$$

as directrix. The z axis is sometimes called *the* axis of the cylinder.

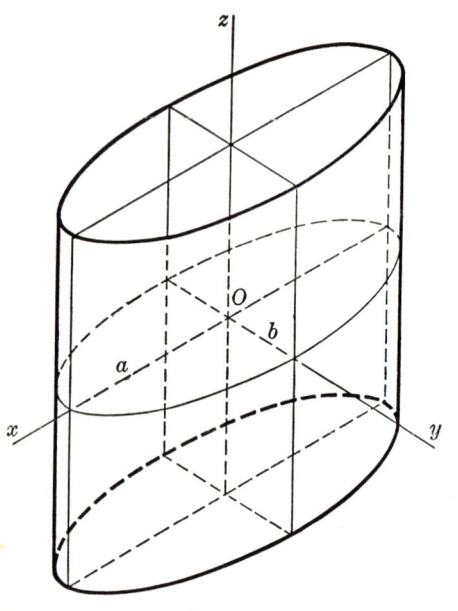

Fig. 37

106. (10) Imaginary elliptic cylinder.

$$(10) \quad \frac{x^2}{a^2} + \frac{y^2}{b^2} = -1.$$

This surface is completely imaginary.

107. (11) Hyperbolic cylinder.

$$(11) \quad \frac{x^2}{a^2} - \frac{y^2}{b^2} = -1.$$

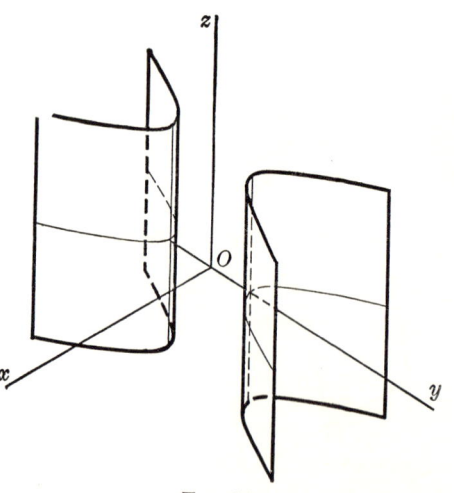

Fig. 38

This surface is symmetrical with respect to the coördinate planes and axes and any point on the z axis. (See Fig. 38.) Since the variable z is missing the surface is a cylinder parallel to the z axis.

108. (12) Real intersecting planes.

(12) $\quad \dfrac{x^2}{a^2} - \dfrac{y^2}{b^2} = 0.$

The surface is reducible, consisting of the two planes

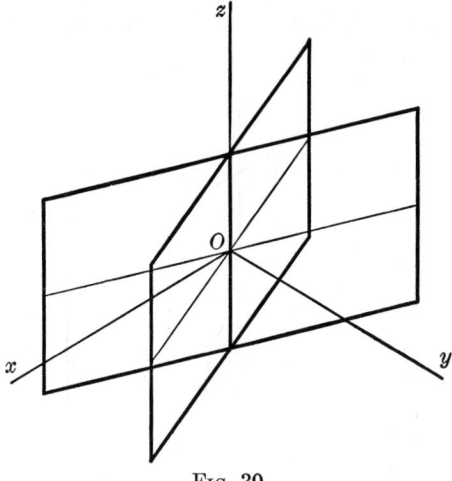

$$bx \pm ay = 0,$$

which intersect in the z axis. (See Fig. 39.)

Fig. 39

109. (13) Imaginary intersecting planes.

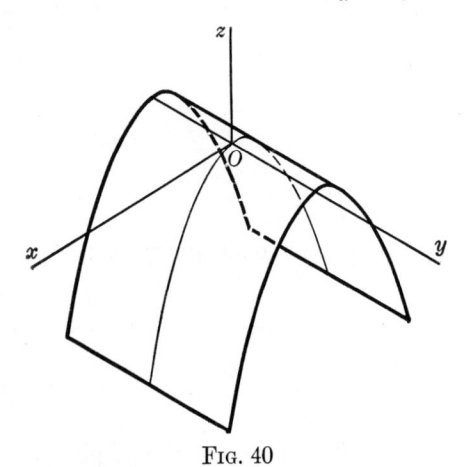

(13) $\quad \dfrac{x^2}{a^2} + \dfrac{y^2}{b^2} = 0.$

The real graph is the z axis. Accordingly, it is sometimes called a *line cylinder*. However, the surface is reducible, consisting of the two imaginary planes,

$$bx \pm iay = 0,$$

which intersect in the z axis.

110. (14) Parabolic cylinder.

(14) $\quad x^2 + 2rz = 0.$

Fig. 40

This surface is symmetrical with respect to the xz plane, the yz plane, and the z axis, and is a cylinder parallel to the y axis. (See Fig. 40.)

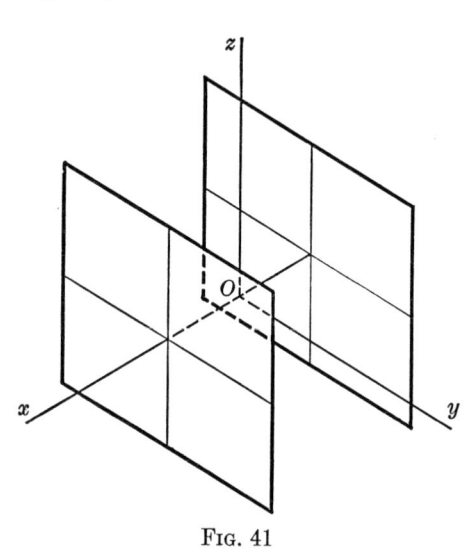

Since the points on the y axis are vertices of (parabolic) sections of (14) by planes parallel to the xz plane, these points are sometimes called **vertices,** and the y axis the **line of vertices,** of the parabolic cylinder.

111. Quadric cylinders. Any quadric surface that is a cylinder is called a **quadric cylinder.** Thus, surfaces (9), (10), (11), and (14) are quadric cylinders. They are the only irreducible quadric cylinders.

Fig. 41

112. (15) Real parallel planes.

(15) $$x^2 = a^2.$$

This surface is reducible, consisting of the distinct parallel planes

$$x = a, \quad x = -a.$$

(See Fig. 41.)

113. (16) Imaginary parallel planes.

(16) $$x^2 = -a^2.$$

This surface is reducible, consisting of the two imaginary planes

$$x = ia, \quad x = -ia.$$

114. (17) Coincident planes.

(17) $$x^2 = 0.$$

This surface is reducible, consisting of one factor (the yz plane) with multiplicity 2.

115. Equations almost in canonical form.

The surface

(A) $$x^2 + 2y^2 - 3z^2 = 12$$

is obviously a hyperboloid of one sheet, since division of each member by 12 reduces the equation to the canonical form (3).

Although the equation

(B) $$x^2 - 3y^2 + 2z^2 = 12$$

cannot be reduced to canonical form by division, its graph is clearly related to that of equation (A). In fact, it should be easy to see that the two surfaces are identical except for position; that is, the surfaces are *congruent*. The rôles of the y and z axes are interchanged.

The two surfaces

(C) $$x^2 + y^2 + 2z = 0 \quad \text{and} \quad x^2 + y^2 - 2z = 0$$

are also congruent. In this case the rôles of the positively directed z axis and the negatively directed z axis are interchanged.

As suggested by these examples, our classification of surfaces can easily be extended to include the graphs of many equations which differ only slightly from equations in canonical form. Of special interest are equations obtained by sequences of transformations of the following two types:

(i) Interchanging two variables;
(ii) Replacing a variable by its negative.

Regarding such equations we have the theorem:

THEOREM. *If any second degree equation in canonical form is transformed into a new equation by a sequence of transformations of types (i) and (ii), the graphs of these two equations are congruent.*

The following list contains those equations that are related in this way to the canonical equations (3), (4), (5), (7), and (8):

[3] $\quad \dfrac{x^2}{a^2} - \dfrac{y^2}{b^2} + \dfrac{z^2}{c^2} = 1, \qquad -\dfrac{x^2}{a^2} + \dfrac{y^2}{b^2} + \dfrac{z^2}{c^2} = 1;$

[4] $\quad \dfrac{x^2}{a^2} - \dfrac{y^2}{b^2} + \dfrac{z^2}{c^2} = -1, \qquad -\dfrac{x^2}{a^2} + \dfrac{y^2}{b^2} + \dfrac{z^2}{c^2} = -1;$

[5] $\quad \dfrac{x^2}{a^2} - \dfrac{y^2}{b^2} + \dfrac{z^2}{c^2} = 0, \qquad -\dfrac{x^2}{a^2} + \dfrac{y^2}{b^2} + \dfrac{z^2}{c^2} = 0;$

[7] $\quad \dfrac{x^2}{a^2} + \dfrac{y^2}{b^2} - 2z = 0, \quad \dfrac{x^2}{a^2} + \dfrac{z^2}{c^2} \pm 2y = 0, \quad \dfrac{y^2}{b^2} + \dfrac{z^2}{c^2} \pm 2x = 0;$

[8] $\quad \dfrac{x^2}{a^2} - \dfrac{y^2}{b^2} - 2z = 0, \quad \dfrac{x^2}{a^2} - \dfrac{z^2}{c^2} \pm 2y = 0, \quad \dfrac{y^2}{b^2} - \dfrac{z^2}{c^2} \pm 2x = 0.$

For the sake of the student who wishes a more general statement than that given in the Theorem above, let us observe first that the two surfaces

(D) $\qquad\qquad z = x^3 + 2y^3 \quad$ and $\quad z = 2x^3 + y^3$

are *not* congruent, although either equation is obtained from the other by an interchange of x and y.

To treat the case of the general algebraic surface, we think of relabeling the coördinate axes instead of moving the surface itself. Interchanging the letters associated with any two axes, or reversing the direction on any single axis, changes the orientation of the coördinate system from a right-handed system to a left-handed system or the reverse. If two such transformations are applied, the orientation is preserved. Since any right-handed system of axes can be moved into coincidence with any other right-handed system, it follows that if *any* equation is transformed into a new equation by an *even* number of transformations of the types (i) and (ii), the graphs of these two equations are congruent.

The reason that it is unnecessary in the Theorem above to specify that an even number of transformations be made is that in a second degree equation in canonical form there is always a variable that appears to the second degree only (there is a plane of symmetry) and that the additional transformation of replacing this variable by its negative can always be included to make a total of an even number of transformations. (The surfaces (D) do not have planes of symmetry.)

116. Exercises.

1. Discuss the sections of (9)–(17) by planes parallel to the coördinate planes.

2. Discuss the surface (9) as a surface of revolution.

3. Discuss the symmetry properties of (12), (15), and (17).

4. Discuss the asymptotic relation between (11) and (12).

In Exercises **5-28**, name the surface. If the surface is a surface of revolution, name the axis of revolution. Draw a figure.

5. $3x^2 + y^2 + 4z^2 = 0.$
6. $x^2 + 9z^2 = 9.$
7. $y^2 - 9z^2 = 81.$
8. $4y^2 + 5 = 0.$
9. $9x^2 - 4y^2 + 9z^2 = 36.$
10. $x^2 + 4y^2 = 2z.$
11. $5y^2 = 8.$
12. $4x^2 = y^2 + z^2.$
13. $y^2 + 2z^2 + 4 = 0.$
14. $x^2 + z^2 = y.$
15. $4x^2 + 4y^2 + z^2 = 16.$
16. $4x = 4y^2 - z^2.$
17. $3x^2 + 4y^2 + z^2 = -12.$
18. $z^2 = 0.$
19. $4y^2 + 9z^2 = 0.$
20. $4x^2 + 4y^2 + 4z^2 = 25.$
21. $x^2 - 4y^2 - 4z^2 = 4.$
22. $x^2 = 4y.$
23. $x^2 - 16y^2 = 0.$
24. $x = (y - 3z)(y + 3z).$
25. $4y^2 + 4z^2 = 9.$
26. $x^2 + y^2 + z^2 = x.$
27. $x^2 + y^2 + z^2 + 2xy + 2xz + 2yz = 1.$
28. $y = z^2 - 4z + 3.$

In Exercises **29-36**, write an equation in one of the seventeen canonical forms of this chapter whose graph satisfies the given conditions.

29. Ellipsoid with axes 8, 6, and 4, two of whose intercepts are (2, 0, 0) and (0, 0, −4).

30. One-sheeted hyperboloid of revolution containing the points (5, 0, 0) and (5, 5, 2).

31. Two-sheeted hyperboloid of revolution with equal transverse and conjugate axes, tangent to the planes $z = \pm 3$.

32. Right circular cone containing the line $x = y = z$.

33. Elliptic paraboloid containing the ellipse

$$9x^2 + y^2 = 144, \quad z = -2.$$

34. Hyperbolic paraboloid. The projection on the yz plane of the section by the plane $x = 2$ is the parabola

$$y^2 = 50z + 1, \quad x = 0.$$

35. One-sheeted hyperboloid of revolution containing the line

$$x = 1, \quad y = z.$$

36. Parabolic cylinder containing the point (12, 10, −24).

37. Let C be the circle of radius 3 lying in the plane $x = 2$ with center on the x axis. Write

(a) the equation of the right circular cylinder containing C;

(b) the equation of the cone with vertex at the origin containing C;

(c) the equation of the elliptic paraboloid of revolution with vertex at the origin containing C;

⋆ (d) the equation of the family of spheres containing C.

In Exercises **38-41**, write an equation of the described locus of a point P. Name the locus.

38. The point P is equidistant from the point (0, 0, −1) and the plane $z = 1$.

39. The point P is equidistant from the z axis and the xy plane.

40. The distance of P from the origin is equal to twice its distance from the xy plane.

41. The sum of the squares of the distances of P from the origin and the xy plane is equal to 4.

In Exercises **42** and **43**, determine the nature of the described locus of a point P. Name the surface.

42. The sum of the squares of the distances of P from a line and a fixed point on the line is constant.

43. The sum of the squares of the distances of P from a line and a plane perpendicular to the line is constant.

44. A line segment parallel to the xy plane extends from a point on the line $x = 3$, $y = 0$ to a point on the sphere $(x + 3)^2 + y^2 + z^2 = 4$. Find the locus of the midpoint of this segment.

* 45. Find the locus of the center of a sphere that passes through the point $(3, 0, 0)$ and is tangent to the sphere $(x + 3)^2 + y^2 + z^2 = 4$.

* 46. Find the equation in canonical form of the cone with vertex at the origin that is tangent to the sphere $x^2 + y^2 + (z - 5)^2 = 9$. (Cf. Exs. **36** and **37**, § **156**.)

* 47. Prove that the locus of a point the sum of whose distances from two distinct fixed points is constant (greater than the distance between the two fixed points) is a prolate spheroid. *Suggestion:* Let the two fixed points be $(c, 0, 0)$ and $(-c, 0, 0)$, let the constant sum be $2a$, and let $b^2 = a^2 - c^2$.

* 48. Prove that the locus of a point the absolute value of the difference of whose distances from two distinct fixed points is constant is a two-sheeted hyperboloid of revolution.

* 49. Prove that the surface generated by revolving one line about a distinct fixed line, the two lines being thought of as secured rigidly with relation to each other, is (*i*) a right circular cylinder if the lines are parallel and (*ii*) a one-sheeted hyperboloid of revolution if the two lines are skew but not perpendicular. *Suggestion:* Let the fixed line be the z axis and the moving line have the initial position $x = a$, $y = kz$. The equation of this surface is $x^2 + y^2 - k^2z^2 = a^2$.

* 50. Describe as a locus the graph of the equation

$$(x - x_1)^2 + (y - y_1)^2 + (z - z_1)^2 = (ax + by + cz + d)^2.$$

Then name the surface given by this equation, considering the cases where $ax_1 + by_1 + cz_1 + d$ is equal to zero or different from zero, and the cases where $a^2 + b^2 + c^2 > 1$, $=1$, or <1. *Suggestion:* By choosing an appropriate coördinate system, consider the related equation

$$x^2 + y^2 + z^2 = k^2x^2,$$

if the point (x_1, y_1, z_1) is in the plane $ax + by + cz + d = 0$; otherwise

$$(x + p)^2 + y^2 + z^2 = (x - p)^2$$

if $a^2 + b^2 + c^2 = 1$,

$$(x - k^2p)^2 + y^2 + z^2 = k^2(x - p)^2$$

if $a^2 + b^2 + c^2 \neq 1$.

CHAPTER VI

THE GENERAL EQUATION OF THE SECOND DEGREE

117. Introduction. The general equation of the second degree in the three variables x, y, and z (see § 65) can be written in the form

(1) $\quad f(x, y, z) \equiv ax^2 + by^2 + cz^2 + 2fyz + 2gxz + 2hxy$
$\qquad + 2px + 2qy + 2rz + d = 0.$

Throughout the remainder of this book the notation $f(x, y, z)$ will be used exclusively to represent the left member of this equation, and it will be assumed that the coefficients are real.

The function $f(x, y, z)$ can also be written in the form

(2) $\quad f(x, y, z) \equiv \left\{ \begin{array}{l} axx + hxy + gxz + px \\ + hyx + byy + fyz + qy \\ + gzx + fzy + czz + rz \\ + px + qy + rz + d \end{array} \right\},$

or in the form

(3) $\quad f(x, y, z) \equiv (ax + hy + gz + p)x + (hx + by + fz + q)y$
$\qquad + (gx + fy + cz + r)z + (px + qy + rz + d).$

Associated with $f(x, y, z)$ are two symmetric† matrices,

$$e \equiv \begin{pmatrix} a & h & g \\ h & b & f \\ g & f & c \end{pmatrix}, \qquad E \equiv \begin{pmatrix} a & h & g & p \\ h & b & f & q \\ g & f & c & r \\ p & q & r & d \end{pmatrix},$$

which play an important rôle in the analysis of the general equation of the second degree.

The determinant of the matrix E will be denoted by Δ, and the cofactor of each element of Δ by the corresponding capital letter. For example, the determinant of e is denoted by D, since this determinant is the cofactor of d in Δ. The surface (1) is called **singular** or **non-singular** according as E is singular or non-singular. The ellipsoids, hyperboloids, and paraboloids are non-singular. The other quadrics are singular.

† A matrix is **symmetric** if and only if corresponding rows and columns are identical.

132   THE GENERAL EQUATION   [Ch. VI

Since frequent reference will be made to the matrices e and E for second degree equations in canonical form, we list below the matrices E for the seventeen canonical equations, indicating elements only by the symbols $+$, $-$, and 0:

$$\begin{pmatrix} + & 0 & 0 & 0 \\ 0 & + & 0 & 0 \\ 0 & 0 & + & 0 \\ 0 & 0 & 0 & - \end{pmatrix} \begin{pmatrix} + & 0 & 0 & 0 \\ 0 & + & 0 & 0 \\ 0 & 0 & + & 0 \\ 0 & 0 & 0 & + \end{pmatrix} \begin{pmatrix} + & 0 & 0 & 0 \\ 0 & + & 0 & 0 \\ 0 & 0 & - & 0 \\ 0 & 0 & 0 & - \end{pmatrix} \begin{pmatrix} + & 0 & 0 & 0 \\ 0 & + & 0 & 0 \\ 0 & 0 & - & 0 \\ 0 & 0 & 0 & + \end{pmatrix}$$

1. Real ellipsoid   2. Imaginary ellipsoid   3. Hyperboloid of one sheet   4. Hyperboloid of two sheets

$$\begin{pmatrix} + & 0 & 0 & 0 \\ 0 & + & 0 & 0 \\ 0 & 0 & - & 0 \\ 0 & 0 & 0 & 0 \end{pmatrix} \begin{pmatrix} + & 0 & 0 & 0 \\ 0 & + & 0 & 0 \\ 0 & 0 & + & 0 \\ 0 & 0 & 0 & 0 \end{pmatrix} \begin{pmatrix} + & 0 & 0 & 0 \\ 0 & + & 0 & 0 \\ 0 & 0 & 0 & + \\ 0 & 0 & + & 0 \end{pmatrix} \begin{pmatrix} + & 0 & 0 & 0 \\ 0 & - & 0 & 0 \\ 0 & 0 & 0 & + \\ 0 & 0 & + & 0 \end{pmatrix}$$

5. Real quadric cone   6. Imaginary quadric cone   7. Elliptic paraboloid   8. Hyperbolic paraboloid

$$\begin{pmatrix} + & 0 & 0 & 0 \\ 0 & + & 0 & 0 \\ 0 & 0 & 0 & 0 \\ 0 & 0 & 0 & - \end{pmatrix} \begin{pmatrix} + & 0 & 0 & 0 \\ 0 & + & 0 & 0 \\ 0 & 0 & 0 & 0 \\ 0 & 0 & 0 & + \end{pmatrix} \begin{pmatrix} + & 0 & 0 & 0 \\ 0 & - & 0 & 0 \\ 0 & 0 & 0 & 0 \\ 0 & 0 & 0 & + \end{pmatrix} \begin{pmatrix} + & 0 & 0 & 0 \\ 0 & - & 0 & 0 \\ 0 & 0 & 0 & 0 \\ 0 & 0 & 0 & 0 \end{pmatrix}$$

9. Real elliptic cylinder   10. Imaginary elliptic cylinder   11. Hyperbolic cylinder   12. Real intersecting planes

$$\begin{pmatrix} + & 0 & 0 & 0 \\ 0 & + & 0 & 0 \\ 0 & 0 & 0 & 0 \\ 0 & 0 & 0 & 0 \end{pmatrix} \begin{pmatrix} + & 0 & 0 & 0 \\ 0 & 0 & 0 & 0 \\ 0 & 0 & 0 & + \\ 0 & 0 & + & 0 \end{pmatrix} \begin{pmatrix} + & 0 & 0 & 0 \\ 0 & 0 & 0 & 0 \\ 0 & 0 & 0 & 0 \\ 0 & 0 & 0 & \mp \end{pmatrix} \begin{pmatrix} + & 0 & 0 & 0 \\ 0 & 0 & 0 & 0 \\ 0 & 0 & 0 & 0 \\ 0 & 0 & 0 & 0 \end{pmatrix}$$

13. Imaginary intersecting planes   14. Parabolic cylinder   15, 16. Parallel planes   17. Coincident planes

A notation that will be used in Chapter IX to facilitate certain proofs should be mentioned now, if only to underline the fact that the matrices e and E are related to the function $f(x, y, z)$ in a natural way. If the standard a_{ij} notation for matrices is used and if the variables are written x_1, x_2, and x_3, then the homogeneous second degree part of $f(x, y, z)$ can be written more concisely in the form

$$\sum_{i,j=1}^{3} a_{ij} x_i x_j.$$

This type of expression is called a **quadratic form**, and is discussed again in Chapter IX. Furthermore, if an artificial fourth coördinate $x_4 = 1$ is introduced (this may remind some readers of homogeneous coördinates), the complete function $f(x_1, x_2, x_3)$ can be written as the quadratic form

$$\sum_{i,j=1}^{4} a_{ij} x_i x_j.$$

118. Intersections of quadrics and lines. The parametric equations of the line through the point $P_0(x_0, y_0, z_0)$ with direction cosines λ, μ, ν are

(1) $\qquad x = x_0 + \lambda t, \qquad y = y_0 + \mu t, \qquad z = z_0 + \nu t,$

where the parameter t is the directed distance of the point $P(x, y, z)$ from the point P_0. The points, real or imaginary, where this line meets the quadric surface $f(x, y, z) = 0$ correspond to the roots of the following equation in t:

(2) $\qquad f(x_0 + \lambda t, \quad y_0 + \mu t, \quad z_0 + \nu t) = 0.$

In expanded form this equation is

(3) $\qquad e(\lambda, \mu, \nu) t^2 + 2\phi(x_0, y_0, z_0, \lambda, \mu, \nu) t + f(x_0, y_0, z_0) = 0,$

where

(4) $\qquad e(\lambda, \mu, \nu) \equiv a\lambda^2 + b\mu^2 + c\nu^2 + 2f\mu\nu + 2g\lambda\nu + 2h\lambda\mu$

$$\equiv \left\{ \begin{array}{l} a\lambda\lambda + h\lambda\mu + g\lambda\nu \\ + h\mu\lambda + b\mu\mu + f\mu\nu \\ + g\nu\lambda + f\nu\mu + c\nu\nu \end{array} \right\}$$

$\equiv (a\lambda + h\mu + g\nu)\lambda + (h\lambda + b\mu + f\nu)\mu + (g\lambda + f\mu + c\nu)\nu,$

and

(5) $\quad \phi(x_0, y_0, z_0, \lambda, \mu, \nu) \equiv (ax_0 + hy_0 + gz_0 + p)\lambda$
$\qquad\qquad + (hx_0 + by_0 + fz_0 + q)\mu + (gx_0 + fy_0 + cz_0 + r)\nu$
$\qquad\qquad \equiv (a\lambda + h\mu + g\nu)x_0 + (h\lambda + b\mu + f\nu)y_0$
$\qquad\qquad\qquad + (g\lambda + f\mu + c\nu)z_0 + (p\lambda + q\mu + r\nu).$

There are four cases to consider:

(i) $e(\lambda, \mu, \nu) \neq 0$. The equation (3) is a quadratic equation with real coefficients, whose roots are either real or conjugate imaginary numbers. The line (1) meets the surface $f(x, y, z) = 0$ in *two* points, real or imaginary, which correspond to the two roots of (3). The line is said to determine a **chord** of the quadric surface, even if the two points of intersection are coincident. The midpoint of this chord corresponds to the arithmetic mean of the two roots of (3), and is always real. (Why?)

(ii) $e(\lambda, \mu, \nu) = 0$, $\phi(x_0, y_0, z_0, \lambda, \mu, \nu) \neq 0$. The equation (3) is linear, and the line meets the quadric surface in one real point.

(iii) $e(\lambda, \mu, \nu) = 0$, $\phi(x_0, y_0, z_0, \lambda, \mu, \nu) = 0$, $f(x_0, y_0, z_0) \neq 0$. The line and the quadric surface have no points in common, real or imaginary.

(iv) $e(\lambda, \mu, \nu) = 0$, $\phi(x_0, y_0, z_0, \lambda, \mu, \nu) = 0$, $f(x_0, y_0, z_0) = 0$. The line lies entirely in the quadric surface, and is called a **ruling** of the surface.

As with the vanishing of a determinant (see the footnote on page 63), the vanishing of $e(\lambda, \mu, \nu)$ can be thought of as exceptional, so that in the absence of other information one can expect a straight line to meet a quadric surface in exactly two points.

EXAMPLE. Find the points where the line
$$\frac{x-1}{2} = \frac{y+2}{1} = \frac{z-1}{-3}$$
meets the surface
$$x^2 + 6y^2 + 4yz + 2xy + 8y + z - 2 = 0.$$

Solution. We write the equations of the line in parametric form, using the direction numbers 2, 1, -3 instead of direction cosines:
$$x = 1 + 2t, \quad y = -2 + t, \quad z = 1 - 3t,$$
and substitute these expressions for x, y, and z in the equation of the surface. The quadratic equation obtained reduces to $2t^2 + 7t - 4 = 0$, whose roots are $\frac{1}{2}$ and -4. Therefore the points of intersection are $(2, -\frac{3}{2}, -\frac{1}{2})$ and $(-7, -6, 13)$.

119. Degenerate sections of quadrics. It will be proved in Chapter VII that any non-vacuous section of a quadric surface by a plane that is not a factor of the surface is a conic section or a line. The following special case of this theorem will be useful later in this chapter.

THEOREM. *If a plane Π and a quadric surface S have two distinct lines in common, then either they have no other points in common or every point of Π lies in S.*

Proof. Let Λ_1 and Λ_2 be two distinct lines lying in Π and S, let P_0 be any point of Π and S not on Λ_1 or Λ_2, and let Λ be any line in Π through P_0 not parallel to Λ_1 or Λ_2, or passing through their point of intersection. By § **118**, since Λ has three points in common with S (what are these points?), it lies entirely in S. By a repetition of this argument, we see that every point of Π lies in S.

120. Exercises.

In Exercises **1-4**, find the points where the line

$$\frac{x-1}{2} = \frac{y+2}{1} = \frac{z-1}{-3}$$

meets the given quadric surface.

1. $7y^2 + 2yz + 9x - 45 = 0.$
2. $6y^2 + 2yz + 6x - 36 = 0.$
3. $5y^2 + 2yz + 3x - 20 = 0.$
4. $6y^2 + 2yz + 5x + 13 = 0.$

In Exercises **5-8**, find the points where the line

$$\frac{x-2}{-1} = \frac{y}{3} = \frac{z-1}{2}$$

meets the given quadric surface.

5. $y^2 - z^2 + 2xz + 10x + 4y - 28 = 0.$
6. $x^2 + z^2 - 4xy + 8y - 5 = 0.$
7. $2y^2 - 4z^2 + xz - x + 4y + 4 = 0.$
8. $2y^2 - 5z^2 - xz - 8x + 5y - 7 = 0.$

121. Diametral planes.

Let $f(x, y, z) = 0$ be a quadric surface and let (λ, μ, ν) be a direction for which $e(\lambda, \mu, \nu) \neq 0$. As we observed in § **118**, any line with the direction (λ, μ, ν) determines a chord of $f(x, y, z) = 0$ which has a real midpoint. Let this midpoint be the fixed point $P_0(x_0, y_0, z_0)$ of § **118**. Then the sum of the two roots of equation (3), § **118**, is zero, since these roots (at least if they are real) are equal in magnitude and opposite in sign. (What happens if the roots are imaginary?) Conversely, if the sum of these roots is zero, P_0 is the midpoint of the chord. But a necessary and sufficient condition for the sum of the roots of a quadratic equation to vanish is that the coefficient of the linear term vanish. Therefore a necessary and sufficient condition for P_0 to be the midpoint of the chord under consideration is that $\phi(x_0, y_0, z_0, \lambda, \mu, \nu)$ vanish. Dropping subscripts, we can phrase this result as follows:

THEOREM. *The locus of midpoints of chords of the quadric surface $f(x, y, z) = 0$ with direction cosines λ, μ, ν is the plane*

(1) $\quad (a\lambda + h\mu + g\nu)x + (h\lambda + b\mu + f\nu)y$
$\quad\quad + (g\lambda + f\mu + c\nu)z + (p\lambda + q\mu + r\nu) = 0.$

In equation (1) *the direction cosines λ, μ, ν can be replaced by any set of direction numbers l, m, n of the family of chords.*

DEFINITION. *A plane that bisects a system of parallel chords of a quadric surface is called a **diametral plane** of the surface.*

The term *diametral plane* will be used only for the case where $e(\lambda, \mu, \nu) \neq 0$, that is, where there are actual chords. Whether $e(\lambda, \mu, \nu)$ vanishes or not, a plane whose equation has the form (1) is said to be **conjugate** to the direction (λ, μ, ν), or to any line having the direction (λ, μ, ν), with respect to the quadric surface $f(x, y, z) = 0$. By referring to the third form for $e(\lambda, \mu, \nu)$, equation (4), § **118**, it is easy to see that the plane conjugate to the direction (λ, μ, ν) is a diametral plane if and only if it is not parallel to (λ, μ, ν), and that if $e(\lambda, \mu, \nu) \neq 0$ the coefficients of x, y, and z in equation (1) cannot all vanish simultaneously (Ex. **10**, §**123**). As shown in Example 2, below, it is possible to have a direction without a conjugate plane.

NOTE. The equation of the plane conjugate to the direction (λ, μ, ν) can be written $\phi(x, y, z, \lambda, \mu, \nu) = 0$.

EXAMPLE 1. Find the diametral plane of the sphere
$$x^2 + y^2 + z^2 - 4x + 6y - 2z + d = 0$$
that bisects chords with direction numbers 2, −2, 1.

Solution. Direct substitution in equation (1), with the direction numbers 2, −2, 1 instead of direction cosines, gives the equation $2x - 2y + z - 11 = 0$. Notice that (*i*) this plane is independent of d, and therefore is as well defined if the sphere is imaginary as it is if the sphere is real; (*ii*) the plane passes through the center $(2, -3, 1)$ of the sphere and is perpendicular to the family of chords.

EXAMPLE 2. Examine equation (1) for the hyperbolic cylinder $x^2 - y^2 = 1$ and a direction having the given direction numbers: (*a*) (1, 0, 0); (*b*) (1, 1, 0); (*c*) (0, 0, 1).

Solution. (*a*) The plane (1) is the yz plane and is a diametral plane since $e(\lambda, \mu, \nu) = 1$. (*b*) The plane (1) is $x = y$. This is not a diametral plane since $e(\lambda, \mu, \nu) = 0$. It is parallel to the given direction. (*c*) Equation (1) reduces to $0 = 0$. There is no plane conjugate to the z axis.

122. Centers. A center of an arbitrary set of points was defined in § **73** as a point of symmetry of the set. After it has been proved that every equation of the second degree can be reduced to canonical form, it will be evident by individual examination of the seventeen canonical equations that for a *quadric surface* the following definition is equivalent:

DEFINITION. *A **center** of a quadric surface is a point with the property that any line through the point (i) determines a chord of the surface whose midpoint is the given point, or (ii) has no point in common with the surface, or (iii) lies entirely in the surface.*

The point $P_0(x_0, y_0, z_0)$ is therefore a center of the surface $f(x, y, z) = 0$ if and only if $\phi(x_0, y_0, z_0, \lambda, \mu, \nu)$ vanishes for every set of direction cosines λ, μ, ν; that is, if and only if, in the first form for $\phi(x_0, y_0, z_0, \lambda, \mu, \nu)$, (5), § **118**, the coefficients of λ, μ, and ν vanish. We therefore have the theorem:

Theorem I. *The centers of the quadric surface $f(x, y, z) = 0$ are the points whose coördinates are solutions of the system*

(1)
$$\begin{aligned} ax + hy + gz + p &= 0, \\ hx + by + fz + q &= 0, \\ gx + fy + cz + r &= 0. \end{aligned}$$

These equations are easily remembered, since their augmented matrix is the matrix of the first three rows of E.

Note. The system (1) is equivalent to the system
$$\frac{\partial f}{\partial x} = \frac{\partial f}{\partial y} = \frac{\partial f}{\partial z} = 0.$$

With the notation introduced in § **117**, if $D \neq 0$, the system (1) has a unique solution; there is exactly one center, whose coördinates can be written $\dfrac{P}{D}, \dfrac{Q}{D}, \dfrac{R}{D}$ (Ex. **13**, § **123**); the surface is called a **central quadric**.

If $D = 0$, the system (1) may be consistent or inconsistent. If $D = 0$ and the system (1) is consistent, the centers form a line if the rank of the matrix e is 2, and a plane if the rank is 1.

In general, we can state that *if there is at least one center, then there is exactly one center, a line of centers, or a plane of centers according as the rank of the matrix e is* 3, 2, *or* 1.

By examining the canonical forms for the equations of the quadrics, we see that the only central quadrics are the ellipsoids, hyperboloids, and cones; that the paraboloids and parabolic cylinder have no centers; that the elliptic and hyperbolic cylinders and intersecting planes have a line of centers; and that parallel or coincident planes have a plane of centers. These centers are all real, even when the surfaces are imaginary.

By comparing the equations (1) for centers, equation (1), § **121**, for a diametral plane, and the two forms (5), § **118**, for $\phi(x_0, y_0, z_0, \lambda, \mu, \nu)$, we immediately have the theorem:

Theorem II. *Any center lies in every diametral plane.*

Remark. This theorem can be put in the stronger form: *any center lies in every plane that is conjugate to some direction,* since the details of the proof of the above theorem do not depend on the non-vanishing of $e(\lambda, \mu, \nu)$.

EXAMPLE 1. Find the center of the quadric surface
$$2x^2 + 3y^2 - z^2 + 6xy - 2xz + 8yz - 2x - 28y + 18z + 7 = 0.$$
Solution. The three equations defining the center are
$$2x + 3y - z - 1 = 0,$$
$$3x + 3y + 4z - 14 = 0,$$
$$-x + 4y - z + 9 = 0.$$
The determinant $D \neq 0$ and the only solution is $x = 3$, $y = -1$, $z = 2$. The surface is therefore a central quadric with center $(3, -1, 2)$.

EXAMPLE 2. Find the centers of the quadric surface
$$3x^2 - 20z^2 - 2xy + 4xz + 4yz + 6y - 48z = 0.$$
Solution. The three equations defining the centers are
$$3x - y + 2z = 0,$$
$$-x + 2z + 3 = 0,$$
$$x + y - 10z - 12 = 0.$$
These three planes have the following line of centers in common:
$$\frac{x-1}{2} = \frac{y-1}{8} = \frac{z+1}{1}.$$

EXAMPLE 3. Find the centers of the quadric surface
$$y^2 + 4z^2 + 4yz + 2x + 6y + 2z = 0.$$
Solution. The first equation of system (1) is $1 = 0$. The system is therefore inconsistent and there are no centers.

123. Exercises.

In Exercises **1-4**, find the diametral plane of the given surface that bisects chords having the given direction numbers.

1. $3x^2 - y^2 + 4z^2 + 5 = 0$; 2, 4, -3.
2. $5x^2 + 4xy - 6yz + 2x + 7 = 0$; 1, 0, 2.
3. $xy + xz + yz = 1$; 1, 1, 1.
4. $z^2 + 2yx + 6x - 12z - 3 = 0$; 3, -1, 2.

In Exercises **5-8**, find the centers of the given surface, if there are any. If there is more than one center give the equations of the line of centers or the equation of the plane of centers.

5. $4x^2 + y^2 - 4xy + 8yz + 24x - 20y - 48z + 7 = 0.$
6. $x^2 + 4y^2 + 4xy + 6xz + 12yz - 8x - 16y - 6z = 0.$
7. $x^2 + 4xy + 4y^2 + 4x + 8y + 3 = 0.$
8. $xy + yz + x + 2z = 0.$

9. If $f(x, y, z) \equiv x^2 + y^2 - 2z^2 - 1$, show that the plane conjugate to the line $x = y = z$ is $x + y - 2z = 0$. Is this a diametral plane?

10. Prove that if $e(\lambda, \mu, \nu) \neq 0$, the coefficients of x, y, and z in equation (1), § **121**, cannot all vanish simultaneously. *Suggestion:* See the third form for $e(\lambda, \mu, \nu)$, (4), § **118**.

11. Prove that there exist direction cosines for which the coefficients of x, y, and z in equation (1), § **121**, all vanish if and only if $D = 0$, that is, if and only if the quadric is not a central quadric. (See Ex. **10**.)

12. Prove that for a central quadric and a given plane through the center, there is, except for sense, exactly one direction to which the plane is conjugate. (Cf. Exs. **10** and **11**.)

13. Prove that the coördinates of the center of a central quadric are $\dfrac{P}{D}$, $\dfrac{Q}{D}$, and $\dfrac{R}{D}$, as stated in § **122**.

14. Prove that $e(1, 0, 0) = a$, $e(0, 1, 0) = b$, and $e(0, 0, 1) = c$. Assuming that a, b, and c are all different from zero, prove that the three planes defined by system (1), § **122**, are the diametral planes that bisect chords parallel to the coördinate axes, and that any diametral plane is a linear combination of these three.

* Supplementary Exercises

In Exercises **15-22**, prove that the described set of planes is the set of all diametral planes of the given quadric surface, whose equation is assumed to be in the canonical form given in Chapter V. (For the hyperboloids and real quadric cone, see Ex. **13**, § **128**.)

15. Real ellipsoid; planes through the origin.
16. Elliptic paraboloid; planes parallel to the z axis.
17. Hyperbolic paraboloid; planes parallel to the z axis but not parallel to either line of the trace of the surface in the xy plane.
18. Real elliptic cylinder; planes containing the z axis.
19. Hyperbolic cylinder; planes containing the z axis but not asymptotic to the cylinder.
20. Real intersecting planes; planes containing the z axis but different from either of the given planes.
21. Parabolic cylinder; planes parallel to the yz plane.
22. Parallel or coincident planes; the yz plane.

23. Let Π_1 and Π_2 be diametral planes conjugate to the lines Λ_1 and Λ_2, respectively. Prove that if Λ_2 is parallel to Π_1, Λ_1 is parallel to Π_2. Two diametral planes that stand in this relation are called **conjugate**. Prove that two conjugate diametral planes cannot be parallel. Their line of intersection is called a **diameter**.

24. Three diametral planes are said to be **conjugate** if they have one and only one point in common and each is conjugate to the line of intersection of the other two. The three lines of intersection are called **conjugate diameters**. Examining the seventeen fundamental quadrics in turn, prove (*i*) that for each of the central quadrics the coördinate planes form a system of conjugate diametral planes and (*ii*) that for any one of the other quadrics there is no set of three conjugate diametral planes. (See Ex. 23.) *Suggestion for part* (*ii*): See Exs. **16-22.**

124. Singular points and regular points.

DEFINITION. A **singular point** *of a quadric surface is a point of the surface that is a center.* A **regular point** *of a quadric surface is a point of the surface that is not a center.*

125. Exercises.

1. By examining the seventeen canonical equations of Chapter V, show that the following are the only examples of singular points for quadric surfaces: (*i*) the vertex of a cone, (*ii*) the points on the line of intersection of a pair of intersecting planes, and (*iii*) the points of a pair of coincident planes.

*** 2.** Prove that the singular points of the surface $f(x, y, z) = 0$ are those points whose coördinates are the solutions of the system of four linear equations in three variables that have E as their augmented matrix. Prove that the surface $f(x, y, z) = 0$ has a singular point if and only if the ranks of e and E are equal, and show that this statement is consistent with that of Exercise **1.**

*** 3.** Discuss the following statements: (*i*) One can expect a quadric surface to be non-singular. (*ii*) One can expect a quadric surface to be a central quadric. (*iii*) One can expect a quadric surface to be an ellipsoid or a hyperboloid. (*iv*) One can expect a quadric surface to be irreducible. (*v*) One can expect a straight line to meet a quadric surface in two distinct points. (*vi*) One can expect a conjugate plane to be a diametral plane. (*vii*) One can expect a point on a quadric surface to be a regular point. (See the footnote on page 63.)

126. Tangent planes and normal lines.

DEFINITION. *A line is a* **tangent line** *of a quadric surface at a* **regular point** P_0 *of the surface if and only if it meets the surface in the two coincident points* P_0 *and* P_0. *A plane is a* **tangent plane** *of a quadric surface* **at a regular point** P_0 *of the surface if and only if it contains all of the tangent lines and rulings of the surface that pass through* P_0. *A line is a* **normal line** *of a quadric surface* **at a regular point** P_0 *of the surface if and only if it is normal to a tangent plane of the surface at* P_0. *A line (or plane) is a* **tangent line** (*or* **plane**) *of a quadric surface if and only if it is a tangent line (or plane) of the surface at some regular point of the surface. A line is a* **normal line** *of a quadric surface if and only if it is a normal line of the surface at some regular point of the surface.*

§ 126] TANGENT PLANES AND NORMAL LINES

THEOREM. *If $P_0(x_0, y_0, z_0)$ is a regular point of the quadric surface $f(x, y, z) = 0$, there are just one tangent plane and one normal line of the surface at the point P_0. Every line through P_0 that lies in the tangent plane of the surface at P_0 is either a tangent line or a ruling of the surface. The equation of the tangent plane can be written in any of the four forms:*

(1) $(ax_0 + hy_0 + gz_0 + p)(x - x_0) + (hx_0 + by_0 + fz_0 + q)(y - y_0)$
$+ (gx_0 + fy_0 + cz_0 + r)(z - z_0) = 0,$

(2) $\dfrac{\partial f}{\partial x}(x_0, y_0, z_0)(x - x_0) + \dfrac{\partial f}{\partial y}(x_0, y_0, z_0)(y - y_0)$
$+ \dfrac{\partial f}{\partial z}(x_0, y_0, z_0)(z - z_0) = 0,$

(3) $(ax_0 + hy_0 + gz_0 + p)x + (hx_0 + by_0 + fz_0 + q)y$
$+ (gx_0 + fy_0 + cz_0 + r)z + (px_0 + qy_0 + rz_0 + d) = 0,$

(4) $ax_0 x + by_0 y + cz_0 z + f(y_0 z + z_0 y) + g(x_0 z + z_0 x) + h(x_0 y + y_0 x)$
$+ p(x_0 + x) + q(y_0 + y) + r(z_0 + z) + d = 0.$

Equations of the normal line are

(5) $\dfrac{x - x_0}{ax_0 + hy_0 + gz_0 + p} = \dfrac{y - y_0}{hx_0 + by_0 + fz_0 + q}$
$= \dfrac{z - z_0}{gx_0 + fy_0 + cz_0 + r},$

(6) $\dfrac{x - x_0}{\dfrac{\partial f}{\partial x}(x_0, y_0, z_0)} = \dfrac{y - y_0}{\dfrac{\partial f}{\partial y}(x_0, y_0, z_0)} = \dfrac{z - z_0}{\dfrac{\partial f}{\partial z}(x_0, y_0, z_0)}.$

Proof. Since P_0 is not a center, the coefficients of x, y, and z in equation (1) are not all zero, and therefore form a set of direction numbers. A line through P_0 with direction (λ, μ, ν) is a tangent line or a ruling of the surface $f(x, y, z) = 0$ if and only if $\phi(x_0, y_0, z_0, \lambda, \mu, \nu) = 0$; that is (cf. (5), § **118**), if and only if it lies in the plane (1). Therefore the plane (1) is a tangent plane and the only tangent plane of the quadric surface $f(x, y, z) = 0$ at the point P_0. The only remaining part of the proof that is not trivial is showing the equivalence of equations (1) and (3), but this equivalence is an immediate consequence of the expansion of $f(x, y, z)$ given by equation (3), § 117.

For many students equation (4) will provide the simplest method of obtaining the equation of the tangent plane, since it has the same form as that customarily used for getting the equation of a tangent line of a conic. According to this method, the equation of the tangent plane can be written down directly from the equation of the quadric by making replacements of the same type as those used in plane analytic geometry, typified as follows:

$$x^2 \to x_0 x,$$
$$2xy \to x_0 y + y_0 x,$$
$$2x \to x_0 + x.$$

This same method can be adapted to the problem of finding the equations of the normal line.

EXAMPLE. Write the equations of the tangent plane and normal line of the surface
$$x^2 - 2y^2 - 3xz + yz + 4x - 7z + 15 = 0$$
at the point $(3, -1, 2)$.

Solution. Multiplying by 2 to avoid fractions, and making the necessary replacements, we have for the tangent plane

$$2 \cdot 3x - 4(-1)y - 3(2x + 3z) + (2y - z) + 4(x + 3) - 7(z + 2) + 30 = 0,$$
or
$$4x + 6y - 17z + 28 = 0.$$

The normal line is therefore

$$\frac{x-3}{4} = \frac{y+1}{6} = \frac{z-2}{-17}.$$

127. Normal planes and tangent lines. If P is a point on the curve of intersection C of two quadric surfaces S_1 and S_2, and if S_1 and S_2 have distinct normal lines Λ_1 and Λ_2, and therefore distinct tangent planes Π_1 and Π_2, at P, then the plane containing Λ_1 and Λ_2 is called the **normal plane** of C at P and the line of intersection of Π_1 and Π_2 is called the **tangent line** of C at P.

EXAMPLE. Find the equations of the normal plane and tangent line of the curve of intersection of the quadrics

$$5x^2 + y^2 + 4z^2 + 2xy + 2xz - 6y - 14z + 26 = 0,$$
$$4x^2 - 5z^2 + 6xy + 4yz - 16x - 4z - 2 = 0$$

at the point $(-1, 5, 2)$.

Solution. According to equation (3), § **126**, the normals to the two surfaces at the point $(-1, 5, 2)$ have direction numbers 2, 1, 0 and 3, 1, -2, and therefore the direction perpendicular to these two normals has direction numbers 2, -4, 1. The equation of the normal plane is thus $2(x + 1) - 4(y - 5) + (z - 2) = 0$, and the equations of the tangent line are

$$\frac{x+1}{2} = \frac{y-5}{-4} = \frac{z-2}{1}.$$

128. Exercises.

In Exercises **1-4**, write the equations of the tangent plane and normal line of the given quadric surface at the specified point.

1. $5x^2 + y^2 + 2z^2 = 49$; $(1, -6, 2)$.
2. $2x^2 + z^2 + 6xy - 2xz + 10y + 12z - 25 = 0$; $(1, 2, -1)$.
3. $2xy + 3xz - 4yz = 6$; $(2, 0, 1)$.
4. $x^2 + 6yz + 4x - 2y + 14z = 0$; $(0, 0, 0)$.

In Exercises **5** and **6**, find the point of the given quadric surface at which the given plane is tangent.

5. $3x^2 - y^2 + 2z^2 + 4 = 0$; $3x - 5y - 6z + 4 = 0$.
6. $xy + 2yz + 3x + 4z + 2 = 0$; $2y - z + 7 = 0$.

7. Prove that if $f(x, y, z) = 0$ is a real sphere, the plane (3), § **126**, is perpendicular to the line segment joining the center of the sphere and the point P_0.

In Exercises **8** and **9,** find the equations of the normal plane and tangent line of the curve of intersection of the given quadric surfaces at the specified point.

8. $x^2 + y^2 + 2z^2 = 7$, $xy = 2$; $(1, 2, 1)$.
9. $x^2 - 2y^2 + 2z = 0$, $8x^2 + y^2 - z^2 + 8 = 0$; $(2, 3, 7)$.

10. Define *normal plane* and *tangent line* for the section of a quadric surface by a plane. Prove that the tangent line of the trace of a quadric surface in the xy plane is identical with the tangent line as defined in plane analytic geometry.

11. Find the equations of the normal plane and tangent line of the section of the surface

$$xy + xz + yz = -3$$

by the plane

$$3x + y - 4z = 3$$

at the point $(-1, 2, -1)$. (See Ex. **10**.)

*Supplementary Exercises

12. Prove that the lines

$$\Lambda_1 : \frac{x}{1} = \frac{y}{0} = \frac{z}{1}, \quad \Lambda_2 : \frac{x}{1} = \frac{y-1}{0} = \frac{z}{1}$$

are rulings of the cone and hyperboloid of one sheet

$$S_1 : x^2 + y^2 - z^2 = 0, \quad S_2 : x^2 + y^2 - z^2 = 1,$$

respectively. Prove that as a point P moves along the line Λ_1 the tangent plane of S_1 at P remains fixed. Prove that as a point P moves along the line Λ_2 the tangent plane of S_2 at P revolves about the line Λ_2, taking all possible positions except that parallel to the y axis. (Cf. Ex. **22**, § **133**.)

13. Prove that the diametral planes of the hyperboloid of one sheet, the hyperboloid of two sheets, or the real quadric cone, whose equations in canonical form are

(1) $\quad \dfrac{x^2}{a^2} + \dfrac{y^2}{b^2} - \dfrac{z^2}{c^2} = 1, \quad \dfrac{x^2}{a^2} + \dfrac{y^2}{b^2} - \dfrac{z^2}{c^2} = -1, \quad \dfrac{x^2}{a^2} + \dfrac{y^2}{b^2} - \dfrac{z^2}{c^2} = 0,$

respectively, consist of all planes through the origin that are not tangent to the cone. (Cf. Exs. **15-22**, § **123**.)

14. For the hyperboloid of one sheet and the real quadric cone, equations (1), prove that any plane tangent to the cone cuts the hyperboloid in two parallel lines. *Suggestion:* Let the plane be tangent to the cone at the point (λ, μ, ν), where $\lambda^2 + \mu^2 + \nu^2 = 1$. This plane, the xy plane, and the hyperboloid have in common the two points

$$\left(\dfrac{\pm a^2 \mu}{\sqrt{\lambda^2 b^2 + \mu^2 a^2}}, \dfrac{\mp b^2 \lambda}{\sqrt{\lambda^2 b^2 + \mu^2 a^2}}, 0 \right).$$

The two lines through these points having the direction (λ, μ, ν) lie in the hyperboloid. By § **119**, the section consists only of these two lines.

15. Let S be the ellipsoid or hyperboloid

$$\dfrac{x^2}{p} + \dfrac{y^2}{q} + \dfrac{z^2}{r} = 1,$$

and for a given direction (λ, μ, ν) introduce the notation:

$$A \equiv \lambda^2 p + \mu^2 q + \nu^2 r, \quad B \equiv \sqrt{|A|}.$$

Prove that there exist real planes with normal direction (λ, μ, ν) tangent to S if and only if $A > 0$, that the equations of these tangent planes are

(2) $\quad \lambda x + \mu y + \nu z = \pm B,$

and that their points of tangency are

$$\left(\pm \dfrac{p\lambda}{B}, \pm \dfrac{q\mu}{B}, \pm \dfrac{r\nu}{B} \right).$$

In Exercises **16-18**, find equations of the planes which are tangent to the given quadric surface and whose normals have the given set of direction numbers. Find the points of tangency. (See Ex. **15**.)

16. $\dfrac{x^2}{5} + \dfrac{y^2}{1} + \dfrac{z^2}{4} = 1;\ 2,\ 5,\ 1.$

17. $\dfrac{x^2}{12} + \dfrac{y^2}{8} - \dfrac{z^2}{4} = 1;\ 1,\ 1,\ -1.$

18. $\dfrac{x^2}{25} - \dfrac{y^2}{16} - \dfrac{z^2}{9} = 1;\ 2,\ -1,\ -4.$

19. Let S be the elliptic or hyperbolic paraboloid

$$\frac{x^2}{p} + \frac{y^2}{q} + 2z = 0.$$

Prove that there exists a (real) plane with normal direction (λ, μ, ν) tangent to S if and only if $\nu \neq 0$, that the equation of this tangent plane is

$$\lambda x + \mu y + \nu z = \frac{p\lambda^2 + q\mu^2}{2\nu},$$

and that its point of tangency is

$$\left(\frac{p\lambda}{\nu}, \frac{q\mu}{\nu}, -\frac{p\lambda^2 + q\mu^2}{2\nu^2}\right).$$

In Exercises **20** and **21**, find the equation of the plane which is tangent to the given quadric surface and whose normals have the given set of direction numbers. Find the point of tangency. (See Ex. **19**.)

20. $\dfrac{x^2}{2} + \dfrac{y^2}{3} + 2z = 0;\ 5,\ -2,\ 1.$

21. $\dfrac{x^2}{6} - \dfrac{y^2}{8} + 2z = 0;\ 4,\ 3,\ -2.$

22. Let P be any point of the curve of intersection of a quadric surface S and a plane Π, whose equations are $S = 0$ and $\Pi = 0$, respectively, and let T be the quadric surface whose equation is

$$S + k\Pi^2 = 0,$$

where k is any real constant. Assuming that P is a regular point of S, prove that P is a regular point of T, and that the two quadrics S and T have a common tangent plane at P.

23. Let P be a point of the ellipsoid or hyperboloid

$$\frac{x^2}{p} + \frac{y^2}{q} + \frac{z^2}{r} = 1,$$

not in the xy plane, and let Q be the point where the normal to the quadric at P meets the xy plane. Discuss the locus of the midpoint M of PQ. Show that, if r is not equal to $2p$ or to $2q$, M lies on a quadric surface of the same type as the original. What is the locus if $p = q = 1, r = 2$?

24. The family of quadric surfaces

(3) $$\frac{x^2}{a^2 - t} + \frac{y^2}{b^2 - t} + \frac{z^2}{c^2 - t} = 1,$$

where $a > b > c > 0$ and t is a parameter, is called a **family of confocal quadrics**.

(a) Prove that the traces of this family in any coördinate plane are a family of confocal conics.

(b) Determine the values of t for which the surfaces are (i) ellipsoids; (ii) hyperboloids of one sheet; (iii) hyperboloids of two sheets.

(c) Prove that through any point not on a coördinate plane there pass three surfaces of the family (3), one of each type mentioned in part (b).

(d) Prove that the three surfaces of part (c) are mutually perpendicular; that is, that their tangent planes (or normals) are mutually perpendicular.

Suggestions: (c) Transform equation (3) to the form $f(t) = 0$, where $f(t)$ is a third degree polynomial in t, and consider the signs of $f(-\infty)$, $f(c^2)$, $f(b^2)$, and $f(a^2)$. (d) Let (x_1, y_1, z_1) be any point not on a coördinate plane, and let t_1 and t_2 be two distinct roots of (3). If the left member of equation (3) is denoted by $g(x, y, z, t)$, then $g(x_1, y_1, z_1, t_1) - g(x_1, y_1, z_1, t_2) = 1 - 1 = 0$, and this fact can be expressed by the equation

$$\frac{x_1}{a^2 - t_1} \cdot \frac{x_1}{a^2 - t_2} + \frac{y_1}{b^2 - t_1} \cdot \frac{y_1}{b^2 - t_2} + \frac{z_1}{c^2 - t_1} \cdot \frac{z_1}{c^2 - t_2} = 0.$$

25. Discuss the section of a quadric surface by a diametral plane as the locus of the points of tangency of a system of parallel lines. (Cf. Ex. **34, § 156.**)

* **129. Poles and polar planes.** If $P_0(x_0, y_0, z_0)$ is any point that is not a center of the quadric surface $f(x, y, z) = 0$, equation (3), **§ 126,**

(1) $\quad (ax_0 + hy_0 + gz_0 + p)x + (hx_0 + by_0 + fz_0 + q)y$
$\quad\quad + (gx_0 + fy_0 + cz_0 + r)z + (px_0 + qy_0 + rz_0 + d) = 0,$

is the equation of a plane, whether P_0 lies on the surface or not. This plane is called the **polar plane** of P_0 with respect to the quadric $f(x, y, z) = 0$, and P_0 is called a **pole** of the plane with respect to the quadric.

Notice that equations (1) and (2), **§ 126,** are always equivalent and that equations (3) and (4), **§ 126,** are always equivalent, but that all four equations are equivalent if and only if P_0 is on the quadric. Therefore equations (1) and (2), **§ 126,** are to be used only for tangent planes, whereas equations (3) and (4), **§ 126,** may be used for polar planes in general. A tangent plane is a special case of a polar plane when P_0 is on the quadric.

For convenience, we repeat equation (4), **§ 126:**

(2) $\quad ax_0x + by_0y + cz_0z + f(y_0z + z_0y) + g(x_0z + z_0x) + h(x_0y + y_0x)$
$\quad\quad + p(x_0 + x) + q(y_0 + y) + r(z_0 + z) + d = 0.$

Because of the symmetrical relationship between (x, y, z) and (x_0, y_0, z_0) in this equation, we have only to substitute (i) x_1, y_1, z_1, x_2, y_2, and z_2 for x, y, z, x_0, y_0, and z_0, respectively, and (ii) x_2, y_2, z_2, x_1, y_1, and z_1 for x, y, z, x_0, y_0, and z_0, respectively, to prove the theorem:

§ 129] POLES AND POLAR PLANES

Theorem I. *If neither P_1 nor P_2 is a center of a quadric surface S, and if P_1 lies in the polar plane of P_2, then P_2 lies in the polar plane of P_1.*

As we shall see in the Exercises of § **130**, a plane may be the polar plane of more than one point. That is, the pole of a plane is not necessarily uniquely determined. The basic theorem on existence and uniqueness of poles is the following:

Theorem II. *A necessary and sufficient condition that a plane Π have one and only one pole with respect to a quadric surface S is that S be non-singular and Π not be the plane conjugate to any direction.*

Proof. Let the quadric S and the plane Π have equations $f(x, y, z) = 0$ and $\alpha x + \beta y + \gamma z + \delta = 0$, respectively. Then Π is the polar plane of the point $P_0(x_0, y_0, z_0)$ if and only if the coefficients of x, y, and z and the constant term of equation (1) are proportional to α, β, γ, and δ; that is, if and only if the system of equations

$$(3) \quad \begin{aligned} ax + hy + gz + k\alpha + p &= 0, \\ hx + by + fz + k\beta + q &= 0, \\ gx + fy + cz + k\gamma + r &= 0, \\ px + qy + rz + k\delta + d &= 0 \end{aligned}$$

in the unknowns x, y, z, and k is satisfied when $x = x_0$, $y = y_0$, $z = z_0$, and k has some value different from zero. The plane Π has exactly one pole if and only if the system (3) has exactly one solution (x, y, z, k), where $k \neq 0$. Letting the determinant of the coefficients of system (3) be denoted by T, and recalling that the determinant of the matrix E is denoted by Δ, we see that the system (3) has exactly one solution if and only if $T \neq 0$ and that if $T \neq 0$, the value of k given by Cramer's rule is $-\Delta/T$, which vanishes if and only if S is singular. It remains to be proved that if S is non-singular, the determinant T vanishes if and only if the plane Π is conjugate to some direction. But Π is conjugate to some direction if and only if the system

$$(4) \quad \begin{aligned} al + hm + gn + \alpha &= 0, \\ hl + bm + fn + \beta &= 0, \\ gl + fm + cn + \gamma &= 0, \\ pl + qm + rn + \delta &= 0 \end{aligned}$$

has a non-trivial solution (l, m, n). Since $\Delta \neq 0$, the rank of the coefficient matrix of system (4) is 3, and therefore a (non-trivial) solution of (4) exists if and only if the rank of the augmented matrix is 3; that is, if and only if $T = 0$.

130. Exercises.

In Exercises **1-4**, find the polar plane of the given point with respect to the given quadric surface.

1. $(2, -1, 6)$; $x^2 + 3y^2 - 2z^2 + 4 = 0$.
2. $(3, 0, -4)$; $xy - xz + 2yz = 5$.
3. $(3, 2, 7)$; $2x^2 + y^2 + 3z^2 + 10xy - 2xz + 8y - 8 = 0$.
4. $(1, 1, 1)$; $3x^2 - y^2 + 2z^2 + 4xy - 10xz - 2y + 6z = 0$.

In Exercises **5** and **6**, find the pole of the given plane with respect to the given quadric surface.

5. $5x - 4y - 3z = 15$; $5x^2 + y^2 - 6z^2 = 30$.
6. $4x + y + 2z + 2 = 0$; $x^2 + 4xy + 4yz - 8z = 0$.

7. The poles of the plane $2x - y + 2z + 2 = 0$ with respect to the quadric $x^2 + 2yz - 4z = 0$ lie on a line. Write the equations of this line.

8. Prove that if the plane Π_1 has a pole lying in the plane Π_2, then every pole of Π_2 lies in Π_1.

9. Prove that if P is any point that is not a center of the quadric surface S and if Π is the polar plane of P with respect to S, then the following four statements are equivalent (that is, each implies the other three):

(*i*) P lies in S.
(*ii*) P lies in Π.
(*iii*) Π is tangent to S at P.
(*iv*) Π is tangent to S.

10. Prove that if P_1 is a regular point of a quadric surface, if P_2 lies in the tangent plane of the surface at P_1, and if P_2 is not a center of the surface, then P_1 lies in the polar plane of P_2.

11. Let C be the section of a non-singular quadric surface S by a plane Π that is not conjugate to any direction. Prove that the planes tangent to S at points of C pass through the pole of Π with respect to S, and have no other points in common. (See Ex. **10**. Cf. Ex. **36, § 156**.)

* Supplementary Exercises

12. Prove that the polar plane of a point with respect to a non-singular quadric does not pass through a center of that quadric.

13. Prove that a plane cannot have *more* than one pole with respect to a non-singular quadric.

14. If $f(x, y, z) = 0$ is a non-singular quadric and if $\alpha x + \beta y + \gamma z + \delta = 0$ is a plane that is not conjugate to any direction, prove that its pole with respect to the quadric is

$$\left(\frac{A\alpha + H\beta + G\gamma + P\delta}{P\alpha + Q\beta + R\gamma + D\delta},\ \frac{H\alpha + B\beta + F\gamma + Q\delta}{P\alpha + Q\beta + R\gamma + D\delta},\ \frac{G\alpha + F\beta + C\gamma + R\delta}{P\alpha + Q\beta + R\gamma + D\delta} \right),$$

where the notation is that introduced in **§ 117**.

15. Prove that the polar plane of any point with respect to a quadric surface passes through all singular points of the surface.

In Exercises **16-24**, each second degree equation is assumed to be in the canonical form given in Chapter V.

16. For an ellipsoid or a hyperboloid of one or two sheets, prove that a plane has a pole if and only if it does not pass through the origin.

17. For an elliptic or hyperbolic paraboloid, prove that a plane has a pole if and only if it is not parallel to the z axis.

18. For a quadric cone, prove that a plane has a pole if and only if it passes through the origin, and that any such plane has a line of poles through the origin (excluding the origin itself).

19. For an elliptic or hyperbolic cylinder, prove that a plane has a pole if and only if it is parallel to the z axis but does not contain it, and that any such plane has a line of poles parallel to and distinct from the z axis.

20. For a pair of intersecting planes, prove that a plane has a pole if and only if it contains the z axis, and that any such plane has a plane of poles containing the z axis (excluding the points on the z axis itself).

21. Give an analysis of poles and polar planes, similar to those of Exercises **16-20**, for a parabolic cylinder, a pair of distinct parallel planes, and a pair of coincident planes.

22. For an ellipsoid or a hyperboloid of one or two sheets, prove that if Λ is any line through the origin, the polar planes of points of Λ different from the origin consist of all planes parallel to and distinct from the plane conjugate to Λ.

23. For an elliptic or hyperbolic paraboloid, prove that if Λ is any line parallel to the z axis, the polar planes of points of Λ constitute a family of parallel planes.

24. If S is an ellipsoid or a hyperboloid of one or two sheets and Λ any line not passing through the origin, or if S is an elliptic or hyperbolic paraboloid and Λ any line not parallel to the z axis, prove that the polar planes of points of Λ consist of all planes of a pencil except for the plane that is conjugate to Λ. The axis of this pencil is called the **polar line** of Λ (with respect to S).

25. Prove that if S is a non-singular quadric (with an equation in canonical form) and if Λ_1 is any line that has a polar line Λ_2, then Λ_1 is the polar line of Λ_2. The two lines Λ_1 and Λ_2 are called **conjugate polar lines** (with respect to S). (See Ex. **24**.)

26. Prove that if S is a non-singular quadric (with an equation in canonical form) and if Λ_1 and Λ_2 are conjugate polar lines, then Λ_1 and Λ_2 are identical if and only if they are a ruling of S, and Λ_1 and Λ_2 are distinct and intersecting if and only if they are both tangent to S at their point of intersection. (See Exs. **24** and **25**.)

27. Let S be a non-singular quadric (with an equation in canonical form) and let Λ_{12} and Λ_{34} be any skew conjugate polar lines. Let P_1 be any point on Λ_{12} whose polar plane Π_1 intersects Λ_{12} in some point P_2, and let P_3 be any point on Λ_{34} whose polar plane Π_3 intersects Λ_{34} in some point P_4. Prove that the points P_1, P_2, P_3, and P_4 are the vertices of a tetrahedron each face of which is the polar plane of the opposite vertex and each edge of which is the polar line of the opposite edge. (See Fig. 42.) This tetrahedron is called **self-polar** (with respect to S). (See Exs. **24-26**.)

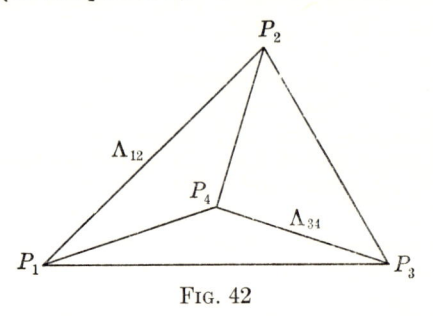

FIG. 42

28. Let Π be a plane with equation $\alpha x + \beta y + \gamma z + \delta = 0$ and let S be a non-singular quadric surface with equation $f(x, y, z) = 0$. Prove that the vanishing of the determinant

$$\begin{vmatrix} a & h & g & p & \alpha \\ h & b & f & q & \beta \\ g & f & c & r & \gamma \\ p & q & r & d & \delta \\ \alpha & \beta & \gamma & \delta & 0 \end{vmatrix}$$

is a necessary and sufficient condition for the plane Π to be either (*i*) tangent to S or (*ii*) conjugate to some direction with respect to S without being a diametral plane of S.

29. Discuss the following statements: (*i*) A point can be expected to have a polar plane with respect to a quadric surface. (*ii*) A plane can be expected to have one and only one pole with respect to a quadric surface. (See the footnote on page 63, and Ex. **3**, § **125**.)

131. Ruled surfaces.

DEFINITION. *A real surface is a **ruled surface** if and only if through every real point of the surface passes a real ruling.*

NOTE. It is important to specify that the rulings to be considered be real. A real sphere, for example, is not a ruled surface, although through each point of the surface pass two (imaginary) rulings. (Two such rulings of the unit sphere with center at the origin are the lines

$$x = t, y = \pm it, \quad z = 1.)$$

Among the seventeen surfaces discussed in Chapter **V**, it is obvious that the real cone, the real cylinders, and the real reducible quadrics are ruled surfaces. There are two other ruled quadrics, which we shall consider now.

I. *Hyperboloid of one sheet.* Consider the three equations

(1) $$\left(\frac{x}{a} + \frac{z}{c}\right)\left(\frac{x}{a} - \frac{z}{c}\right) = \left(1 + \frac{y}{b}\right)\left(1 - \frac{y}{b}\right),$$

(2) $$p\left(\frac{x}{a} + \frac{z}{c}\right) = q\left(1 + \frac{y}{b}\right),$$

(3) $$q\left(\frac{x}{a} - \frac{z}{c}\right) = p\left(1 - \frac{y}{b}\right),$$

where p and q are real numbers that are not both zero. Equation (1) is that of the hyperboloid and equations (2) and (3) are those of two intersecting planes. We now make two statements which can be readily verified:

(*i*) Any solution of the system consisting of the two equations (2) and (3) is a solution of equation (1). (*ii*) Any solution of equation (1) is a solution of a system of two equations of the form (2) and (3), where p and q are uniquely determined except for a factor of proportionality.

As a consequence we can conclude that the two planes (2) and (3) always intersect in a ruling of the hyperboloid, and through every point of the hyperboloid passes one and only one such ruling. If p and q are considered as parameters, equations (2) and (3) thus provide a family of rulings which can be thought of as *generating* the hyperboloid (1).

Similarly, the system of two equations

(4) $p'\left(\frac{x}{a} + \frac{z}{c}\right) = q'\left(1 - \frac{y}{b}\right)$

and

(5) $q'\left(\frac{x}{a} - \frac{z}{c}\right) = p'\left(1 + \frac{y}{b}\right),$

where p' and q' are real numbers that are not both zero, defines a family of rulings. It can be shown (show it!) that no ruling is a member of both families, and therefore that through each point passes exactly one ruling from each family. (See Fig. 43.)

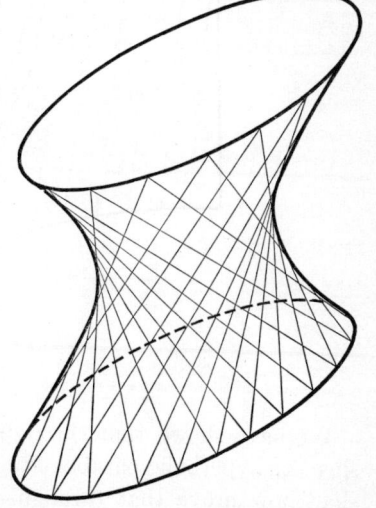

Fig. 43

II. *Hyperbolic paraboloid.* This surface, whose equation can be written

(6) $$\left(\frac{x}{a} + \frac{y}{b}\right)\left(\frac{x}{a} - \frac{y}{b}\right) = -2z,$$

also has two families of rulings, whose equations can be written

(7) $\quad p\left(\dfrac{x}{a} + \dfrac{y}{b}\right) = -2z,\qquad$ (8) $\quad \dfrac{x}{a} - \dfrac{y}{b} = p,$

and

(9) $\quad p'\left(\dfrac{x}{a} - \dfrac{y}{b}\right) = -2z,\qquad$ (10) $\quad \dfrac{x}{a} + \dfrac{y}{b} = p',$

where p and p' are parameters. As with the hyperboloid of one sheet, these families of lines are mutually exclusive, and through each point of the surface passes exactly one ruling of each family (prove these statements). (See Fig. 44.)

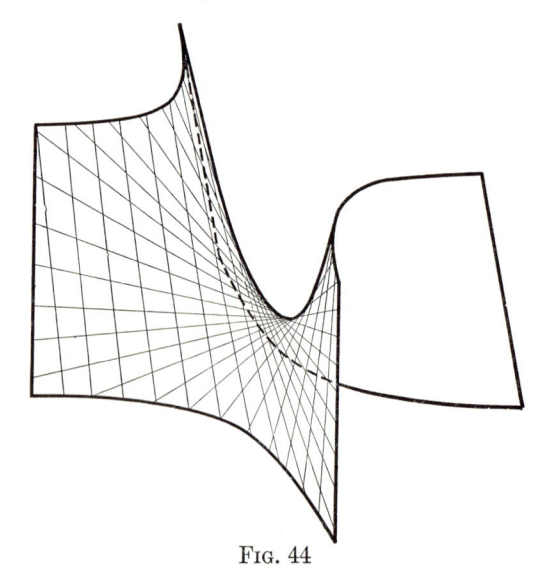

Fig. 44

We have shown that through each point of a hyperboloid of one sheet or hyperbolic paraboloid pass *at least* two distinct rulings. We shall now prove that through each point pass *exactly* two distinct rulings.

§ 131] RULED SURFACES

Theorem. *Through each point of a hyperboloid of one sheet or hyperbolic paraboloid pass exactly two rulings, which are the section of the quadric by the tangent plane at the point.*

Proof. The tangent plane at any point of a quadric of either type contains every ruling through that point, and therefore the two rulings obtained in this section. Since the tangent plane does not lie entirely in the surface, and since this plane and the surface have two distinct lines in common, by the Theorem of § **119** they have no other points in common. There are therefore just two rulings through the point.

As a consequence of this theorem, the hyperboloid of one sheet and the hyperbolic paraboloid are called **doubly ruled** surfaces.

Example 1. Find the rulings of the hyperboloid of one sheet

$$\frac{x^2}{25} + \frac{y^2}{1} - \frac{z^2}{4} = 1$$

that pass through the point $(5, 3, 6)$.

Solution. To find one ruling we substitute $x = 5$, $y = 3$, and $z = 6$ in equation (2) or (3) to determine values of p and q. Accordingly we let $p = q = 1$ and have the planes

$$2x - 10y + 5z = 10 \quad \text{and} \quad 2x + 10y - 5z = 10,$$

whose line of intersection has direction numbers 0, 1, 2. The planes (4) and (5) pass through the point $(5, 3, 6)$ if $p' = 1$, $q' = -2$, in which case a set of direction numbers for the line of intersection is 15, 4, 10. The rulings are

$$\frac{x-5}{0} = \frac{y-3}{1} = \frac{z-6}{2} \quad \text{and} \quad \frac{x-5}{15} = \frac{y-3}{4} = \frac{z-6}{10}.$$

A check on our work is given by means of the equation of the tangent plane at the point $(5, 3, 6)$:

$$2x + 30y - 15z = 10.$$

The normal to this plane is perpendicular to both rulings and therefore the plane contains them both.

Example 2. Prove that neither family of rulings of a hyperboloid of one sheet contains two distinct parallel lines.

Solution. Considering only the family of rulings given by equations (2) and (3), we note that the two rulings corresponding to p and q and to p_1 and q_1 have direction numbers

$$a(q^2 - p^2),\ 2bpq,\ c(p^2 + q^2) \quad \text{and} \quad a(q_1^2 - p_1^2),\ 2bp_1q_1,\ c(p_1^2 + q_1^2).$$

Assuming that these lines are parallel, we wish to show that they are identical. For simplicity we shall assume furthermore that p_1 and q_1 have been multiplied by a suitable common factor of such a magnitude that the two sets of direction numbers given above are not only proportional, but identical. It follows easily then that $p^2 = p_1^2$, $q^2 = q_1^2$. Finally, since $pq = p_1q_1$, either $p = p_1$ and $q = q_1$ or $p = -p_1$ and $q = -q_1$, and the two rulings are identical.

*132. Further discussion of ruled surfaces.

The method of § **131** might well be called a "trick method," since it is applicable only to two particular quadric surfaces whose equations are written in canonical form. It is our purpose in this section to present briefly a treatment of rulings of quadric surfaces based on first principles, which is applicable to any equation, whether it is in canonical form or not.

We recall (§ **118**) that the line through the point $P_0(x_0, y_0, z_0)$ with direction (λ, μ, ν) is a ruling of the surface $f(x, y, z) = 0$ if and only if the following system is simultaneously satisfied:

(1)
$$\begin{aligned} e(\lambda, \mu, \nu) &= 0, \\ \phi(x_0, y_0, z_0, \lambda, \mu, \nu) &= 0, \\ f(x_0, y_0, z_0) &= 0. \end{aligned}$$

This system of equations can be used as the basis of an analytical proof that the ruled quadrics and their rulings are precisely those enumerated in § **131**. For example, the ellipsoid is not a ruled surface since, for any real direction (λ, μ, ν),

$$e(\lambda, \mu, \nu) = \frac{\lambda^2}{a^2} + \frac{\mu^2}{b^2} + \frac{\nu^2}{c^2} \neq 0.$$

We shall illustrate the use of system (1) by proving the following theorem concerning a hyperboloid of one sheet and its asymptotic cone (with equations in canonical form):

Theorem. *Corresponding to any ruling of a hyperboloid of one sheet, there is a parallel ruling of the asymptotic cone. Conversely, corresponding to any ruling of the asymptotic cone, there is a parallel ruling from each family of rulings of the hyperboloid.*

Proof. The systems (1) for the hyperboloid and cone are

(2)
$$\begin{aligned} \frac{\lambda^2}{a^2} + \frac{\mu^2}{b^2} - \frac{\nu^2}{c^2} &= 0, \\ \frac{x_0\lambda}{a^2} + \frac{y_0\mu}{b^2} - \frac{z_0\nu}{c^2} &= 0, \\ \frac{x_0^2}{a^2} + \frac{y_0^2}{b^2} - \frac{z_0^2}{c^2} &= 1; \end{aligned}$$

(3)
$$\begin{aligned} \frac{\lambda^2}{a^2} + \frac{\mu^2}{b^2} - \frac{\nu^2}{c^2} &= 0, \\ \frac{x_0\lambda}{a^2} + \frac{y_0\mu}{b^2} - \frac{z_0\nu}{c^2} &= 0, \\ \frac{x_0^2}{a^2} + \frac{y_0^2}{b^2} - \frac{z_0^2}{c^2} &= 0. \end{aligned}$$

The first part of the theorem is easily established, for if (λ, μ, ν) is the direction of any ruling of the hyperboloid, and therefore satisfies the first of equations (2), the line

$$x = \lambda t, \qquad y = \mu t, \qquad z = \nu t$$

is a ruling of the cone, as is seen by direct substitution in system (3). Conversely, let (λ, μ, ν) be the direction of any ruling of the cone, and let $P_1(x_1, y_1, z_1)$ and $P_2(x_2, y_2, z_2)$ be the points where the line $\dfrac{\lambda}{a^2}x + \dfrac{\mu}{b^2}y = 0$, $z = 0$ meets the ellipse $\dfrac{x^2}{a^2} + \dfrac{y^2}{b^2} = 1$, $z = 0$. The two lines

$$x = x_1 + \lambda t, \quad y = y_1 + \mu t, \quad z = \nu t,$$
$$x = x_2 + \lambda t, \quad y = y_2 + \mu t, \quad z = \nu t$$

are rulings of the hyperboloid, as is seen by direct substitution in system (2). Since neither family of rulings of a hyperboloid of one sheet contains two parallel lines (Example 2, § **131**), one of these two rulings must belong to one family, and the other ruling to the other family.

EXAMPLE *1.* Find the locus of the point on the hyperbolic paraboloid

$$\frac{x^2}{4} - y^2 + 2z = 0$$

at which the two rulings are perpendicular.

Solution. The first two equations of system (1) for this paraboloid are

$$\frac{\lambda^2}{4} - \mu^2 = 0, \quad \frac{x_0\lambda}{4} - y_0\mu + \nu = 0,$$

whence $\lambda = \pm 2\mu$, $\nu = y_0\mu \mp \dfrac{x_0}{2}\mu$, and

$$\lambda : \mu : \nu = 2 : \pm 1 : -\frac{x_0}{2} \pm y_0.$$

These two directions are perpendicular if and only if

$$4 - 1 + \frac{x_0^2}{4} - y_0^2 = 3 - 2z = 0,$$

and the locus is the hyperbola

$$z = \frac{3}{2}, \quad \frac{x^2}{4} - y^2 + 3 = 0.$$

EXAMPLE 2. Investigate the surface

$$xy + xz + yz + 1 = 0$$

for rulings at the point $(1, 0, -1)$.

Solution. For convenience we multiply by 2 to avoid fractions. The matrix E is

$$\begin{pmatrix} 0 & 1 & 1 & 0 \\ 1 & 0 & 1 & 0 \\ 1 & 1 & 0 & 0 \\ 0 & 0 & 0 & 2 \end{pmatrix},$$

and the three equations (1) defining rulings are

$$\lambda\mu + \lambda\nu + \mu\nu = 0,$$
$$(y_0 + z_0)\lambda + (x_0 + z_0)\mu + (x_0 + y_0)\nu = 0,$$
$$x_0 y_0 + x_0 z_0 + y_0 z_0 + 1 = 0.$$

The coördinates $(1, 0, -1)$ satisfy the third equation, and the second becomes $\lambda = \nu$. One solution is therefore $\lambda = \nu = 0$, $\mu = 1$. Another is obtained by substituting λ for ν in the first equation and dividing by λ. This gives $\mu + \lambda + \mu = 0$. Then $\lambda : \mu : \nu = 2 : -1 : 2$. There are thus two rulings through the given point, the first one parallel to the y axis:

$$\frac{x-1}{0} = \frac{y}{1} = \frac{z+1}{0} \quad \text{and} \quad \frac{x-1}{2} = \frac{y}{-1} = \frac{z+1}{2}.$$

The equation of the tangent plane at the given point is $x - z = 2$, which clearly contains the two rulings. Since there are just two rulings through the point $(1, 0, -1)$, the surface must be either a hyperboloid of one sheet or a hyperbolic paraboloid. Since $D \neq 0$, it is a central quadric and therefore a hyperboloid of one sheet. A more complete and straightforward analysis of this surface can be given by the methods of Chapter VIII (Ex. **19**, § **156**.)

133. Exercises.

In Exercises **1-4**, write the equations of the rulings of the given surface through the specified point.

1. $x^2 + y^2 - z^2 = 1$; $(1, 1, 1)$.
2. $\dfrac{x^2}{9} + \dfrac{y^2}{4} - \dfrac{z^2}{1} = 1$; $(12, -14, 8)$.
3. $\dfrac{x^2}{4} - \dfrac{y^2}{9} + 2z = 0$; $(-6, 15, 8)$.
* 4. $xy + 2xz + 3yz + y + z + 2 = 0$; $(1, -1, 1)$.

In Exercises **5** and **6**, write the equations of the ruling (or rulings) of the given surface having the specified set of direction numbers.

5. $\dfrac{x^2}{16} + \dfrac{y^2}{4} - z^2 = 1$; $16, -6, 5$.
6. $\dfrac{x^2}{25} - y^2 + 2z = 0$; $5, -1, -6$.

In Exercises **7-10**, find the equation of the ruled surface generated by the given family of lines, k, k_1, and k_2 being parameters.

7. $k_1(x - 3y) = k_2 z$, $k_2(x + 3y) = k_1 z$.
8. $k_1(x - 3y) = k_2(1 - 2z)$, $k_2(x + 3y) = k_1(1 + 2z)$.
9. $k(x - 3y) = z$, $x + 3y = 4k$.
10. $k_1(y + z + 1) = k_2(x + z)$, $k_2(y - z - 1) = k_1(x - z)$.

In Exercises **11** and **12**, find the equation of the hyperboloid of one sheet generated by revolving the given line about the specified axis. (Cf. Ex. **49**, § **116**.)

11. $x = 3$, $z = 2y$; z axis. **12.** $z = 2$, $y = 3x$; x axis.

*Supplementary Exercises

13. Prove that the rulings of a hyperbolic paraboloid, with an equation in canonical form, through the point (x_0, y_0, z_0) have direction numbers $a, \pm b, -\dfrac{x_0}{a} \pm \dfrac{y_0}{b}$. Use this result to check Exercise **3**.

14. Prove that no two distinct rulings of a hyperbolic paraboloid are parallel. (See Ex. **13**.)

15. Prove that the locus of points of a hyperbolic paraboloid at which the rulings are perpendicular is a hyperbola.

16. Prove that for a hyperboloid of one sheet or a hyperbolic paraboloid two rulings are skew if and only if they belong to the same family.

17. For a real quadric cone, prove that no three distinct rulings are parallel to the same plane.

18. For a hyperboloid of one sheet, prove that no three distinct rulings of one family are parallel to the same plane. *Suggestion:* See Ex. **17** and the Theorem of § **132**.

19. For a hyperbolic paraboloid, prove that all of the rulings of either family are parallel to one plane. More specifically, if the equation is in canonical form and if Λ_1 and Λ_2 denote the rulings through the origin, prove that all of the rulings of one family are parallel to the plane containing Λ_1 and the z axis, and that all of the rulings of the other family are parallel to the plane containing Λ_2 and the z axis. (See Ex. **13**.)

20. If Λ_1 and Λ_2 are any two lines of one family of rulings of a hyperbolic paraboloid, and if L_1, L_2, and L_3 are any three lines of the other family, meeting Λ_1 and Λ_2 in the points P_1, P_2, and P_3 and Q_1, Q_2, and Q_3, respectively, prove that the segments P_1P_2 and P_2P_3 are proportional to the segments Q_1Q_2 and Q_2Q_3. *Suggestion:* Use the fact (Ex. **19**) that L_1, L_2, and L_3 are all parallel to one plane.

21. Prove that the directions of the two rulings of a hyperboloid of one sheet, with an equation in canonical form, through the point (x_0, y_0, z_0) satisfy the two systems of equations

$$\pm \frac{bc}{a}\lambda + z_0\mu - y_0\nu = 0,$$

$$-z_0\lambda \pm \frac{ac}{b}\mu + x_0\nu = 0,$$

$$-y_0\lambda + x_0\mu \pm \frac{ab}{c}\nu = 0.$$

Hence prove that $\lambda : \mu : \nu = x_0z_0 \pm \dfrac{ac}{b}y_0 : y_0z_0 \mp \dfrac{bc}{a}x_0 : z_0^2 + c^2$.

158   THE GENERAL EQUATION   [Ch. VI]

Suggestion: Since $\dfrac{\lambda}{a^2}\lambda + \dfrac{\mu}{b^2}\mu - \dfrac{\nu}{c^2}\nu = 0$ and $\dfrac{x_0}{a^2}\lambda + \dfrac{y_0}{b^2}\mu - \dfrac{z_0}{c^2}\nu = 0$,

$$\lambda : \mu : \nu = \frac{y_0\nu - z_0\mu}{b^2c^2} : \frac{z_0\lambda - x_0\nu}{a^2c^2} : \frac{y_0\lambda - x_0\mu}{a^2b^2}.$$

Letting the constant of proportionality be k, we have
$$b^2c^2k\lambda + z_0\mu - y_0\nu = 0,$$
$$-z_0\lambda + a^2c^2k\mu + x_0\nu = 0,$$
$$-y_0\lambda + x_0\mu + a^2b^2k\nu = 0,$$
a homogeneous system of equations whose determinant vanishes if and only if $k = \pm 1/abc$.

22. Prove that as a point P moves along a ruling Λ of a hyperboloid of one sheet or a hyperbolic paraboloid, the tangent plane at P revolves about Λ, assuming every position except that of the plane conjugate to Λ. (Cf. Ex. 12, § 128, and Exs. 24-26, § 130.)

23. Using the notation introduced in § **117**, writing $f(x, y, z)$ in the form
$$f(x_1, x_2, x_3) \equiv \sum_{i,j=1}^{4} a_{ij}x_ix_j,$$
where $x_1 = x$, $x_2 = y$, $x_3 = z$, $x_4 = 1$, and furthermore, letting $\lambda_1 = \lambda$, $\lambda_2 = \mu$, $\lambda_3 = \nu$, $\lambda_4 = 0$, prove each of the following statements:

(*i*) $e(\lambda_1, \lambda_2, \lambda_3) = \sum_{i,j=1}^{3} a_{ij}\lambda_i\lambda_j.$

(*ii*) $\phi(x_1, x_2, x_3, \lambda_1, \lambda_2, \lambda_3) = \sum_{i,j=1}^{4} a_{ij}x_i\lambda_j = \sum_{i,j=1}^{4} a_{ij}\lambda_ix_j.$

(*iii*) The equation of the diametral plane conjugate to the direction $(\lambda_1, \lambda_2, \lambda_3)$ is
$$\sum_{i,j=1}^{4} a_{ij}\lambda_ix_j = 0.$$

(*iv*) The centers of the surface $f(x_1, x_2, x_3) = 0$ are the points whose coordinates satisfy the system of equations
$$\sum_{j=1}^{4} a_{ij}x_j = 0, \quad i = 1, 2, 3.$$

(*v*) The singular points of the surface $f(x_1, x_2, x_3) = 0$ are the points whose coördinates satisfy the system of equations
$$\sum_{j=1}^{4} a_{ij}x_j = 0, \quad i = 1, 2, 3, 4.$$

(*vi*) The polar plane of the point (y_1, y_2, y_3) with respect to the surface $f(x_1, x_2, x_3) = 0$ is
$$\sum_{i,j=1}^{4} a_{ij}y_ix_j = 0.$$

CHAPTER VII

COÖRDINATE AND POINT TRANSFORMATIONS

134. Introduction. The student is already familiar with transformations of coördinates in plane analytic geometry. A conic section may have a much simpler equation with respect to one set of axes than it has with respect to some other set. We speak of *simplifying* an equation, meaning that we choose a new system of axes in an appropriate manner. This is accomplished by means of two fundamental types of transformations, *translations* and *rotations*, the rotations being taken about the origin.

In a similar manner and for similar reasons we shall be interested in simplifying equations in three variables by means of two basic transformations, translations and rotations.

Consider two rectangular coördinate systems, as indicated in Fig. 45, and assume for simplicity that both are right-handed. Any

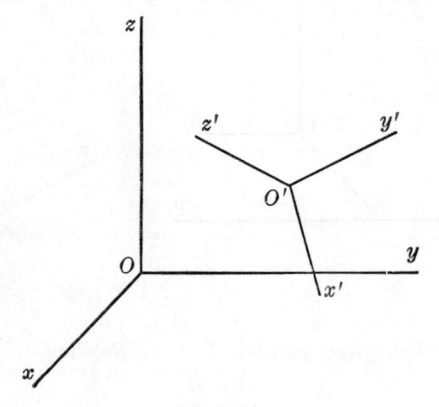

Fig. 45

point in space has two sets of coördinates, one for each set of axes. The basic problem is to find a relationship between these two sets of coördinates that will give the coördinates of any point in either system in terms of its coördinates in the other system. We shall obtain this relationship by resolving the general problem into two simpler parts.

As indicated in Fig. 46, let a third system be introduced in such a way that its origin coincides with the origin of one coördinate system

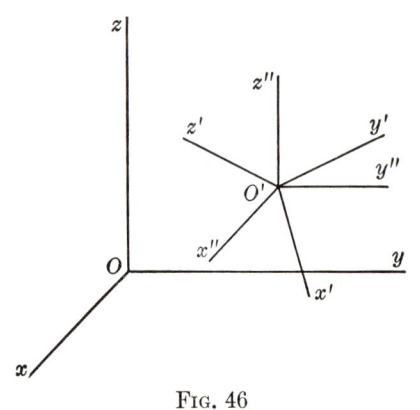

FIG. 46

and its axes are respectively parallel to those of the other. In this way the problem is reduced to two simpler ones associated with the coördinate systems illustrated in Fig. 47 and appropriately labeled.

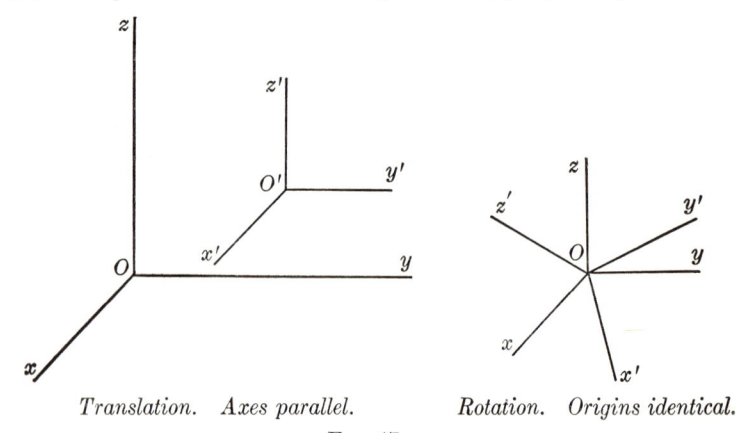

Translation. Axes parallel.   *Rotation. Origins identical.*

FIG. 47

Once the relationship between the primed and unprimed coördinates has been obtained for each of these two special types of transformations, the more general problem is resolved. For example (see Fig. 46), if x', y', and z' are expressed in terms of x'', y'', and z'', and the latter are expressed in terms of x, y, and z, the primed coördinates are immediately expressible in terms of the unprimed coördinates.

135. Translations. Inspection of Fig. 48 indicates that if the coördinates of O' and P in the xyz system are (x_0, y_0, z_0) and (x, y, z),

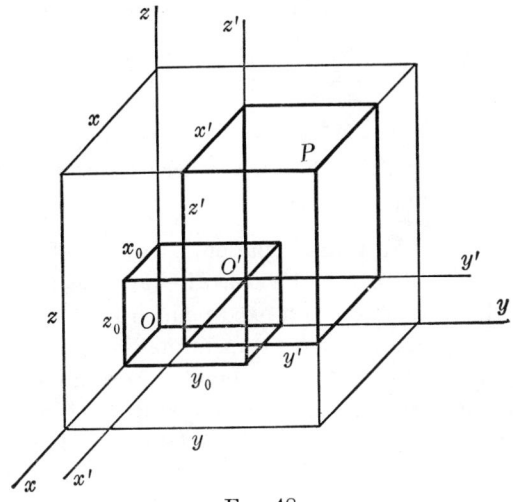

FIG. 48

respectively, and the coördinates of P in the $x'y'z'$ system are (x', y', z'), then the two sets of coördinates of P are related by the equations

(1) $\begin{aligned} x &= x' + x_0, \\ y &= y' + y_0, \\ z &= z' + z_0; \end{aligned}$
(2) $\begin{aligned} x' &= x - x_0, \\ y' &= y - y_0, \\ z' &= z - z_0. \end{aligned}$

The student should convince himself that these relations hold regardless of the signs of the numbers involved.

System (2) transforms the coördinates of a point in the xyz system into the coördinates of the point in the $x'y'z'$ system, while system (1) transforms the coördinates in the reverse direction. For example, if the new origin O' is the point $(3, -4, 2)$, the point with xyz coördinates $(5, 2, -3)$ has $x'y'z'$ coördinates $(5 - 3, 2 + 4, -3 - 2)$ or $(2, 6, -5)$, and the point with $x'y'z'$ coördinates $(4, 0, -1)$ has xyz coördinates $(4 + 3, -4, -1 + 2)$ or $(7, -4, 1)$.

Systems (1) and (2) can also be used to transform an equation in the variables x, y, and z (or x', y', and z') into an equation in the variables x', y', and z' (or x, y, and z). However, in this case it is system (1) that applies to the transformation from the xyz system to the $x'y'z'$ system and system (2) that applies to the transformation

in the reverse direction. For example, the sphere whose xyz equation is
$$x^2 + y^2 + z^2 - 6x + 8y - 4z + 20 = 0$$
has an $x'y'z'$ equation obtained by substituting in this equation $x = x' + 3, y = y' - 4, z = z' + 2$:
$$x'^2 + y'^2 + z'^2 = 9.$$
The equation in the $x'y'z'$ system is simpler because the origin of this system is the center of the sphere.

136. Two aspects of a transformation. Any transformation of the type discussed in § **134** can be regarded in two ways: (*i*) as a transformation of the coördinates of a fixed point from one system of axes to another system of axes; (*ii*) as a transformation in a fixed coördinate system from one point to a different point. For example, the particular translation considered in § **135** could be alternatively regarded as a shifting of the entire space which carries the point $(3, -4, 2)$ into the origin, the point $(5, 2, -3)$ into the point $(2, 6, -5)$, and the sphere of radius 3 with center at $(3, -4, 2)$ into the sphere of radius 3 with center at the origin. That is, we can think either of holding the space fixed while moving the axes, or of holding the axes fixed while moving the space. The equation of a quadric surface may be regarded as simplified either by moving the axes into an advantageous position with respect to the fixed surface, or by moving the surface into a suitable position with respect to fixed axes.

Both aspects of a system of transformation equations are important, and the student should spend some time thinking about the subtle distinction between them.

The fact, as expressed in § **134,** that any coördinate transformation from one rectangular coördinate system to another can be achieved by means of a translation and a rotation with fixed origin can be stated in terms of point transformations as follows: *Any rigid motion is the result of a translation followed by a rotation with fixed origin.* Furthermore, by introducing in Fig. 45 a third coördinate system with origin O and axes parallel to the primed axes, we can establish the following fact (do it): *Any rigid motion is the result of a rotation with fixed origin followed by a translation.* (Cf. Ex. **18**, § **140**.)

* **137. Symbolic notation for transformations.** It is sometimes convenient to represent a point transformation, or **mapping,** symbolically. For example, the translation that carries the point $(3, -4, 2)$ into

the origin, and therefore the point $P(x, y, z)$ into the point $Q\ (x - 3, y + 4, z - 2)$, can be represented by the equation
$$Q = T(P)$$
or
$$T((x, y, z)) = (x - 3, y + 4, z - 2).$$

With this notation it is natural to regard Q as a single-valued function of P, in the sense that if any point P is given, a point Q is uniquely determined.

NOTE. We shall consider in this book only single-valued transformations that are defined on the entire three dimensional Euclidean space, although many of the results which we shall obtain are valid in much greater generality.

The transformation that carries every point into itself is called the **identity transformation**, and is denoted by I. Thus $I(P) = P$ for every point P. The identity transformation can be thought of as a translation where $x_0 = y_0 = z_0 = 0$.

Consider two point transformations, T and U, successively applied. That is, let the point P be transformed by the transformation T into a point Q, and then let Q be transformed by the transformation U into a point R. Symbolically, we may write
$$Q = T(P), \qquad R = U(Q) = U(T(P)).$$

The point P is thus carried into the point R in two stages, but could be thought of as being carried into R by one new transformation V, written
$$V = UT,$$
called the **product** of the transformations U and T, and defined by the equation
$$V(P) = U(T(P)).$$

Notice that the operations are not applied in the order in which they are normally read on a printed page, but in the right-to-left direction.†

An example might clarify this idea of *multiplying* transformations. Let T and U be the translations:
$$T((x, y, z)) = (x - 3, y + 4, z - 2)$$
$$U((x, y, z)) = (x + 7, y - 5, z + 2).$$
Then the transformation $V = UT$ is
$$V((x, y, z)) = (x - 3 + 7, y + 4 - 5, z - 2 + 2)$$
$$= (x + 4, y - 1, z),$$

† Some authors (with good reason) use the other order, the transformations being applied in the left-to-right direction.

and the transformation $W = TU$ is

$$W((x, y, z)) = (x + 7 - 3, y - 5 + 4, z + 2 - 2)$$
$$= (x + 4, y - 1, z).$$

When, as in this example, the result of applying two transformations, T and U, is independent of the order of application, we say that the two transformations **commute,** or that the **commutative law** holds, and write

$$UT = TU.$$

It can easily be proved that the commutative law holds for any two translations, but we shall see later that it fails for rotations.

It is an important fact that for any transformations, T, U, and V, the **associative law** holds:

$$(TU)V = T(UV).$$

This is a direct consequence of the definition of the product of two transformations. Proof is requested formally in Exercise **15,** § **140.**

The identity transformation behaves like the number 1 when multiplied by a transformation:

$$IT = TI = T.$$

Two transformations, T and U, whose product in either order is the identity:

$$TU = UT = I,$$

are called **inverse transformations.** (Cf. Exs. **16** and **17,** § **140.**) Each is said to be the **inverse** of the other. The inverse of T is usually written T^{-1}, and satisfies the relations

$$TT^{-1} = T^{-1}T = I.$$

The inverse of the translation

$$T((x, y, z)) = (x - 3, y + 4, z - 2)$$

is thus the translation

$$T^{-1}((x, y, z)) = (x + 3, y - 4, z + 2).$$

In general, the two translations defined by equations (1) and (2) § **135,** considered in this way, are inverse transformations. If

$$T((x, y, z)) = (x - x_0, y - y_0, z - z_0)$$

then

$$T^{-1}((x, y, z)) = (x + x_0, y + y_0, z + z_0).$$

The statements of the last paragraph of § 136 take the following symbolic form: *If V is any rigid motion, there exist translations, T_1 and T_2, and rotations with fixed origin, U_1 and U_2, such that*

$$V = T_1 U_1 = U_2 T_2.$$

In other words, any rigid motion can be factored as the product of a translation and a rotation with fixed origin, or as the product of a rotation with fixed origin and a translation. (Cf. Ex. **18**, § **140**, Ex. **25**, § **147**.)

EXAMPLE. Prove by symbolic methods that if a transformation T has an inverse, this inverse is unique.

Solution. Assume that U and V are any inverses of T; that is, that

$$TU = UT = TV = VT = I.$$

Then
$$U(TV) = UI = U$$
and
$$(UT)V = IV = V.$$

But by the associative law, $U(TV) = (UT)V$. Therefore $U = V$.

138. Invariance. Most of the concepts that have been discussed in this book have been thought of as having meaning independent of any coördinate system. For example, two points have a certain distance between them and, although we have a formula for this distance in terms of their coördinates in any given coördinate system, we have confidence that this formula will always give the same value. An analytic proof that this is true is requested in Exercise **21**, § **140**, when the transformation is a translation, and in Exercise **36**, § **147**, when the transformation is a rotation with fixed origin. We say that the distance between two points, as given by the formula, is **invariant** under translations and rotations, according to the following definition:

DEFINITION. *Any property or quantity that is left unchanged by a transformation is said to be **invariant** under (or with respect to) the transformation, and is called **an invariant** under (or with respect to) the transformation.*

In contrast with the notion of distance between two points, certain ideas have been *defined* in terms of a particular coördinate system. Among these are *algebraic surface, reducible algebraic surface, singular quadric surface,* and *polar plane.* In these cases it is essential for logical completeness to prove that these concepts have the same meaning in one coördinate system as in any other. In Chapter IX we shall prove that Δ (§ **117**) is an invariant under any change of coördinates, and that therefore a quadric surface is either always singular (in all coördinate systems) or never singular (in any coördinate system). (Cf. Example 3, § **139.**) Proofs of invariance of certain properties or quantities with respect to translations are requested in some of the Exercises of § **140**. Proofs of invariance with respect to rotations are given or requested in later sections of the book.

EXAMPLE. Prove analytically that the distance between a plane and a point, as given by formula (1), § **32**, is invariant under translation.

Solution. Under the translation formulas of § **135** the point (x_1, y_1, z_1) has coördinates
$$(x_1', y_1', z_1') = (x_1 - x_0, y_1 - y_0, z_1 - z_0)$$
in the new coördinate system, and the plane $ax + by + cz + d = 0$ has the new equation
$$a(x' + x_0) + b(y' + y_0) + c(z' + z_0) + d = 0.$$
The distance formula applied in the new coördinate system gives the distance
$$\frac{|a(x_1' + x_0) + b(y_1' + y_0) + c(z_1' + z_0) + d|}{\sqrt{a^2 + b^2 + c^2}}.$$
But this is equal to
$$\frac{|ax_1 + by_1 + cz_1 + d|}{\sqrt{a^2 + b^2 + c^2}},$$
the distance provided by the formula in the original coördinate system.

139. Simplification of equations by translations. It is the purpose of this section to explore the possibilities of simplifying an equation of a quadric surface by means of a translation, thought of either as a shifting of the coördinate axes to a set of parallel axes, or as a translation of the surface itself to a new position.

First let us observe the result of substituting
$$x = x' + x_0, \qquad y = y' + y_0, \qquad z = z' + z_0$$
in the general equation of the second degree, $f(x, y, z) = 0$. The new equation,

(1) $$\begin{aligned}F(x', y', z') &\equiv f(x' + x_0, y' + y_0, z' + z_0) \\ &\equiv ax'^2 + by'^2 + cz'^2 + 2fy'z' + 2gx'z' + 2hx'y' \\ &\quad + 2(ax_0 + hy_0 + gz_0 + p)x' + 2(hx_0 + by_0 + fz_0 + q)y' \\ &\quad + 2(gx_0 + fy_0 + cz_0 + r)z' + f(x_0, y_0, z_0) = 0,\end{aligned}$$

is also of the second degree. In fact, the coefficients of the second degree terms are identical with those of the original equation. We therefore have the theorem:

Theorem I. *For any quadric surface the coefficients of the second degree terms, and therefore the matrix e, are invariant under translation.*

This theorem tells us that whatever simplification of a second degree equation is to be effected by a translation must occur in the terms of degree less than the second. A natural question to ask is the following: "Can the terms of the first degree be eliminated by means of a translation?" The answer is, "Sometimes they can and sometimes they cannot." For example, the method of completing squares to reduce the equation of a sphere to the form

$$(x - x_0)^2 + (y - y_0)^2 + (z - z_0)^2 = R^2,$$

(§ 81), can be interpreted as a means of determining the origin of a new coördinate system with respect to which the sphere has an equation without first degree terms:

$$x'^2 + y'^2 + z'^2 = R^2.$$

However, no translation can eliminate the first degree term from the equation of the parabolic cylinder

$$x^2 = y.$$

Suppose the question is asked, "Under what circumstances can the terms of the first degree be eliminated by means of a translation?" The answer is contained in the theorem:

Theorem II. *The terms of the first degree of an equation of a quadric surface can be eliminated by means of a translation if and only if the surface has a center, in which case the first degree terms are eliminated if and only if the new origin is a center.*

Proof. From the form of the three equations which the coördinates of a center must satisfy ((1), § 122), it is clear that the coefficients of x', y', and z' in equation (1) all vanish if and only if (x_0, y_0, z_0) is a center.

Since, under a translation to a set of axes with origin at a center of a quadric surface, the second degree terms remain unchanged, while the first degree terms are eliminated, there is some interest in focusing our attention on the new constant term, which we shall denote by d'.

From the expanded form of equation (1) we see that $d' = f(x_0, y_0, z_0)$. In the expansion

$$f(x, y, z) \equiv (ax+hy+gz+p)x + (hx+by+fz+q)y \\ + (gx+fy+cz+r)z + (px+qy+rz+d),$$

each of the first three terms vanishes at a center. Therefore, if the new origin (x_0, y_0, z_0) is a center, the new constant term has the value

(2) $\qquad d' = f(x_0, y_0, z_0) = px_0 + qy_0 + rz_0 + d.$

If a translation is made to a center, (x_0, y_0, z_0), of the quadric surface, the transformed equation therefore becomes

(3) $\qquad ax'^2 + by'^2 + cz'^2 + 2fy'z' + 2gx'z' + 2hx'y' \\ + (px_0 + qy_0 + rz_0 + d) = 0.$

If there is more than one center, it is immaterial which center is chosen as the new origin, by virtue of the theorem:

THEOREM III. *The equation obtained from that of a given quadric surface by a translation to a center of the surface as the new origin is independent of the center chosen.*

Proof. If (x_0, y_0, z_0) and (x_1, y_1, z_1) are both centers of the surface $f(x, y, z) = 0$, the following two systems of equations are simultaneously satisfied ((1), § **122**):

$$(ax_0 + hy_0 + gz_0 + p)x_1 = 0, \qquad (ax_1 + hy_1 + gz_1 + p)x_0 = 0,$$
$$(hx_0 + by_0 + fz_0 + q)y_1 = 0, \qquad (hx_1 + by_1 + fz_1 + q)y_0 = 0,$$
$$(gx_0 + fy_0 + cz_0 + r)z_1 = 0; \qquad (gx_1 + fy_1 + cz_1 + r)z_0 = 0.$$

Equating the sums of the left members of the two systems of equations, and subtracting identical terms from the two members of the resulting equation, we have

$$px_1 + qy_1 + rz_1 = px_0 + qy_0 + rz_0,$$

and therefore

$$px_1 + qy_1 + rz_1 + d = px_0 + qy_0 + rz_0 + d.$$

By equation (3), the proof is complete.

EXAMPLE 1. Find a center of the surface

$$3x^2 - y^2 - 2z^2 - 4xy + 2xz - 6yz + 44x - 6y + 38z + 86 = 0,$$

and transform the equation by means of a translation to new axes with this center as origin.

§ 139] SIMPLIFICATION BY TRANSLATION

Solution. The matrix E of the given equation is

$$\begin{pmatrix} 3 & -2 & 1 & 22 \\ -2 & -1 & -3 & -3 \\ 1 & -3 & -2 & 19 \\ 22 & -3 & 19 & 86 \end{pmatrix}.$$

Equations (1), § 122, for the coördinates of the centers are obtained from the first three rows of this matrix. It is readily seen that the centers lie on the line whose parametric equations are

$$x = -4 + t, \quad y = 5 + t, \quad z = -t.$$

A translation to axes meeting at any point on this line will remove the first degree terms from the equation of the surface and give a new constant term obtained from the last row of E:

$$d' = 22(-4 + t) - 3(5 + t) + 19(-t) + 86 = -17.$$

Therefore the transformed equation is

$$3x'^2 - y'^2 - 2z'^2 - 4x'y' + 2x'z' - 6y'z' = 17.$$

The procedure in case there is no center is indicated to a certain extent in the solution of the following example:

EXAMPLE 2. Reduce the following equation to canonical form by means of a translation, and identify the surface:

$$2x^2 - 3y^2 - 20x - 6y + 12z + 131 = 0.$$

Solution. Using the standard method of completing squares, we can write the given equation in the form

$$2(x^2 - 10x + 25) - 3(y^2 + 2y + 1) + 12z + 131 = 47$$

or

$$2(x - 5)^2 - 3(y + 1)^2 + 12(z + 7) = 0.$$

Translating the axes to a new set with origin $(5, -1, -7)$, we have the equation in canonical form:

$$\frac{x'^2}{3} - \frac{y'^2}{2} + 2z' = 0.$$

The surface is a hyperbolic paraboloid with vertex at the point $(5, -1, -7)$.

EXAMPLE 3. Prove that Δ (§ 117) is invariant under translation.

Solution. According to the expanded form of $F(x', y', z')$, (1), the value of the transformed fourth order determinant Δ' is

$$\begin{vmatrix} a & h & g & ax + hy + gz + p \\ h & b & f & hx + by + fz + q \\ g & f & c & gx + fy + cz + r \\ ax + hy + gz + p & hx + by + fz + q & gx + fy + cz + r & f(x, y, z) \end{vmatrix},$$

where subscripts have been omitted to save space. Add to the fourth row the result of multiplying the first three rows by $-x$, $-y$, and $-z$, respectively. Then perform similar operations on the columns. The resulting determinant is the same as the original Δ.

140. Exercises.

1. Write the equations for the translation to a new coördinate system whose origin has coördinates $(8, -3, 9)$ in the original system.

In Exercises 2-7, find a center of the given surface and transform the equation by means of a translation to new axes with this center as origin.

2. $x^2 + y^2 + z^2 - 6x + 2y + 10z + 19 = 0$.
3. $4x^2 - y^2 + 2z^2 + 16x - 12y = 25$.
4. $x^2 + 3z^2 - 6xy - 26x + 6y + 18z + 63 = 0$.
5. $2y^2 - 2z^2 + 2xy - 2xz - 2x + 2y - 6z + 11 = 0$.
6. $2x^2 + 3y^2 + 3z^2 - 2xy + 2xz + 4yz - 28x - 6y - 34z + 132 = 0$.
7. $4x^2 + y^2 + 9z^2 - 4xy + 12xz - 6yz - 8x + 4y - 12z - 8 = 0$.

In Exercises **8** and **9**, reduce the given equation to canonical form by means of a translation, and identify the surface.

8. $2x^2 + y^2 - 8x - 22y + 8z - 7 = 0$.
9. $x^2 + 26x + 7y = 64$.

In Exercises **10** and **11**, simplify the given equation by means of a translation and identify the surface.

10. $2x^2 + 3y^2 + z^2 - 24x + 6y + 16z + 141 = 0$.
11. $5x^2 - y^2 + 6z^2 + 10x + 16y - 24z - 35 = 0$.

12. Prove that by means of a translation the equation of any plane can be reduced to the form
$$ax + by + cz = 0.$$

13. Prove that by means of a translation the equation of any line can be reduced to the form
$$\frac{x}{l} = \frac{y}{m} = \frac{z}{n}.$$

14. Prove that by means of a translation the equation of any quadric surface can be reduced to a form where the constant term is zero.

★ 15. Prove the associative law for multiplication of transformations.

★ 16. Prove that a transformation T has an inverse (that is single-valued) if and only if the following two conditions hold:

(i) $T(P_1) = T(P_2)$ implies $P_1 = P_2$.
(ii) If Q is any point, there is a point P such that $Q = T(P)$.

Such a transformation is called **one-to-one** (or **one-one**).

★ 17. Show that the following is an example of two transformations T and U for which $UT = I$ and $TU \neq I$:

$$T((x, y, z)) = (e^x, e^y, e^z);$$
$$U((x, y, z)) = (\ln x, \ln y, \ln z) \text{ if } x > 0, y > 0, z > 0,$$
$$= (x, y, z) \text{ otherwise.}$$

Notice that T violates the second condition, and U the first condition, of Exercise **16**.

* **18.** Prove that any rigid motion V can be represented in the forms
$$V = T_1 U = U T_2,$$
where T_1 and T_2 are translations and U is a rotation with fixed origin. (See §§ **136, 137.** Cf. Ex. **25,** § **147.**)

19. Prove that the rank of E is invariant under translation. *Suggestion:* Show that the steps used in the solution of Example 3, § **139,** will also furnish the proof in this case.

20. Use the invariance of D and Δ with respect to a translation (why is D invariant?) to prove that if the new origin is the center of a central quadric, then the new constant term in the equation of the quadric is
$$d' = \frac{\Delta}{D}.$$

In Exercises **21-28,** prove that the given property or quantity is invariant under translation; that is, that it has the same meaning in any two coördinate systems that are related by a translation.

21. The distance between two points (x_1, y_1, z_1) and (x_2, y_2, z_2), as given by the formula
$$d = \sqrt{(x_2 - x_1)^2 + (y_2 - y_1)^2 + (z_2 - z_1)^2}.$$

22. The value of the determinant
$$\begin{vmatrix} x_1 & y_1 & z_1 & 1 \\ x_2 & y_2 & z_2 & 1 \\ x_3 & y_3 & z_3 & 1 \\ x_4 & y_4 & z_4 & 1 \end{vmatrix},$$
where (x_i, y_i, z_i), $i = 1, 2, 3, 4$, are any four points.

23. The algebraic character of a surface.

24. The reducibility of an algebraic surface.

25. The degree of an algebraic surface.

* **26.** The property of a point's being a center of a given quadric surface.

* **27.** The property of a plane's being the diametral plane of a given quadric surface conjugate to the direction (λ, μ, ν).

* **28.** The property of a plane's being the polar plane of a particular point with respect to a given quadric surface.

* **29.** Suppose one attempts to define a plane, to be labeled the "pohlar plane" of the point (x_0, y_0, z_0) with respect to the quadric surface $f(x, y, z) = 0$, by means of the equation
$$p(x - x_0) + q(y - y_0) + r(z - z_0) = 0.$$

Prove that the property of a plane's being the "pohlar plane" of a particular point with respect to a given quadric surface is not invariant under translation, and therefore is a meaningless concept when the coördinate system is unspecified. (Cf. Ex. **41,** § **147.**)

141. Rotations. Two rectangular coördinate systems having the same origin O are shown in Fig. 49. With respect to the $x'y'z'$ system let the direction cosines of the x, y, and z axes be $(\lambda_1, \mu_1, \nu_1)$, $(\lambda_2, \mu_2, \nu_2)$, and $(\lambda_3, \mu_3, \nu_3)$, respectively. Then with respect to the xyz system the direction cosines of the x', y', and z' axes are $(\lambda_1, \lambda_2, \lambda_3)$, (μ_1, μ_2, μ_3), and (ν_1, ν_2, ν_3), respectively.

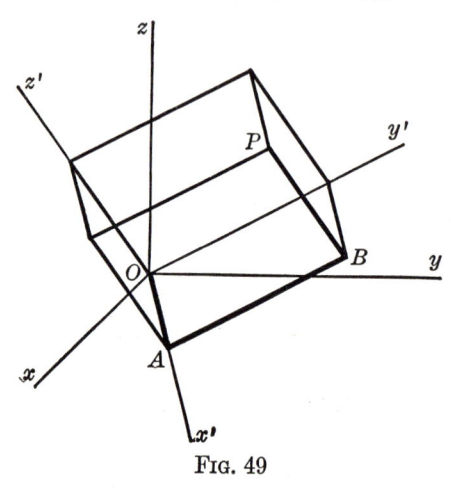

Let P be an arbitrary point (other than the origin and let its coördinates in the two systems be (x, y, z) and (x', y', z'). Relations between the primed and unprimed coördinates are most easily obtained by projections. We shall first project the directed line segment

FIG. 49

OP on the x axis. The result is x and is the same as the projection of the broken line segment $OABP$ on the x axis, or the sum of the individual projections (see § 25). The projection of OA on the x axis is $OA \cdot \cos(xOx') = x'\lambda_1$, the projection of AB on the x axis is $AB \cdot \cos(xOy') = y'\mu_1$, and the projection of BP on the x axis is $BP \cdot \cos(xOz') = z'\nu_1$. Combining these projections, we have the equation $x = \lambda_1 x' + \mu_1 y' + \nu_1 z'$. Projections of this same broken line segment on the y axis and the z axis give two more equations of a similar nature.

In the same way, the segment OP can be projected on the primed axes to give x', y', and z' in terms of x, y, and z. The broken line segment from O to P will in this case consist of segments parallel to the unprimed axes. The complete collection of six equations follows:

(1) $\begin{aligned} x &= \lambda_1 x' + \mu_1 y' + \nu_1 z', \\ y &= \lambda_2 x' + \mu_2 y' + \nu_2 z', \\ z &= \lambda_3 x' + \mu_3 y' + \nu_3 z'; \end{aligned}$ (2) $\begin{aligned} x' &= \lambda_1 x + \lambda_2 y + \lambda_3 z, \\ y' &= \mu_1 x + \mu_2 y + \mu_3 z, \\ z' &= \nu_1 x + \nu_2 y + \nu_3 z. \end{aligned}$

As in the case of translations, each of these two sets of equations can be obtained by solving the other set for the three unknowns. (Cf. Ex. **18**, § 147.)

The transformation equations (1) and (2) can be easily remembered by the device:

	x'	y'	z'
x	λ_1	μ_1	ν_1
y	λ_2	μ_2	ν_2
z	λ_3	μ_3	ν_3

Since the elements of the rows (or columns) of the **rotation matrix**

$$(3) \qquad A = \begin{pmatrix} \lambda_1 & \mu_1 & \nu_1 \\ \lambda_2 & \mu_2 & \nu_2 \\ \lambda_3 & \mu_3 & \nu_3 \end{pmatrix}$$

are direction cosines of perpendicular directions, we have immediately the following 12 relations between the elements of A:

$$(4) \quad \begin{aligned} \lambda_1^2 + \mu_1^2 + \nu_1^2 &= 1, \\ \lambda_2^2 + \mu_2^2 + \nu_2^2 &= 1, \\ \lambda_3^2 + \mu_3^2 + \nu_3^2 &= 1; \end{aligned} \qquad \begin{aligned} \lambda_1^2 + \lambda_2^2 + \lambda_3^2 &= 1, \\ \mu_1^2 + \mu_2^2 + \mu_3^2 &= 1, \\ \nu_1^2 + \nu_2^2 + \nu_3^2 &= 1; \end{aligned}$$

$$(5) \quad \begin{aligned} \lambda_1\lambda_2 + \mu_1\mu_2 + \nu_1\nu_2 &= 0, \\ \lambda_1\lambda_3 + \mu_1\mu_3 + \nu_1\nu_3 &= 0, \\ \lambda_2\lambda_3 + \mu_2\mu_3 + \nu_2\nu_3 &= 0; \end{aligned} \qquad \begin{aligned} \lambda_1\mu_1 + \lambda_2\mu_2 + \lambda_3\mu_3 &= 0, \\ \lambda_1\nu_1 + \lambda_2\nu_2 + \lambda_3\nu_3 &= 0, \\ \mu_1\nu_1 + \mu_2\nu_2 + \mu_3\nu_3 &= 0. \end{aligned}$$

This list of relations will be expanded to twenty-two in **§ 142**.

* A rotation, like a translation, can be considered either as a coördinate transformation or as a point transformation. Products of rotations are defined as in the case of any transformations. That multiplication of rotations is not commutative is easily verified by a simple example.† However, the associative law holds since it is valid for any combinations of transformations. The transformations defined by equations (1) and (2) can be considered as inverse transformations. (Cf. Ex. **39**, § **147**.) The identity transformation can be regarded as a rotation with matrix

$$(6) \qquad \begin{pmatrix} 1 & 0 & 0 \\ 0 & 1 & 0 \\ 0 & 0 & 1 \end{pmatrix}.$$

Combinations of rotations are treated more completely in Chapter IX by means of matrix algebra.

---

† Place a book (or similar object) in a horizontal plane, and take a convenient system of axes meeting at the center of the book. Now consider the results of applying the following two point transformations, first in one order and then in the reverse order: (*i*) a rotation of 90° about the z axis; (*ii*) a rotation of 90° about the y axis.

142. Determinant of a rotation matrix.

THEOREM. *The determinant D of the rotation matrix A is equal to 1.*

Proof. As shown in § 23, the cofactors of the elements of the first row of D are direction numbers of a line perpendicular to the directions $(\lambda_2, \mu_2, \nu_2)$ and $(\lambda_3, \mu_3, \nu_3)$. Furthermore, since the angle between these directions is 90° and sin (90°) = 1, by the formula of § 21 for $\sin^2 \theta$ the sum of the squares of these cofactors is 1, and the cofactors form a set of direction cosines. Since λ_1, μ_1, ν_1 and their cofactors are direction cosines of the same line they must be either equal or negatives. By expanding the determinant with respect to the elements of the first row we have

$$D = \pm(\lambda_1^2 + \mu_1^2 + \nu_1^2) = \pm 1.$$

We shall now show by a (non-algebraic) continuity argument that this value must be 1.

Consider a rigid system of axes with a fixed origin to be moving continuously from coincidence with the xyz system into coincidence with the $x'y'z'$ system. We can think of time t as a parameter and the entire motion taking place within, say, one minute. The direction cosines of the moving axes, taken with respect to the xyz system, are continuous functions of t, $\lambda_1(t)$ for instance varying continuously from $\lambda_1(0) = 1$ to $\lambda_1(1) = \lambda_1$. Since the determinant D is a polynomial in $\lambda_1, \lambda_2, \cdots, \nu_3$, it becomes a continuous function of t, $D(t)$. Since, as shown in the preceding paragraph, $D(t) = \pm 1$, and is continuous, it can assume no value between -1 and 1, and must therefore be either identically equal to -1 or identically equal to 1. At the beginning of the motion the axes coincide with the xyz axes and therefore $D(0)$ is the determinant

$$\begin{vmatrix} 1 & 0 & 0 \\ 0 & 1 & 0 \\ 0 & 0 & 1 \end{vmatrix} = 1.$$

Consequently $D = D(1) = 1$, as was to be proved.

By a similar argument we could show that a transformation with fixed origin that reverses sense, one system of axes being right-handed and the other left-handed, always has a determinant equal to -1. Such a transformation is sometimes called a **skew-rotation**.

Since $D = 1$, each element in the first row is equal to its cofactor (rather than the negative of its cofactor). It is now readily seen that this is a property of the elements of any row, and that any element of D is equal to its cofactor.

Therefore, in addition to the twelve equations (4) and (5), § **141**, the elements of A are related by ten more equations. One equation is obtained by setting D equal to 1, and the other nine equations are obtained by equating each element to its cofactor. These twenty-two relations are far from independent, as is demonstrated in Chapter IX.

EXAMPLE 1. Write the rotation matrix for the rotation to new axes if the x' and y' axes have direction cosines in the original coördinate system:
$$\tfrac{2}{7}, \tfrac{3}{7}, \tfrac{6}{7} \quad \text{and} \quad \tfrac{6}{7}, \tfrac{2}{7}, -\tfrac{3}{7},$$
respectively.

Solution. Since the z' axis is perpendicular to the x' and y' axes, it has direction cosines $\tfrac{3}{7}, -\tfrac{6}{7}, \tfrac{2}{7}$ or $-\tfrac{3}{7}, \tfrac{6}{7}, -\tfrac{2}{7}$. In order that the determinant of the rotation matrix be 1, it is necessary to choose the second set, the rotation matrix being

$$\begin{pmatrix} \tfrac{2}{7} & \tfrac{6}{7} & -\tfrac{3}{7} \\ \tfrac{3}{7} & \tfrac{2}{7} & \tfrac{6}{7} \\ \tfrac{6}{7} & -\tfrac{3}{7} & -\tfrac{2}{7} \end{pmatrix}.$$

EXAMPLE 2. If the $x'y'z'$ coördinate system is defined as in Example 1, find the equation in x', y', and z' of the surface whose equation in x, y, and z is
$$103x^2 + 125y^2 + 66z^2 - 48xy - 12xz - 60yz - 294 = 0.$$

Solution. If we substitute in this equation
$$x = \tfrac{2}{7}x' + \tfrac{6}{7}y' - \tfrac{3}{7}z',$$
$$y = \tfrac{3}{7}x' + \tfrac{2}{7}y' + \tfrac{6}{7}z',$$
$$z = \tfrac{6}{7}x' - \tfrac{3}{7}y' - \tfrac{2}{7}z',$$
and collect terms, the new equation reduces to
$$x'^2 + 2y'^2 + 3z'^2 = 6.$$
The surface is an ellipsoid. The student is not urged to carry through the substitutions (cf. Ex. **11**, § **147**).

★ 143. Axis of a rotation. Considered as a point transformation, the rotation with matrix

$$\begin{pmatrix} \cos\theta & \sin\theta & 0 \\ -\sin\theta & \cos\theta & 0 \\ 0 & 0 & 1 \end{pmatrix}$$

is a rotation about the z axis through the angle θ. Every point on the z axis is transformed into itself. (Prove these statements.)

More generally, any rotation has such an axis of invariant points, according to the following definition and theorem:

DEFINITION. *Any point that is transformed into itself by a (point) transformation is called a* ***fixed point*** *of the transformation.*

THEOREM. *Any rotation with fixed origin, different from the identity, has a unique line of fixed points passing through the origin, called the* **axis** *of the rotation.*

Proof. The point (x, y, z) is transformed into itself by the rotation T with matrix (3), § **141**, if and only if the following system of equations is satisfied:

(1)
$$\lambda_1 x + \mu_1 y + \nu_1 z = x,$$
$$\lambda_2 x + \mu_2 y + \nu_2 z = y,$$
$$\lambda_3 x + \mu_3 y + \nu_3 z = z.$$

In other words, the fixed points of T are the points whose coördinates satisfy the system of homogeneous linear equations whose coefficient matrix is

$$B = \begin{pmatrix} \lambda_1 - 1 & \mu_1 & \nu_1 \\ \lambda_2 & \mu_2 - 1 & \nu_2 \\ \lambda_3 & \mu_3 & \nu_3 - 1 \end{pmatrix}.$$

According as the rank of B is 0, 1, 2, or 3, the fixed points of T are all points in space, the points on a plane through the origin, the points on a line through the origin, or the origin only. Obviously, the rank of B is 0 if and only if $T = I$, the identity transformation. It remains to be shown that the rank of B cannot be 1 or 3. To this end we write down the cofactor of each element of $|B|$ (the determinant of B), making use of the fact that each element of D, the determinant of the rotation matrix A, is equal to its cofactor. The cofactor of each element of $|B|$ is written in the corresponding position in the following matrix:

$$C = \begin{pmatrix} \lambda_1 - \mu_2 - \nu_3 + 1 & \mu_1 + \lambda_2 & \nu_1 + \lambda_3 \\ \mu_1 + \lambda_2 & -\lambda_1 + \mu_2 - \nu_3 + 1 & \nu_2 + \mu_3 \\ \nu_1 + \lambda_3 & \nu_2 + \mu_3 & -\lambda_1 - \mu_2 + \nu_3 + 1 \end{pmatrix}.$$

To prove that $|B| = 0$, we expand it with respect to the elements of any row or column (with the aid of C), and again use the fact that each element of D is equal to its cofactor.

Finally, if the rank of B is 1, the cofactor of each element of $|B|$ is zero, and therefore every element of C is zero. In particular, the elements on the main diagonal of C vanish, whence we have the system of equations

$$-\lambda_1 + \mu_2 + \nu_3 = 1,$$
$$\lambda_1 - \mu_2 + \nu_3 = 1,$$
$$\lambda_1 + \mu_2 - \nu_3 = 1,$$

whose only solution is $\lambda_1 = \mu_2 = \nu_3 = 1$. But these relations hold only for the identity I, and for the identity the rank of B is zero. With this contradiction the proof is complete.

EXAMPLE. Find the equations of the axis of the rotation of Example 1, § **142**.

Solution. The system of equations (1) is equivalent to the system

$$5x - 6y + 3z = 0,$$
$$3x - 5y + 6z = 0,$$
$$2x - y - 3z = 0.$$

Solving this system, we have the equations of the axis·

$$\frac{x}{3} = \frac{y}{3} = \frac{z}{1}.$$

* **144. Angle of a rotation.** To determine the angle through which points are carried by a rotation, we have only to choose any point whose radius vector is perpendicular to the axis of the rotation, and find the point into which it is carried. This method is illustrated in the following example (for another method, see § **176**):

EXAMPLE. Determine the angle of the rotation of Example 1, § **142**.

Solution. As found in the Example of § **143**, the axis of the rotation has direction numbers 3, 3, 1. The point P (1, -1, 0) has a radius vector perpendicular to the axis of rotation, and is transformed into the point $Q(-\frac{1}{7}, \frac{4}{7}, -\frac{9}{7})$. If O is the origin and θ the angle between the directed segments OP and OQ, then

$$\cos \theta = \frac{1 \cdot (-1) + (-1) \cdot 4 + 0 \cdot (-9)}{\sqrt{2} \cdot \sqrt{98}} = -\frac{5}{14}.$$

Therefore the angle sought is

$$\theta = \text{arc cos } (-\tfrac{5}{14}).$$

145. Invariance of degree.

THEOREM. *Under a coördinate transformation from one rectangular set of axes to another, the algebraic character and the degree of an algebraic surface are invariant.*

Proof can be restricted to a transformation that is either a translation or a rotation with fixed origin. Since the equations defining a transformation of either type are linear in the coördinates, substitution in the given algebraic equation will always yield a new algebraic equation whose degree cannot exceed that of the original equation. We therefore can say that the algebraic character of the surface is preserved, and that the degree of the surface cannot be increased. But, on the other hand, the degree cannot be decreased, for if it could the inverse transformation applied to the new equation would increase the degree to its original value. Therefore the degree is invariant.

One might express a particular case of this theorem by saying, "Once a quadric surface, always a quadric surface."

146. Plane sections of algebraic surfaces.

THEOREM I. *If every point of a plane is on an algebraic surface, the plane is a factor of the surface.*

Proof. Let a rectangular coördinate system be chosen so that the given plane is the xy plane, and let the equation of the surface be $F(x, y, z) = 0$, where $F(x, y, z)$ is a polynomial. Then, by the assumptions of the theorem, $F(x, y, 0) \equiv 0$ for all values of x and y. But $F(x, y, 0)$ is a polynomial in the variables x and y, and if a polynomial in any number of variables vanishes for all values of the variables, every coefficient must also vanish.† This means that every term of $F(x, y, z)$ contains z as a factor, and therefore the surface $F(x, y, z) = 0$ has the plane $z = 0$ as a factor. Since this plane is a factor of the surface in one coördinate system, it is a factor in every coördinate system (why?).

THEOREM II. *Any non-vacuous plane section of an algebraic surface by a plane that is not a factor of the surface is an algebraic plane curve‡ whose degree does not exceed that of the surface.*

Proof. Choose a coördinate system as in the proof of Theorem I, and set $z = 0$.

As a corollary to this theorem we have:

THEOREM III. *Any non-vacuous section of a quadric surface by a plane that is not a factor of the surface is a conic section or a line.*

---

† For a general proof of this fact, see Maxime Bôcher, *Introduction to Higher Algebra* (The Macmillan Company, 1935), Theorem 1, § 2. For the special case of a polynomial in two variables of degree at most 2,

$$ax^2 + 2hxy + by^2 + 2px + 2qy + d,$$

a simple proof is obtained by equating this polynomial to zero, after substituting a few pairs of values for x and y. For example, the substitution $x = 0, y = 0$ gives $d = 0$, after which the substitutions $x = 1, y = 0$ and $x = 2, y = 0$ give $a = p = 0$ and the substitutions $x = 0, y = 1$ and $x = 0, y = 2$ give $b = q = 0$. Finally, the substitution $x = 1, y = 1$ gives $h = 0$.

‡ The definitions of an algebraic plane curve and the degree of an algebraic plane curve are similar to those for algebraic surfaces, and can be supplied by the student.

A particular case of this theorem is the well known fact that any section of a right circular cone is a conic section.

Examples of the two types of non-vacuous sections mentioned in Theorem III are the traces in the xy plane of the sphere $x^2 + y^2 + z^2 = 1$ and the parabolic cylinder $z^2 = y$. The trace in the xy plane of the hyperbolic cylinder $yz = 1$ is vacuous. The student should appreciate that the trace in the xy plane of the parabolic cylinder $y^2 = z$ is *two coincident lines*, not one line, and that the trace in the xy plane of the hyperbolic cylinder $z^2 - y^2 = 1$ is imaginary, not vacuous.

In the sense of the footnote on page 63, in the absence of other information one can expect a section of a quadric surface to be a conic section (in fact, an ellipse or a non-degenerate hyperbola). All other sections can be considered as exceptional.

EXAMPLE. Determine the nature of the section of the surface
$$x^2 + xy - 2yz = 4 \text{ by the plane } x + 2y - 2z = 6.$$

Solution. By a suitable rotation of axes we can choose the z' axis normal to the given plane. That is, we let the direction cosines of the z' axis be $\frac{1}{3}, \frac{2}{3}, -\frac{2}{3}$. Any two lines through the origin that are perpendicular to the z' axis and to each other can be chosen as the x' and y' axes. For example, the x' axis might have direction numbers $(0, 1, 1)$, in which case the y' axis would have direction numbers $(4, -1, 1)$. However, two more suitable directions for the x' and y' axes are $(\frac{2}{3}, \frac{1}{3}, \frac{2}{3})$ and $(\frac{2}{3}, -\frac{2}{3}, -\frac{1}{3})$, respectively. Accordingly, let the rotation matrix be

$$\begin{pmatrix} \frac{2}{3} & \frac{2}{3} & \frac{1}{3} \\ \frac{1}{3} & -\frac{2}{3} & \frac{2}{3} \\ \frac{2}{3} & -\frac{1}{3} & -\frac{2}{3} \end{pmatrix},$$

the transformation equations being

$$3x = 2x' + 2y' + z', \qquad 3x' = 2x + y + 2z,$$
$$3y = x' - 2y' + 2z', \qquad 3y' = 2x - 2y - z,$$
$$3z = 2x' - y' - 2z'; \qquad 3z' = x + 2y - 2z.$$

Substitution in the original equation of the quadric surface and the plane gives the new equations

$$2x'^2 + 4y'^2 + 11z'^2 + 16x'y' + 5x'z' + 2y'z' = 36,$$

and

$$z' = 2,$$

respectively. To find the equation of the section, we substitute $z' = 2$ in the new equation, obtaining

$$x'^2 + 8x'y' + 2y'^2 + 5x' + 2y' + 4 = 0.$$

The section is therefore a (non-degenerate) hyperbola.

147. Exercises.

In Exercises **1-4**, supply the missing elements of the rotation matrix.

1. $\begin{pmatrix} \frac{3}{7} & -\frac{6}{7} & \frac{2}{7} \\ \frac{6}{7} & \frac{2}{7} & -\frac{3}{7} \\ & & \end{pmatrix}.$
2. $\begin{pmatrix} & \frac{6}{19} & \frac{15}{19} \\ & \frac{10}{19} & \frac{6}{19} \\ & \frac{15}{19} & \end{pmatrix}.$

3. $\begin{pmatrix} \frac{1}{3} & & \frac{2}{3} \\ & \frac{1}{3} & \frac{2}{3} \end{pmatrix}$. 4. $\begin{pmatrix} \frac{1}{\sqrt{2}} & -\frac{1}{\sqrt{2}} \\ \frac{1}{\sqrt{3}} & \end{pmatrix}$.

In Exercises 5 and 6, write the rotation equations for a transformation of coördinates to a new set of axes, two of which have the given directions in the original coördinate system.

5. $x' : \left(\frac{1}{\sqrt{3}}, \frac{1}{\sqrt{3}}, \frac{1}{\sqrt{3}}\right); z' : \left(\frac{1}{\sqrt{2}}, 0, -\frac{1}{\sqrt{2}}\right)$.

6. $y' : \left(\frac{1}{\sqrt{14}}, \frac{2}{\sqrt{14}}, \frac{3}{\sqrt{14}}\right); z' : \left(\frac{2}{\sqrt{5}}, -\frac{1}{\sqrt{5}}, 0\right)$.

In Exercises 7 and 8, transform the given equation to one in x', y', and z' according to the given rotation matrix.

7. $x + y = 8$;

$$\begin{pmatrix} \frac{2}{3} & \frac{1}{3} & -\frac{2}{3} \\ \frac{1}{3} & \frac{2}{3} & \frac{2}{3} \\ \frac{2}{3} & -\frac{2}{3} & \frac{1}{3} \end{pmatrix}$$

8. $6x + 8y - 9z + 5 = 0$;

$$\begin{pmatrix} \frac{1}{\sqrt{6}} & -\frac{1}{\sqrt{6}} & \frac{2}{\sqrt{6}} \\ \frac{1}{\sqrt{2}} & \frac{1}{\sqrt{2}} & 0 \\ -\frac{1}{\sqrt{3}} & \frac{1}{\sqrt{3}} & \frac{1}{\sqrt{3}} \end{pmatrix}.$$

9. Transform the equation

$$2y^2 + z^2 - 4xz - 4yz = 4$$

by the rotation of Exercise 7 and identify the surface.

10. Transform the equation

$$xy - \sqrt{2}\, yz = 3$$

by the rotation of Exercise 8 and identify the surface.

11. Verify the result of Example 2, §142, by applying the inverse transformation to the new equation.

12. Write down the four possible rotation matrices corresponding to transformations of coördinates where the new coördinate planes are as follows:

$x'y' : 2x - 2y + z = 0;\; x'z' : 2x + y - 2z = 0;\; y'z' : x + 2y + 2z = 0$.

In Exercises 13 and 14, transform the given equation to one in x', y' and z', if the new coördinate planes are those specified.

13. $x^2 - z^2 - 4xy - 4yz + 3 = 0$; planes of Exercise 12.
14. $x^2 + y^2 + z^2 = 25$;

$x'y' : 3x + 12y + 4z = 0;\; x'z' : 12x - 4y + 3z = 0;\; y'z' : 4x + 3y - 12z = 0$.

15. Prove that the equation of a sphere with center at the origin is invariant under any rotation with fixed origin. Actually make the substitutions and carry through the transformation.

16. Verify the twenty-two relations of §§ **141** and **142** for the rotation matrices of Exercises **7** and **8**.

17. Show that the points in the first octant of the $x'y'z'$ coördinate system of Example 1, § **142**, are those whose xyz coördinates satisfy the inequalities:

$$2x + 3y + 6z > 0$$
$$6x + 2y - 3z > 0$$
$$-3x + 6y - 2z > 0.$$

18. Obtain system (2), § **141**, by solving system (1), § **141**, for x', y', and z'.

19. Write the matrix for the rotation for which the y' and z' axes are coincident with the y and z axes, respectively, but are oppositely directed, and for which the x' axis is identical with the x axis.

20. Write the matrix for the rotation for which the x' and y' axes are identical with the z and y axes, respectively.

21. Show that any relabeling of the axes or changing of direction on the axes that preserves right-handedness is equivalent to a suitable rotation with the origin fixed.

In Exercises **22** and **23**, write the matrix for a rotation that will reduce the given equation to canonical form.

22. $y^2 - z^2 + 1 = 0$.   **23.** $y^2 - z^2 + 2x = 0$.

24. If the axes of a new system of coördinates meet at the point $(4, -3, 1)$ and have directions as follows:

$$x' : (\tfrac{2}{3}, \tfrac{1}{3}, \tfrac{2}{3}); \quad y' : (-\tfrac{2}{3}, \tfrac{2}{3}, \tfrac{1}{3}); \quad z' : (-\tfrac{1}{3}, -\tfrac{2}{3}, \tfrac{2}{3}),$$

express the coördinates of each system in terms of those of the other.

★ Supplementary Exercises

25. Prove that any rigid motion can be represented in the form

$$T((x, y, z)) = (\lambda_1 x + \mu_1 y + \nu_1 z + a,\ \lambda_2 x + \mu_2 y + \nu_2 z + b,$$
$$\lambda_3 x + \mu_3 y + \nu_3 z + c).$$

26. A transformation T that is equal to its inverse (that is, whose square T^2 is the identity I) is called **involutory** or **an involution**. Prove that a rotation with fixed origin is an involution if and only if its matrix is symmetric. Show that the following is the matrix of an involutory rotation:

$$\begin{pmatrix} -\tfrac{1}{3} & -\tfrac{2}{3} & -\tfrac{2}{3} \\ -\tfrac{2}{3} & -\tfrac{1}{3} & \tfrac{2}{3} \\ -\tfrac{2}{3} & \tfrac{2}{3} & -\tfrac{1}{3} \end{pmatrix}.$$

In Exercises **27** and **28**, determine the nature of the section of the given surface by the given plane.

27. $z^2 - 2xy + x + y - 1 = 0;\ x + y = 2.$

Suggestion: Let $z' = \dfrac{x+y}{\sqrt{2}}$, $y' = z$.

28. $z^2 - 4xy = 10$; $x + y + z = 4$.

Suggestion: Let $z' = \dfrac{x+y+z}{\sqrt{3}}$, $y' = \dfrac{-x+y}{\sqrt{2}}$.

In Exercises **29** and **30**, find the equations of the axis of the rotation whose matrix is given.

29. $\begin{pmatrix} -\frac{2}{3} & \frac{1}{3} & \frac{2}{3} \\ -\frac{1}{3} & \frac{2}{3} & -\frac{2}{3} \\ -\frac{2}{3} & -\frac{2}{3} & -\frac{1}{3} \end{pmatrix}$.

30. $\begin{pmatrix} \frac{3}{7} & \frac{2}{7} & -\frac{6}{7} \\ -\frac{6}{7} & \frac{3}{7} & -\frac{2}{7} \\ \frac{2}{7} & \frac{6}{7} & \frac{3}{7} \end{pmatrix}$.

31. Find the angle of the rotation of Exercise **29**.

32. Find the angle of the rotation of Exercise **30**.

33. Show that the angle of the rotation of Exercise **26** is 180°. More generally, prove that the angle of any involutory rotation different from the identity is 180°.

34. Show that the angle of each rotation of Exercise **22** is 120°.

35. Show that the angle of each rotation of Exercise **23** is 90°, 120°, or 180°.

In Exercises **36-38**, prove that the given property or quantity is invariant with respect to a rotation with fixed origin.

36. The distance between two points, (x_1, y_1, z_1) and (x_2, y_2, z_2), as given by the formula

$$d = \sqrt{(x_2 - x_1)^2 + (y_2 - y_1)^2 + (z_2 - z_1)^2}.$$

37. The distance between a plane and a point, as given by formula (1), §32.

38. The reducibility of an algebraic surface.

39. Prove analytically that if T and U are the (point) rotations having the matrices

$$\begin{pmatrix} \lambda_1 & \mu_1 & \nu_1 \\ \lambda_2 & \mu_2 & \nu_2 \\ \lambda_3 & \mu_3 & \nu_3 \end{pmatrix} \text{ and } \begin{pmatrix} \lambda_1 & \lambda_2 & \lambda_3 \\ \mu_1 & \mu_2 & \mu_3 \\ \nu_1 & \nu_2 & \nu_3 \end{pmatrix},$$

respectively, then T and U are inverse transformations; that is, $TU = UT = I$. Actually carry through the substitutions.

40. Prove analytically that if T is a translation and U a rotation with fixed origin, then $U^{-1}TU$ is a translation. Actually carry through the substitutions.

41. Suppose one attempts to define a plane, to be labeled the "poelar plane" of the point (x_0, y_0, z_0) with respect to the quadric surface $f(x, y, z) = 0$, by means of the equation

$$a(x - x_0) + b(y - y_0) + c(z - z_0) = 0.$$

§ 147] EXERCISES 183

Prove that the property of a plane's being the "poelar plane" of a particular point with respect to a given quadric surface is invariant under translation, but is not invariant under all rotations with fixed origin. Hence show that this is a meaningless concept when the coördinate system is unspecified. (Cf. Ex. **29**, **§ 140**.) *Suggestion for the second part:* Consider a surface with a simple equation, like an ellipsoid, and use a transformation of the type: $x = -z', y = y', z = x'$.

CHAPTER VIII

ANALYSIS OF THE GENERAL EQUATION OF THE SECOND DEGREE

148. Introduction. In Chapter VI the discussion of the general equation of the second degree (with real coefficients) was initiated, and pertinent notation was introduced. It is the purpose of this chapter to extend and complete this discussion with the aid of coördinate and point transformations. The aim will be (i) to show that with an appropriate coördinate system any quadric surface has an equation in one of the seventeen canonical forms, (ii) to develop methods of determining to which canonical form a given equation is reducible, and (iii) to find the new coördinate planes and axes and the simplified equation.

The following discussions apply to the general equation of the second degree with real coefficients,

(1) $\quad f(x, y, z) \equiv ax^2 + by^2 + cz^2 + 2fyz + 2gxz + 2hxy$
$\quad\quad\quad + 2px + 2qy + 2rz + d = 0,$

and the notation is that of Chapter VI. Repeated reference to the seventeen canonical forms will be necessary, and the student is referred to their representation in matrix form at the beginning of Chapter VI.

149. Principal planes. In order to achieve the goal enunciated above we seek first a plane of symmetry of a certain type, called a *principal plane* according to the definition:

DEFINITION. *A principal plane is a diametral plane that is perpendicular to the chords it bisects.*

A necessary and sufficient condition that the diametral plane corresponding to the direction (λ, μ, ν) be perpendicular to this direction is that the coefficients in the equation of this plane, (1), § **121**, be proportional to λ, μ, ν; that is, that there exist a real number k different from zero such that

(1)
$$a\lambda + h\mu + g\nu = k\lambda,$$
$$h\lambda + b\mu + f\nu = k\mu,$$
$$g\lambda + f\mu + c\nu = k\nu.$$

This result is expressed in slightly different form in the theorem:

Theorem I. *The diametral plane*

(2) $(al + hm + gn)x + (hl + bm + fn)y$
       $+ (gl + fm + cn)z + (pl + qm + rn) = 0,$

which bisects chords with direction numbers l, m, n, is perpendicular to these chords if and only if there exists a real number k different from zero such that

(3)
$$(a - k)l + hm + gn = 0,$$
$$hl + (b - k)m + fn = 0,$$
$$gl + fm + (c - k)n = 0.$$

If (3) is considered as a system of equations in the unknowns l, m, and n, k being a constant, a non-trivial solution exists if and only if

(4)
$$\begin{vmatrix} a - k & h & g \\ h & b - k & f \\ g & f & c - k \end{vmatrix} = 0.$$

Equation (4), considered as a cubic equation in k is called the **characteristic equation** of the matrix e. Its roots are called the **characteristic roots** of e. Its left member is called the **characteristic polynomial** of e. Any non-trivial solution, l, m, n, of system (3) corresponding to a characteristic root k is called a **characteristic vector** of e corresponding to k. If a set of direction cosines, λ, μ, ν, is a characteristic vector of e corresponding to a characteristic root k, (λ, μ, ν) is called a **characteristic direction** of e corresponding to k.

In expanded form the characteristic equation (4) will be written with a change in sign:

(5) $$k^3 - Ik^2 + Jk - D = 0,$$

where D is the determinant of e and

$$I = a + b + c,$$
$$J = ab + ac + bc - f^2 - g^2 - h^2.$$

Notice that I is the sum of the elements of the main diagonal of e (or D) and that J is the sum of their cofactors in D. The sum of the elements of the main diagonal of a square matrix is called the **trace** or **spur** of the matrix, and is an important concept in the theory of matrices.

Before any statement regarding existence of principal planes can be made, we must prove the existence of a *real non-zero* solution of the characteristic equation (4).

THEOREM II. *The characteristic roots of a real symmetric matrix are all real. The roots are all zero if and only if the rank of the matrix is zero; that is, if and only if every element of the matrix vanishes. Characteristic directions corresponding to distinct characteristic roots are perpendicular.*

Proof is given only for a third order matrix e, although the theorem holds in general (with appropriate definitions of such concepts as *characteristic direction* and *perpendicularity* in n-dimensional Euclidean space). For the first part, assume that $k = \alpha + i\beta$ is an imaginary characteristic root of the matrix e, and let $l_1 + il_2$, $m_1 + im_2$, $n_1 + in_2$ be a corresponding characteristic vector. Equating reals and imaginaries after substituting in system (3), we have the six equations

$$al_1 + hm_1 + gn_1 = \alpha l_1 - \beta l_2, \quad -l_2$$
$$hl_1 + bm_1 + fn_1 = \alpha m_1 - \beta m_2, \quad -m_2$$
$$gl_1 + fm_1 + cn_1 = \alpha n_1 - \beta n_2, \quad -n_2$$
$$al_2 + hm_2 + gn_2 = \alpha l_2 + \beta l_1, \quad l_1$$
$$hl_2 + bm_2 + fn_2 = \alpha m_2 + \beta m_1, \quad m_1$$
$$gl_2 + fm_2 + cn_2 = \alpha n_2 + \beta n_1. \quad n_1$$

If the two members of each equation are multiplied by the indicated factors and the products added, the resulting equation is

$$\beta(l_1^2 + m_1^2 + n_1^2 + l_2^2 + m_2^2 + n_2^2) = 0.$$

Since β is assumed to be different from zero, $l_1 = \cdots = n_2 = 0$. But this contradicts the assumption that $l_1 + il_2$, $m_1 + im_2$, and $n_1 + in_2$ are not all zero.

Now assume that zero is the only characteristic root of the matrix e. This means that the characteristic equation (5) becomes $(k - 0)(k - 0)(k - 0) \equiv k^3 = 0$, and therefore that $I = J = D = 0$. Therefore $I^2 - 2J = a^2 + b^2 + c^2 + 2f^2 + 2g^2 + 2h^2 = 0$. But this is possible only if every element of e is zero.

Finally, let $(\lambda_1, \mu_1, \nu_1)$ and $(\lambda_2, \mu_2, \nu_2)$ be directions corresponding to distinct characteristic roots k_1 and k_2, respectively. If we write down the system (1) with subscripts 1, multiply the two members of the three equations by λ_2, μ_2, and ν_2, respectively, and add the results, we obtain the relation

$$a\lambda_1\lambda_2 + b\mu_1\mu_2 + c\nu_1\nu_2 + f(\mu_1\nu_2 + \mu_2\nu_1) + g(\lambda_1\nu_2 + \lambda_2\nu_1)$$
$$+ h(\lambda_1\mu_2 + \lambda_2\mu_1) = k_1(\lambda_1\lambda_2 + \mu_1\mu_2 + \nu_1\nu_2).$$

In a similar way we find that the left member of this equation is equal to $k_2(\lambda_1\lambda_2 + \mu_1\mu_2 + \nu_1\nu_2)$, or that

$$(k_1 - k_2)(\lambda_1\lambda_2 + \mu_1\mu_2 + \nu_1\nu_2) = 0.$$

Since the first factor does not vanish, the second must, and perpendicularity is established. This completes the proof of the theorem.

By a process similar to that used in the proof of the last part of Theorem II, multiplication of the equations of system (1) by λ, μ, and ν, respectively, and addition of the results lead to the relation

$$e(\lambda, \mu, \nu) = k(\lambda^2 + \mu^2 + \nu^2) = k.$$

This establishes the theorem:

THEOREM III. *If (λ, μ, ν) is a direction corresponding to a characteristic root k of e, then $e(\lambda, \mu, \nu) = k$.*

This theorem helps explain the significance of the vanishing or non-vanishing of a characteristic root. We recall that the condition for the plane conjugate to the direction (λ, μ, ν) to be a diametral plane is that $e(\lambda, \mu, \nu) \neq 0$. Therefore, if k is a characteristic root and $k \neq 0$, the plane conjugate to a corresponding characteristic direction is a *diametral* plane, and hence a principal plane that is the perpendicular bisector of actual chords of the quadric surface. If k is a characteristic root and $k = 0$, there is no plane conjugate to a corresponding characteristic direction since, by equations (1), the coefficients of x, y, and z in equation (2) all vanish. Therefore principal planes correspond to *non-vanishing* characteristic roots of the matrix e.

Since Theorem II guarantees that the third order matrix of a quadric surface has a real non-vanishing characteristic root, we can infer the following important result:

THEOREM IV. *Every quadric surface has a principal plane.*

EXAMPLE. Find the principal planes of the surface

$$x^2 + y^2 - xy + xz + yz - 6y - 4z = 0.$$

Solution. To avoid fractions we multiply both members of the given equation by 2. From the matrix

$$E = \begin{pmatrix} 2 & -1 & 1 & 0 \\ -1 & 2 & 1 & -6 \\ 1 & 1 & 0 & -4 \\ 0 & -6 & -4 & 0 \end{pmatrix}$$

we find $I = 4$, $J = 1$, and $D = -6$. The characteristic equation (5) is therefore

$$k^3 - 4k^2 + k + 6 = 0,$$

with roots -1, 2, and 3. Corresponding to $k = -1$, the system (3) is

$$3l - m + n = 0,$$
$$-l + 3m + n = 0,$$
$$l + m + n = 0,$$

whence $l : m : n = 1 : 1 : -2$. Substituting $l = 1$, $m = 1$, and $n = -2$ in equation (2) (with the aid of the matrix E, above), we obtain the equation of the principal plane corresponding to $k = -1$:

$$x + y - 2z - 2 = 0.$$

In a similar way we find that the principal planes corresponding to $k = 2$ and $k = 3$ are $x + y + z - 5 = 0$ and $x - y + 2 = 0$, respectively. As a check, we see that the given surface is a central quadric and that its center, $(1,3,1)$, lies in each principal plane. (Cf. Theorem II, § **122**.)

150. Exercises.

In Exercises **1–10**, find the principal planes of the given surface.

1. $x^2 - 3y^2 + 3z^2 + 8yz + 6x + 29 = 0.$
2. $x^2 + y^2 - 8z^2 - 4xy + 4xz + 4yz + 18y - 36z = 0.$
3. $16x^2 + 4y^2 + 9z^2 - 16xy + 24xz - 12yz + 40x - 20y + 30z + 25 = 0.$
4. $x^2 - z^2 + 2xz + 4yz - 6x - 8y - 2z + 9 = 0.$
5. $2x^2 - y^2 - z^2 + 2xy + 2xz + 4yz + 6x + 6y + 3 = 0.$
6. $4y^2 - 2xy - xz + 2yz + x + y - 3z - 10 = 0.$
7. $5x^2 + 3y^2 + 3z^2 + 2xy - 2xz - 2yz - 12 = 0.$
8. $3x^2 + 2xy + 2xz + 4yz - 2x - 14y + 2z - 9 = 0.$
9. $2y^2 - z^2 - 4yz + 4x - 12y - 6 = 0.$
10. $3x^2 + 2y^2 + 4z^2 - 4xy - 4xz + 4x + 4y + 2z - 36 = 0.$

11. For the seventeen quadric surfaces with equations in canonical form, assuming that no two of the numbers a, b, and c are equal, verify that the principal planes are as follows: (1) – (6), $x = 0$, $y = 0$, $z = 0$; (7) – (13), $x = 0$, $y = 0$; (14) – (17), $x = 0$.

151. Reduction to canonical form.
Let coördinate axes be chosen so that the yz plane is a principal plane of the quadric surface $f(x, y, z) = 0$. Then $(1, 0, 0)$ is a characteristic direction. Equations (1), § **149**, take the form

$$a \cdot 1 = k \cdot 1,$$
$$h \cdot 1 = k \cdot 0,$$
$$g \cdot 1 = k \cdot 0,$$

from which it is evident that $g = h = 0$ and that the characteristic root of e to which $(1, 0, 0)$ corresponds is $k = a \neq 0$.

Equation (2), § **149**, for the diametral plane corresponding to the direction $(1, 0, 0)$ now takes the form

$$ax + p = 0.$$

But since this principal plane is assumed to be the yz plane, $p = 0$. If both members of the equation $f(x, y, z) = 0$ are divided by a, which is not zero, the resulting equation has the form

(1) $\qquad x^2 + by^2 + cz^2 + 2fyz + 2qy + 2rz + d = 0.$

The equation of the trace of the surface (1) in the yz plane is obtained by setting $x = 0$. If $f \neq 0$, a rotation about the origin in the yz plane through a suitable angle θ will eliminate the term in yz from the equation of the trace. Therefore the rotation in space about the x axis with matrix

$$\begin{pmatrix} 1 & 0 & 0 \\ 0 & \cos\theta & \sin\theta \\ 0 & -\sin\theta & \cos\theta \end{pmatrix}$$

will eliminate the yz term from equation (1).

The first goal in the reduction of the general equation of the second degree to canonical form has been reached in the theorem:

Theorem I. *With an appropriate choice of coördinate axes any quadric surface has an equation of the form*

(2) $\qquad x^2 + by^2 + cz^2 + 2qy + 2rz + d = 0.$

The remaining steps in the reduction to canonical form involve simplifications by translations, of the type discussed in § **139**. In some cases, of course, it may be necessary to relabel the axes to obtain the proper arrangement of signs or variables. Except for this minor consideration, the final reduction to canonical form is obtained as follows:

If, in equation (2), b and c are both different from zero, a translation, determined by the method of completing squares, will eliminate the first degree terms and give an equation equivalent to one of the first six canonical forms.

If b or c vanishes and the other does not, the equation is reducible by a translation of axes to one of the canonical forms (7)–(13).

If $b = c = 0$ and q and r are not both zero, we can let the plane

$2qy + 2rz + d = 0$ be the new coördinate plane $y' = 0$. The surface in this case is a parabolic cylinder, type (14).

Finally, if $b = c = q = r = 0$, the surface is reducible, of one of the types (15)–(17).

By means of these coördinate transformations we have established the theorem:

THEOREM II. *With an appropriate choice of coördinate axes, any quadric surface has an equation in canonical form.*

152. Equivalence of second degree equations. It was stated in Chapter IV that two algebraic surfaces $f(x, y, z) = 0$ and $g(x, y, z) = 0$ are identical if and only if $f(x, y, z)$ and $g(x, y, z)$ differ at most by constant factors. This fact was proved in Chapter II for planes. We are now in a position to prove it for quadric surfaces. In fact, we shall prove a stronger theorem:

THEOREM. *Two equations of the second degree in the variables x, y, and z, with real coefficients, are equivalent if and only if their graphs are identical.*

Proof. If two equations of the type specified in the theorem are equivalent, their graphs are clearly identical. Assume now that the graphs are identical. Then with *identical* coördinate transformations both equations can be reduced to the form (2), § **151** (why?). What we wish to prove now is that if these two *new* equations have identical solutions, the coefficients are the same. It will then follow by an application of the inverse transformations to the new equations that the original equations were equivalent. Accordingly, let the new equations be written

(1) $\qquad -x^2 = b_1 y^2 + c_1 z^2 + 2q_1 y + 2r_1 z + d_1$

(2) $\qquad -x^2 = b_2 y^2 + c_2 z^2 + 2q_2 y + 2r_2 z + d_2.$

If these equations have the same solutions, the values of x corresponding to any value of y and any value of z will be the same when determined from equation (1) as they are when determined from equation (2). But this means that the right members of these equations are identically equal, for all values of y and z, and therefore that corresponding coefficients are equal.†

---

† In general, if two polynomials in any number of variables are equal for all values of the variables, corresponding coefficients are equal. This is an immediate consequence of the fact that if a polynomial vanishes identically every coefficient is zero, since the difference between two polynomials that are identically

As a consequence of this theorem we can now speak of *the* equation of a quadric surface in the same sense that we speak of *the* equation of a plane: the equation $f(x, y, z) = 0$ is uniquely determined except for a constant factor.

153. Invariants of a quadric surface. An equation of the second degree can be analyzed by the process of actually carrying through the transformations indicated in § **151**. However, the work involved is usually long and tedious and fortunately can be avoided by the use of certain invariant expressions involving the coefficients of the given equation.

There are three types of transformation with respect to which these expressions must be proved to be invariant: (*i*) translations, (*ii*) rotations with fixed origin, and (*iii*) multiplying both members of the given equation by a real factor different from zero. The third type of transformation of an equation will be called a **multiplying** transformation.

The principal theorem on invariants under transformations of types (*i*) and (*ii*), for our purposes, is the following, whose proof is given in Chapter IX:

THEOREM I. *If the second degree equation $f(x, y, z) = 0$ is transformed by means of a translation or a rotation with fixed origin, the following quantities are invariant*:

$$D, \Delta, \rho_3, \rho_4, I, J, k_1, k_2, k_3,$$

where D and Δ are the determinants of the matrices e and E, respectively, ρ_3 and ρ_4 are the ranks of the matrices e and E, respectively, $I = a + b + c$, $J = ab + ac + bc - f^2 - g^2 - h^2$, and k_1, k_2, and k_3 are the characteristic roots of e.

Let us observe the effect on the invariants of Theorem I of a multiplying transformation (*iii*), the factor being s. Each element of the matrices e and E is multiplied by s. This will clearly not affect their ranks, but it will affect their determinants, D being multiplied by s^3 and Δ by s^4. Since s is real, the sign of Δ will remain the same. It is also easy to see that J is multiplied by s^2, and I, k_1, k_2, and k_3 by s. The most useful invariant property to be obtained here is the distribution of signs among the non-vanishing characteristic roots. They either have the same sign or they do not. Combining these results with Theorem I, we have the theorem:

---

equal is a polynomial that is identically zero. For a brief discussion of one part of this question, and an appropriate reference, see the proof of Theorem I, § **146**.

THEOREM II. *When any coördinate or multiplying transformation is applied to a second degree equation, the following are invariant: ρ_3, ρ_4, the sign of Δ, and whether the non-vanishing characteristic roots of e have the same sign.*

These invariants are easily obtained, as will be demonstrated by means of examples. They also determine to a remarkable degree the nature of a quadric surface, as can be seen if we turn to the seventeen canonical forms. Let us consider a few equations in canonical form to illustrate the significance of these invariants:

Real Ellipsoid. The k's are equal to $\frac{1}{a^2}$, $\frac{1}{b^2}$, and $\frac{1}{c^2}$, all positive, and Δ is negative. If two k's are equal the surface is a spheroid; if all are equal it is a sphere.

Elliptic Paraboloid. The k's are equal to $\frac{1}{a^2}$, $\frac{1}{b^2}$, and 0. The ranks are $\rho_3 = 2$, $\rho_4 = 4$. If the two non-zero k's are equal, the surface is a paraboloid of revolution.

Hyperbolic Cylinder. One k is zero and the others have opposite signs. The ranks are $\rho_3 = 2$, $\rho_4 = 3$.

Parabolic Cylinder. Only one k is different from zero. The ranks are $\rho_3 = 1$, $\rho_4 = 3$.

A more complete investigation of the canonical forms leads to the facts assembled in the following table:

Number	Surface	ρ_3	ρ_4	Sign of Δ	k's same sign?
1	Real ellipsoid...........................	3	4	−	yes
2	Imaginary ellipsoid.....................	3	4	+	yes
3	Hyperboloid of one sheet................	3	4	+	no
4	Hyperboloid of two sheets...............	3	4	−	no
5	Real quadric cone......................	3	3		no
6	Imaginary quadric cone.................	3	3		yes
7	Elliptic paraboloid.....................	2	4	−	yes
8	Hyperbolic paraboloid..................	2	4	+	no
9	Real elliptic cylinder...................	2	3		yes
10	Imaginary elliptic cylinder..............	2	3		yes
11	Hyperbolic cylinder....................	2	3		no
12	Real intersecting planes................	2	2		no
13	Imaginary intersecting planes...........	2	2		yes
14	Parabolic cylinder.....................	1	3		
15	Real parallel planes....................	1	2		
16	Imaginary parallel planes...............	1	2		
17	Coincident planes......................	1	1		

The following facts are readily verified in a similar manner:

I. *The number of non-vanishing characteristic roots of e is equal to the rank of e.*

II. *Two non-vanishing characteristic roots of e are equal if and only if the quadric is a surface of revolution.*

EXAMPLE. Identify the surface
$$2x^2 + 4y^2 + z^2 + 2xy + 6yz - 4x - 10 = 0.$$

Solution. From the matrix
$$E = \begin{pmatrix} 2 & 1 & 0 & -2 \\ 1 & 4 & 3 & 0 \\ 0 & 3 & 1 & 0 \\ -2 & 0 & 0 & -10 \end{pmatrix}$$
it is found that $D = -11$, $\Delta = 130$, and the characteristic equation is
$$k^3 - 7k^2 + 4k + 11 = 0.$$
The characteristic roots are irrational, but could be computed to any desired degree of accuracy. It is clear, either from application of Descartes's Rule of Signs or from a rough graph of the left member of the characteristic equation, that two of its roots are positive and one is negative (since all roots are real). The determinant Δ being positive, the surface is a hyperboloid of one sheet.

154. Exercises.

1-10. Identify the surface of the corresponding Exercise of § 150.

In Exercises **11-20**, identify the given surface.

11. $3x^2 + 2z^2 - 4xz + 2y + 19 = 0.$
12. $3x^2 + y^2 + 3z^2 - 4xz + 2yz + 4z + 9 = 0.$
13. $x^2 + 4y^2 + 9z^2 - 4xy + 6xz - 12yz + 2x + 14y + 10 = 0.$
14. $y^2 + 20z^2 + xy - 5xz - 9yz + x = 0.$
15. $xy + 2xz + 3yz - 2x + 6z = 0.$
16. $x^2 + 3y^2 + 2z^2 + 2xy - 2yz + 4x + 10y + 8 = 0.$
17. $x^2 + 3y^2 + 8z^2 - 2xy - 8yz = 0.$
18. $6z^2 + xy + 2xz + 3yz - 4x + 6y = 0.$
19. $z^2 + 2xy - 4xz + 6yz + 4x + 6y + 12 = 0.$
20. $x^2 + 3z^2 - 2xz + 2yz + 2y + 6z + 9 = 0.$

21. Prove analytically that for a sphere any direction is a characteristic direction and hence that any plane through the center is a principal plane. *Suggestion:* Assume that the center is at the origin.

22. Prove analytically that for a quadric surface of revolution any direction perpendicular to the axis of revolution is a characteristic direction and hence that any plane containing the axis of revolution is a principal plane. *Suggestion:* Assume that the axis of revolution is the z axis.

23. Prove analytically that for a quadric cylinder no plane perpendicular to the cylinder is a principal plane. *Suggestion:* Assume that the equation of the cylinder is in canonical form. (Cf. Ex. **11**, § **150**.)

24. For any quadric surface prove that the maximum number of principal planes any two of which are perpendicular is equal to the rank of e, and that these are uniquely determined if and only if the surface is not a surface of revolution. *Suggestion:* Assume that the equation is in canonical form.

155. Complete analysis of a quadric surface. The analysis described in § **153** provides a complete determination of the nature of any quadric surface, with the following two exceptions: (*i*) real and imaginary elliptic cylinders, and (*ii*) real and imaginary parallel planes, the invariants of Theorem II, § **153**, being the same for the two surfaces in each of these two cases.

In order to determine whether an elliptic cylinder or a pair of parallel planes is real or imaginary, or to obtain more information about a surface than its general nature—for example, its position in space, its equation in canonical form, or the equations of its principal planes in the original coördinate system—it is necessary to know more than what is supplied by the table of § **153**. In this section, we shall discuss methods of applying the invariants of Theorem I, § **153**, to this problem.

By a **complete analysis** of a quadric surface $f(x, y, z) = 0$, we shall mean carrying out each of the following directions:

(*i*) Identify the surface; that is, determine to which canonical form the given equation is reducible.

(*ii*) Write the equation in canonical form.

(*iii*) Find the principal planes.

(*iv*) Find the centers.

(*v*) If the surface is a paraboloid, find the vertex and the tangent plane at the vertex.

(*vi*) If the surface is a parabolic cylinder, find the line of vertices and the tangent plane that contains this line.

(*vii*) If the surface is reducible, find its factors.

(*viii*) Check the results by substituting in the equation in canonical form the new coördinates expressed in terms of the original coördinates. The resulting equation should be equivalent to the original.

The first step in a complete analysis of a quadric surface is finding the invariants D, Δ, ρ_3, ρ_4, I, and J. The characteristic roots of e are obtained from equation (5), § 149. Using the non-vanishing characteristic roots of e, we find the corresponding characteristic vectors from system (3), § **149,** and hence the corresponding principal planes from equation (2), § **149.** If two non-vanishing characteristic roots are equal, the corresponding principal planes are not uniquely determined. (See Example 1, below.) Further details, with illustrative examples, are given below:

I. For surfaces (1)–(4), the value of the constant term in the new equation $k_1x'^2 + k_2y'^2 + k_3z'^2 + d' = 0$ must be determined. Since $D = k_1k_2k_3$ and $\Delta = k_1k_2k_3 d'$, d' is equal to the quotient of these invariants: $d' = \Delta/D$.

EXAMPLE 1. Give a complete analysis of the surface
$$x^2 + y^2 + z^2 - xy + xz - yz - 2y - 2z + 2 = 0.$$

Solution. If we transform the matrix E by means of elementary row transformations that exclude (i) interchanging the last row with another and (ii) adding a non-zero multiple of the last row to another, and if we keep account of factors taken from any of the first three rows, we can obtain a great deal of information with little effort. To avoid fractions, we first multiply both sides of the given equation by 2. Then we transform E:

$$\begin{pmatrix} 2 & -1 & 1 & 0 \\ -1 & 2 & -1 & -2 \\ 1 & -1 & 2 & -2 \\ 0 & -2 & -2 & 4 \end{pmatrix} \sim \begin{pmatrix} 0 & 1 & -3 & 4 \\ 0 & 1 & 1 & -4 \\ 1 & -1 & 2 & -2 \\ 0 & -2 & -2 & 4 \end{pmatrix}$$

$$\sim \begin{pmatrix} 0 & 1 & -3 & 4 \\ 0 & 0 & 4 & -8 \\ 1 & 0 & -1 & 2 \\ 0 & 0 & 0 & -4 \end{pmatrix} \sim 4, \begin{pmatrix} 0 & 1 & 0 & -2 \\ 0 & 0 & 1 & -2 \\ 1 & 0 & 0 & 0 \\ 0 & 0 & 0 & -4 \end{pmatrix}.$$

From this last matrix we can read off the following information: $D = 4$, $\Delta = -16$, $\rho_3 = 3$, $\rho_4 = 4$, the center is $(0, 2, 2)$, and $d' = -4$.

Returning to the matrix e, we write down its characteristic equation:
$$k^3 - 6k^2 + 9k - 4 = 0,$$

the roots of which are 1, 1, and 4. Since these have the same sign, the surface is an ellipsoid. Since Δ is negative, the surface is real. Since two characteristic roots are equal the surface is a surface of revolution. Its equation in the new coördinates is $x'^2 + y'^2 + 4z'^2 = 4$ or, in canonical form,

$$\frac{x'^2}{4} + \frac{y'^2}{4} + z'^2 = 1.$$

The surface is an oblate spheroid.

Corresponding to the characteristic root 4, equations (3), § **149**, are
$$-2l - m + n = 0,$$
$$-l - 2m - n = 0,$$
$$l - m - 2n = 0,$$

which give $l : m : n = 1 : -1 : 1$. The corresponding principal plane therefore has the equation $4x - 4y + 4z = 0$ (according to (2), § **149**), or
$$x - y + z = 0.$$

Corresponding to the characteristic root 1, we have the equations

$$l - m + n = 0,$$
$$-l + m - n = 0,$$
$$l - m + n = 0,$$

which are equivalent. For the other two principal planes we can choose any two directions subject to the two conditions: (i) they are perpendicular and (ii) they satisfy these three equivalent equations. Two simple solutions are $l : m : n = 0 : 1 : 1$ and $l : m : n = 2 : 1 : -1$. The corresponding principal planes are $y + z - 4 = 0$ and $2x + y - z = 0$. As expansion will verify, the original equation can be written in the form

$$\left(\frac{y+z-4}{\sqrt{2}}\right)^2 + \left(\frac{2x+y-z}{\sqrt{6}}\right)^2 + 4\left(\frac{x-y+z}{\sqrt{3}}\right)^2 = 4.$$

II. For surfaces (7) and (8), the value of r' in the equation $k_1 x'^2 + k_2 y'^2 + 2r'z' = 0$ is determined by the relation $\Delta = -k_1 k_2 r'^2$. However, since the plane $z' = 0$ is not a principal plane, its equation in the original coördinate system is not immediately obtainable. One method of finding the equation of this plane is to consider it as the tangent plane of the surface at the vertex, the vertex being the point where the axis of symmetry (the line of intersection of the two principal planes) meets the surface. A second method is shown in Example 2.

EXAMPLE 2. Give a complete analysis of the surface

$$x^2 + 3y^2 + z^2 + 2xy + 2xz + 2yz - 2x + 4y + 2z + 12 = 0.$$

Solution. Proceeding as in Example 1, we have

$$\begin{pmatrix} 1 & 1 & 1 & -1 \\ 1 & 3 & 1 & 2 \\ 1 & 1 & 1 & 1 \\ -1 & 2 & 1 & 12 \end{pmatrix} \sim \begin{pmatrix} 1 & 1 & 1 & -1 \\ 0 & 2 & 0 & 3 \\ 0 & 0 & 0 & 2 \\ 0 & 3 & 2 & 11 \end{pmatrix} \sim \begin{pmatrix} 1 & 0 & 0 & 0 \\ 0 & 2 & 0 & 0 \\ 0 & 0 & 0 & 2 \\ 0 & 0 & 2 & 0 \end{pmatrix}.$$

The second transformation involved elementary column transformations, which can be used without loss in results since it is obvious from the third row of the second matrix that there is no center. We know also that $\rho_3 = 2$, $\rho_4 = 4$, $D = 0$, and $\Delta = -8$. The surface is therefore an elliptic paraboloid.

The characteristic equation is

$$k^3 - 5k^2 + 4k = 0,$$

whose roots are 0, 1, 4. Since $\Delta = -k_1 k_2 r'^2$, $r' = \pm\sqrt{2}$, and the equation of the surface can be written $x'^2 + 4y'^2 + 2\sqrt{2}\,z' = 0$. The principal planes corresponding to $k = 1$ and $k = 4$ are $x - y + z - 2 = 0$ and $x + 2y + z + 1 = 0$, respectively. The direction (λ, μ, ν) of their line of intersection is such that $\lambda : \mu : \nu = 1 : 0 : -1$. Therefore the original equation can be written

$$\left(\frac{x-y+z-2}{\sqrt{3}}\right)^2 + 4\left(\frac{x+2y+z+1}{\sqrt{6}}\right)^2 = \pm 2\sqrt{2}\,\frac{x-z+\delta}{\sqrt{2}}.$$

§ 155] COMPLETE ANALYSIS 197

Expansion shows that the $+$ sign is correct and that $\delta = -5$, and therefore $z' = \dfrac{-x + z + 5}{\sqrt{2}}$. The point of intersection of the principal planes and the plane $x - z - 5 = 0$ is the vertex (and the new origin), $(3, -1, -2)$. As a check, we find the equation of the tangent plane at $(3, -1, -2)$. It is $x - z - 5 = 0$.

III. For surfaces (9), (10), (11), (15), or (16), the value of d' is obtained by the method of § 139: if (x_0, y_0, z_0) is any center, the new constant term is

$$d' = f(x_0, y_0, z_0) = px_0 + qy_0 + rz_0 + d.$$

This method of finding d', incidentally, is available whenever there is a center, and may be used with surfaces of types (1)–(4), or as a check on one's work.

EXAMPLE 3. Give a complete analysis of the surface

$$x^2 + 7y^2 + z^2 + 10xy + 2xz + 10yz + 8x + 4y + 8z - 6 = 0.$$

Solution.

$$\begin{pmatrix} 1 & 5 & 1 & 4 \\ 5 & 7 & 5 & 2 \\ 1 & 5 & 1 & 4 \\ 4 & 2 & 4 & -6 \end{pmatrix} \sim 6, \begin{pmatrix} 1 & 5 & 1 & 4 \\ 1 & 2 & 1 & 1 \\ 0 & 0 & 0 & 0 \\ 0 & -6 & 0 & -10 \end{pmatrix} \sim 18, \begin{pmatrix} 0 & 1 & 0 & 1 \\ 1 & 2 & 1 & 1 \\ 0 & 0 & 0 & 0 \\ 0 & -6 & 0 & -10 \end{pmatrix}.$$

The ranks are $\rho_3 = 2$ and $\rho_4 = 3$. The centers form the line $y = -1$, $x + z = 1$, or $x = 1 - t$, $y = -1$, $z = t$. Choose the center $(1, -1, 0)$. Then $d' = 4 - 2 - 6 = -4$. The characteristic equation is

$$k^3 - 9k^2 - 36k = 0,$$

and $k = 0, 12, -3$. Therefore the surface is a hyperbolic cylinder, with the equation $12x'^2 - 3y'^2 = 4$. The original equation can be written

$$12\left(\frac{x + 2y + z + 1}{\sqrt{6}}\right)^2 - 3\left(\frac{x - y + z - 2}{\sqrt{3}}\right)^2 = 4.$$

If the constant term of the original equation were -2, the surface would reduce to the two intersecting planes:

$$[\sqrt{2}(x + 2y + z + 1) + (x - y + z - 2)][\sqrt{2}(x + 2y + z + 1) - (x - y + z - 2)] = 0.$$

IV. For surface (14), the value of r' in the equation $x'^2 + 2r'z' = 0$ is not so readily found. However, we know that in this case there is just one principal plane, the $y'z'$ plane. If the left member of the equation of this plane (in standard form) in the original coördinate system is squared, the result will equal a multiple of $f(x, y, z)$ except for a linear remainder. This remainder is the left side of the equation of the plane normal to the principal plane and tangent to the cylinder along the line of vertices. When the equations of these planes are normalized (written in normal form), the coefficient of z' can be determined.

EXAMPLE 4. Give a complete analysis of the surface

$$4x^2 + 9y^2 + z^2 - 12xy + 4xz - 6yz - 14x + 12y + 8z - 2 = 0.$$

Solution.

$$\begin{pmatrix} 4 & -6 & 2 & -7 \\ -6 & 9 & -3 & 6 \\ 2 & -3 & 1 & 4 \\ -7 & 6 & 4 & -2 \end{pmatrix} \sim \begin{pmatrix} 0 & 0 & 0 & 1 \\ 0 & 0 & 0 & 0 \\ 2 & -3 & 1 & 0 \\ -3 & 0 & 6 & 0 \end{pmatrix}.$$

The ranks are $\rho_3 = 1$ and $\rho_4 = 3$. The surface is therefore a parabolic cylinder. The characteristic equation is $k^3 - 14k^2 = 0$, with roots $k = 0, 0,$ and 14.

The only principal plane is $2x - 3y + z - 2 = 0$. Squaring the left member of this equation, we find that the original equation can be written

$$(2x - 3y + z - 2)^2 = 6(x - 2z + 1),$$

or

$$\left(\frac{2x - 3y + z - 2}{\sqrt{14}}\right)^2 = -2 \cdot \frac{3\sqrt{5}}{14} \cdot \frac{x - 2z + 1}{-\sqrt{5}}.$$

Therefore the constant r' of the equation in canonical form is $\frac{3}{14}\sqrt{5}$.

156. Exercises.

1-10. For the surface of the corresponding Exercise of § **150**, (*a*) write the equation in canonical form; (*b*) complete the analysis. (See Exs. **1-10**, § **154**.)

In Exercises **11** and **12**, identify the given surface.

11. $x^2 + 4y^2 + 2z^2 + 2xz + 4yz + 4x + 4z + 8 = 0$.
12. $3x^2 + 8y^2 + 5z^2 + 6xz + 8yz - 8y - 4z = 0$.

In Exercises **13** and **14**, (*a*) identify the surface and write its equation in canonical form; (*b*) find the principal planes; (*c*) complete the analysis.

13. $x^2 + 2y^2 + 2z^2 + 2xy - 2xz + 2x + 6y + 2z - 13 = 0$.
14. $9x^2 + y^2 + z^2 - 6xy + 6xz - 2yz + 18x - 6y + 6z - 7 = 0$.

In Exercises **15-22**, (*a*) prove that each surface is a surface of revolution, identify the surface, and write its equation in canonical form; (*b*) determine the axis of revolution and complete the analysis.

15. $5x^2 + 2y^2 + 5z^2 - 4xy - 2xz - 4yz + 6x - 12y + 18z - 3 = 0$.
16. $x^2 + y^2 - 2z^2 - 2xy - 4xz - 4yz + 8y + 20z - 6 = 0$.
17. $5x^2 + 2y^2 + 6z^2 - 4xy + 4x - 4y - 22 = 0$.
18. $5x^2 + 5y^2 + 8z^2 - 8xy - 4xz - 4yz + 10x + 10y - 4z - 18 = 0$.
19. $xy + xz + yz + 1 = 0$.

20. $x^2 + y^2 + z^2 + xy + xz - yz + 6x + 6y + 12 = 0$.
21. $x^2 + y^2 + z^2 + xy + xz + yz + 2x - 2y + 2 = 0$.
22. $2x^2 - y^2 + 7z^2 + 4xy + 12xz + 6yz - 24y - 8z - 56 = 0$.

In Exercises **23-26**, (a) prove that each surface is a cylinder, identify the surface, and find a set of direction numbers for its rulings; (b) complete the analysis.

23. $y^2 - 2xy - 2yz + 2y - 4 = 0$.
24. $x^2 + y^2 + 9z^2 + 2xy + 6xz + 6yz + 4x + 6y + 4z - 21 = 0$.
25. $13x^2 + 25y^2 + 7z^2 + 10xy - 8xz + 20yz + 300 = 0$.
26. $9x^2 + 36y^2 + 4z^2 - 36xy + 12xz - 24yz - 16x - 24y - 48z + 56 = 0$.

27. Find the equation of and describe the quadric surface that contains the points in the first octant whose coördinates satisfy the equation $\sqrt{x} + \sqrt{y} = \sqrt{z}$.

⋆ Supplementary Exercises

28. Identify the surface

$$ax^2 - y^2 + z^2 + 2xy - 2xz + 2y + 1 = 0$$

for all possible values of a.

29. Identify the surface

$$ax^2 + y^2 + z^2 + 4xy - 4xz - 2yz - 12x - 6y + 6z + d = 0$$

for all possible values of a and d.

30. Determine the values of a for which the surface

$$ax^2 + 2yz + 3 = 0$$

is a surface of revolution. Find the axis of revolution and identify the surface in each case.

31. Prove that for all values of a different from 1 the surface

$$a(x^2 + y^2 + z^2) + 2xy + 2xz + 2yz + d = 0$$

is a surface of revolution about the line $x = y = z$. Identify the surface for all possible values of a and d.

32. Prove that if $f(x, y, z) = 0$ is a hyperboloid, then there exists a number α such that $f(x, y, z) = \alpha$ is a real cone. Assuming that $\alpha > 0$ and that $f(x, y, z) = 0$ is a hyperboloid of one sheet, prove that $f(x, y, z) = \beta$ is a hyperboloid of one sheet or two sheets according as $\beta < \alpha$ or $\beta > \alpha$.

33. Determine the values of d for which the following surface is (i) a cone, (ii) a hyperboloid of one sheet, (iii) a hyperboloid of two sheets:

$$x^2 + y^2 + 2z^2 - 4xy + 2xz - 2yz - 2x + 4y - 2z + d = 0.$$

(See Ex. **32**.)

* 34. Let $f(x, y, z) = 0$ be a non-singular quadric surface S and let (λ, μ, ν) be a direction for which $e(\lambda, \mu, \nu) \neq 0$. Prove that the quadric surface

$$e(\lambda, \mu, \nu)f(x, y, z) = [\phi(x, y, z, \lambda, \mu, \nu)]^2$$

is the quadric cylinder whose rulings have the direction (λ, μ, ν) and are tangent to the surface S along the section of that surface by the diametral plane conjugate to the direction (λ, μ, ν). This cylinder is called the **tangent cylinder** to the surface S with direction (λ, μ, ν). (Cf. Ex. **25**, § **128**, and Exs. **36** and **47** of this section.) *Suggestions:* Equation (3), § **118**, has equal roots if and only if its discriminant is equal to zero. Also see Exercise **22**, § **128**.

35. Find the equation of the right circular cylinder with radius R whose axis passes through the origin and has the direction (λ, μ, ν). (Cf. Ex. **34**.)

* **36.** Let $f(x, y, z) = 0$ be a non-singular quadric surface S and let $P_0(x_0, y_0, z_0)$ be a point which is not on the surface and whose polar plane intersects the surface in a real curve C. Prove that the quadric surface

$$f(x_0, y_0, z_0)e(x - x_0, y - y_0, z - z_0) = [\phi(x_0, y_0, z_0, x - x_0, y - y_0, z - z_0)]^2$$

is the real quadric cone with vertex at P_0 whose rulings are tangent to the surface S along C. This cone is called the **tangent cone** to the surface S with vertex at the point P_0. (Cf. Ex. **11**, § **130**, and Exs. **34** and **47** of this section.) *Suggestions:* Show that the given equation can be written in the form

$$f(x_0, y_0, z_0)f(x, y, z) = \Pi^2,$$

where Π is the left member of the equation of the polar plane of P_0, (3), § **126**. Also see Exercise **22**, § **128** and the suggestions of Exercise **34** of this section.

37. Find the equation of the cone with vertex at the origin whose rulings are tangent to the sphere

$$(x - a)^2 + (y - b)^2 + (z - c)^2 = R^2,$$

where $R^2 < a^2 + b^2 + c^2$. (Cf. Ex. **36**.)

38. Find the equation of the cone with vertex at the origin that passes through the points of intersection of the quadric surface $e(x, y, z) = 1$ and the plane $\alpha x + \beta y + \gamma z = 1$.

39. Find the equation of the cone with vertex at the origin that passes through the points of intersection of the quadric surface $e(x, y, z) = 1$ and the sphere $k(x^2 + y^2 + z^2) = 1$.

40. Find the locus of the center of the quadric surface

$$ax^2 + by^2 - (a + b)z^2 + 2x + 2y + 2z = 0,$$

where a and b are parameters.

* **41.** Prove that the cylinder that contains the conic section $f(x, y) = 0$, $z = 0$ and whose rulings have the direction (λ, μ, ν), where $\nu \neq 0$, has the equation

$$f\left(x - \frac{\lambda z}{\nu}, y - \frac{\mu z}{\nu}\right) = 0,$$

and is therefore a quadric cylinder. Deduce from this that the (oblique) projection of any conic section on any plane by means of a system of parallel lines not parallel to the plane of the conic section or to the plane onto which it is projected is a conic section. What can be said about the natures of these two conic sections? (Cf. Ex. **43**.)

42. Find the equation of the cylinder that contains the circle $x^2 + y^2 = 1$, $z = 0$ and whose rulings have the direction numbers 1, 1, 1. (See Ex. **41**.)

*****43.** Prove that the cone that contains the conic section $f(x, y) = 0$, $z = 0$ and whose vertex is the point (x_0, y_0, z_0), where $z_0 \neq 0$, has the equation

$$(z - z_0)^2 \cdot f\left(\frac{x_0 z - z_0 x}{z - z_0}, \frac{y_0 z - z_0 y}{z - z_0}\right) = 0,$$

and is therefore a quadric cone. Deduce from this that the projection of any conic section on any plane by means of a system of lines passing through a fixed point not lying in the plane of the conic section or the plane onto which it is projected is a conic section. What can be said about the natures of these two conic sections? (Cf. Ex. **41**.)

44. Find the equation of the cone that contains the circle $x^2 + y^2 = 1$, $z = 0$ and whose vertex is the point $(1, 1, 1)$. (See Ex. **43**.)

45. Prove that the sum of the squares of the reciprocals of the lengths of any three mutually perpendicular diameters of a real ellipsoid is a constant. *Suggestion:* Let $(\lambda_i, \mu_i, \nu_i)$, $i = 1, 2, 3$, be the direction of a diameter, and let $2t_i$ be its length. Then $(\lambda_i t_i, \mu_i t_i, \nu_i t_i)$ is a point on the ellipsoid. Hence, if the equation is in canonical form,

$$\frac{1}{t_i^2} = \frac{\lambda_i^2}{a^2} + \frac{\mu_i^2}{b^2} + \frac{\nu_i^2}{c^2}.$$

(See Exs. **23** and **24**, § **123**.)

46. Prove that the point of intersection of three mutually perpendicular tangent planes of a real ellipsoid, with an equation in canonical form, lies on the sphere

$$x^2 + y^2 + z^2 = a^2 + b^2 + c^2.$$

This sphere is called the **director sphere** of the ellipsoid. *Suggestion:* The equations of the tangent planes perpendicular to the direction $(\lambda_i, \mu_i, \nu_i)$, $i = 1, 2, 3$, are

$$(\lambda_i x + \mu_i y + \nu_i z)^2 = (\lambda_i^2 a^2 + \mu_i^2 b^2 + \nu_i^2 c^2).$$

Expand these three equations, and add the left members and the right members. (See Ex. **15**, § **128**.)

47. Explain how a diametral plane conjugate to the direction (λ, μ, ν) might be thought of as the polar plane of the "point at infinity in the direction (λ, μ, ν)." (Cf. Exs. **34** and **36**.)

48. Prove that if Λ_1 and Λ_2 are any two skew lines, axes can be chosen so that their equations are

$$z = c, \, y = mx \quad \text{and} \quad z = -c, \, y = -mx,$$

where c and m are positive, or

$$\frac{x}{\lambda} = \frac{y}{\mu} = \frac{z-c}{0} \quad \text{and} \quad \frac{x}{\lambda} = \frac{y}{-\mu} = \frac{z+c}{0},$$

where c, λ, and μ are positive and $\lambda^2 + \mu^2 = 1$.

* **49.** Find the locus of a point that is equidistant from two skew lines. (See Ex. **48** and § **44**.)

* **50.** The end-points of a line segment of constant length lie on two skew lines. Find the locus of the midpoint. (See Ex. **48**.)

* **51.** Two planes, Π_1 and Π_2, are mutually perpendicular and contain the non-perpendicular skew lines Λ_1 and Λ_2, respectively. Find the surface generated by their line of intersection. (See Ex. **48**.)

* **52.** Prove that the surface generated by a line that is coplanar with each of three fixed pairwise skew lines that are not all parallel to any plane is a hyperboloid of one sheet containing the three given lines as rulings. *Suggestion:* Observe first that through any point on any of the given lines there is precisely one line that is coplanar with each of the other two lines. Let two of the skew lines be the second pair of Exercise **48**, and let the third given line and the variable line Λ have equations

$$\frac{x-a}{\lambda_1} = \frac{y-b}{\mu_1} = \frac{z}{\nu_1} \quad \text{and} \quad \frac{x-x_0}{\lambda_2} = \frac{y-y_0}{\mu_2} = \frac{z-z_0}{\nu_2},$$

respectively, where $\nu_1 \neq 0$. The line Λ is coplanar with each of the three given lines if and only if the following three determinants vanish (see Ex. **9**, § **61**):

$$\begin{vmatrix} x_0 & y_0 & z_0 - c \\ \lambda_2 & \mu_2 & \nu_2 \\ \lambda & \mu & 0 \end{vmatrix}, \quad \begin{vmatrix} x_0 & y_0 & z_0 + c \\ \lambda_2 & \mu_2 & \nu_2 \\ \lambda & -\mu & 0 \end{vmatrix}, \quad \begin{vmatrix} x_0 - a & y_0 - b & z_0 \\ \lambda_2 & \mu_2 & \nu_2 \\ \lambda_1 & \mu_1 & \nu_1 \end{vmatrix}.$$

The vanishing of these determinants gives three linear equations in λ_2, μ_2, and ν_2 which have non-trivial solutions if and only if the determinant of their coefficients,

$$\begin{vmatrix} \mu(z-c) & -\lambda(z-c) & \lambda y - \mu x \\ -\mu(z+c) & -\lambda(z+c) & \lambda y + \mu x \\ \mu_1 z - \nu_1(y-b) & \nu_1(x-a) - \lambda_1 z & \lambda_1(y-b) - \mu_1(x-a) \end{vmatrix},$$

where subscripts on x, y, and z have been dropped, vanishes. Show that the third degree terms in the expansion of this determinant cancel, and therefore that the resulting surface is a quadric. Finally, using the initial observation, show that the surface contains the three given lines as rulings and is thus a hyperboloid of one sheet, since that is the only quadric that contains three skew lines that are not parallel to any plane. (See Exs. **16, 18, 19,** § **133**.)

* **53.** Prove that the surface generated by a line that intersects three fixed pairwise skew lines that are parallel to some plane is a hyperbolic paraboloid containing the three given lines as rulings. (See Ex. **52**.)

* **54.** Prove that the surface generated by a line that is parallel to a fixed plane Π and intersects two skew lines neither of which is parallel to Π is a hyperbolic paraboloid containing the two skew lines as rulings. (Cf. Exs. 52 and 53.)

55. Let P_1 and P_2 be any two distinct points on a line Λ_1 and let Q_1 and Q_2 be any two distinct points on a line Λ_2, where Λ_1 and Λ_2 are skew. Prove that the surface generated by a line intersecting Λ_1 and Λ_2 in points that divide the segments P_1P_2 and Q_1Q_2 in the same ratio is a hyperbolic paraboloid. (Cf. Ex. **20**, § **133**.) *Suggestion:* Show that the moving line is always parallel to a fixed plane, and hence that the locus is the same as that of Exercise **54**.

* **56.** A (primed) coördinate system is moving relative to a fixed (unprimed) coördinate system in such a way that the x' axis crosses the x axis, the y' axis crosses the y axis, and the z' axis crosses the z axis at a fixed point different from the origin. Find the locus of the origin of the moving system.

57. Use Theorem II, § **153**, to prove that a quadric surface is a cone with vertex at the origin if and only if its equation has the form $e(x, y, z) = 0$ and $D \neq 0$.

58. Use Theorem II, § **153**, to prove that a quadric surface whose equation has the form $e(x, y, z) = 0$ is reducible if and only if $D = 0$.

59. Use Theorem II, § **153**, to prove that the quadric surface $f(x, y, z) = 0$ is reducible if and only if the rank of E does not exceed 2, and that if the surface is reducible, its factors are distinct or identical according as the rank of E is 2 or 1.

60. Prove that a reducible polynomial of the second degree in the variables x, y, z has essentially only one decomposition into irreducible factors; that is, prove that the factors of any two such decompositions differ only by constant factors. (See § **65**.)

61. Prove that a quadric surface is a real cone with each coördinate axis as a ruling if and only if it has an equation of the form

$$fyz + gxz + hxy = 0.$$

Find the equation of the cone obtained by revolving a coördinate axis about the line $x = y = z$.

62. Obtain the answer to the second part of Exercise **61** by transforming the equation $x'^2 + y'^2 = 2z'^2$ by means of a rotation, the z' axis being the line $x = y = z$.

63. Apply Descartes's Rule of Signs to the characteristic equation of the matrix e to prove that if $J = 0$, then I and D cannot have the same sign, and that if $I = 0$, then $J < 0$.

64. Prove that for any values of a, b, and c the surfaces

$$a(x+y)^2 + b(x+z)^2 + c(y+z)^2 = 1$$

and

$$ax^2 + by^2 + cz^2 = 1$$

are of the same type, that is, belong to *one* of the fundamental seventeen classifications of quadric surfaces. *Suggestion:* Apply Descartes's Rule of Signs to the characteristic equations.

65. Prove that if $f(x, y, z) = 0$ is a paraboloid or an irreducible quadric cylinder, and if (λ, μ, ν) is the direction of the axis (if the surface is a paraboloid) or of the rulings (if the surface is a cylinder). then

$$a\lambda + h\mu + g\nu = 0$$
$$h\lambda + b\mu + f\nu = 0$$
$$g\lambda + f\mu + c\nu = 0.$$

Suggestion: What is the characteristic direction corresponding to $k = 0$?

66. Prove that the plane $\lambda x + \mu y + \nu z + \delta = 0$ is a plane of symmetry but not a principal plane for the quadric surface $f(x, y, z) = 0$ if and only if

$$a\lambda + h\mu + g\nu = 0$$
$$h\lambda + b\mu + f\nu = 0$$
$$g\lambda + f\mu + c\nu = 0$$
$$p\lambda + q\mu + r\nu = 0.$$

Suggestion: Investigate the vanishing of $\phi(x, y, z, \lambda, \mu, \nu)$ identically in x, y, and z.

67. Prove that if two quadric surfaces have one plane section in common, any other points belonging to both surfaces must lie in another plane. *Suggestion:* Let the given plane of common section be the xy plane. Then the equations of the two surfaces can be written

$$ax^2 + by^2 + 2hxy + 2px + 2qy + d + z(\alpha_i x + \beta_i y + \gamma_i z + \delta_i) = 0, \ i = 1, 2.$$

68. Prove that for the quadric surface $f(x, y, z) = 0$ the quantity $p^2 + q^2 + r^2$ is invariant under any rotation with fixed origin. Actually make the substitutions.

69. Prove that if three distinct chords of a quadric surface have the same midpoint, then the three chords either lie in a plane or meet in a center. *Suggestion:* Let the common midpoint be the origin.

70. Let P be a regular point of the quadric surface S, and let Π be a plane through P. Prove that Π is the tangent plane of S at P if and only if the section of S by Π is reducible to two lines (real or imaginary, distinct or identical) through P. *Suggestion:* Let P be the origin and Π the xy plane.

71. Prove that an elliptic paraboloid has no hyperbolic sections and that a hyperbolic paraboloid has no elliptic sections. *Suggestion for the first part:* The curve of intersection of the elliptic paraboloid, with an equation in canonical form, and the plane $z = \alpha x + \beta y + \gamma$ projects into a curve in the xy plane whose equations are

$$z = 0, \ \frac{x^2}{a^2} + \frac{y^2}{b^2} + 2\alpha x + 2\beta y + 2\gamma = 0.$$

The section is therefore an ellipse (real, point, or imaginary). What is the section if the plane is parallel to the z axis? (See Ex. **41.**)

72. Discuss sections of a quadric surface by parallel planes. Prove that these sections are similar in the following sense: (*i*) if one section is a hyperbola (degenerate or not), then every section is a hyperbola (degenerate or not), the asymptotes of any section being obtained by translating those of any other; (*ii*) if one section is an ellipse (real, point, or imaginary), then every section is an ellipse (real, point, or imaginary), any two real sections being similar and having parallel major and minor axes; (*iii*) if one section is a parabola (degenerate or not), then every section is a parabola (degenerate or not), any two sections having parallel axes; (*iv*) if one section is a line, then every non-vacuous section is a line. *Suggestion:* Let the family of planes be parallel to the xy plane, and let axes be chosen so that $h = 0$, the equation of the surface having the form

(1) $\qquad ax^2 + by^2 + 2(gz + p)x + 2(fz + q)y + (cz^2 + 2rz + d) = 0.$

73. Prove that if the sections of a quadric surface by parallel planes are central conics, the centers of these conics lie on a line. *Suggestion:* In equation (1), Ex. **72**, give z a constant value. The coördinates of the center of the section are

$$\left(-\frac{gz + p}{a},\ -\frac{fz + q}{b},\ z\right).$$

Therefore the coördinates of the center satisfy the equations

$$\frac{ax + p}{g} = \frac{by + q}{f} = -z.$$

***74.** Prove that a necessary and sufficient condition for sections of the (non-spherical) quadric surface $f(x, y, z) = 0$ normal to the direction (λ, μ, ν) to be circular is that the plane $\lambda x + \mu y + \nu z = 0$ be a factor of the quadric surface $e(x, y, z) - k(x^2 + y^2 + z^2) = 0$ for some non-vanishing characteristic root k of the matrix e. *Suggestion:* Consider a rotation of coördinate axes, with $x' = \lambda_1 x + \mu_1 y + \nu_1 z$, $y' = \lambda_2 x + \mu_2 y + \nu_2 z$, $z' = \lambda x + \mu y + \nu z$. Show that the sections by the planes $z' =$ a constant are circles if and only if the coefficient of x'^2 in the equation $f(\lambda_1 x' + \lambda_2 y' + \lambda z',\ \mu_1 x' + \mu_2 y' + \mu z',\ \nu_1 x' + \nu_2 y' + \nu z') = 0$ is a non-vanishing constant k for every direction $(\lambda_1, \mu_1, \nu_1)$ perpendicular to (λ, μ, ν). After showing that this coefficient is $e(\lambda_1, \mu_1, \nu_1)$, establish the following necessary and sufficient condition: $e(\lambda_1, \mu_1, \nu_1) - k(\lambda_1^2 + \mu_1^2 + \nu_1^2) = 0$ for all $(\lambda_1, \mu_1, \nu_1)$ perpendicular to (λ, μ, ν). From this deduce the necessary and sufficient condition: $e(x, y, z) - k(x^2 + y^2 + z^2) = 0$ for all $x, y,$ and z such that $\lambda x + \mu y + \nu z = 0$. Finally, show that the condition of Exercise **58** applied to the reducibility of the surface $e(x, y, z) - k(x^2 + y^2 + z^2) = 0$ is equivalent to the statement that k is a characteristic root of e.

75. Prove that for an ellipsoid, a hyperboloid, a cone, an elliptic paraboloid, or an elliptic cylinder, that is not a surface of revolution there are just two families of parallel circular sections. Prove that hyperbolic paraboloids and hyperbolic cylinders have no circular sections. (Cf. Ex. **71**.) *Partial solution for a real ellipsoid:* Let the equation be $k_1 x^2 + k_2 y^2 + k_3 z^2 = 1$, where $0 < k_1 < k_2 < k_3$. The equation $e(x, y, z) - k(x^2 + y^2 + z^2) = 0$ is $(k_2 - k_1)y^2 + (k_3 - k_1)z^2 = 0$ if $k = k_1$, $(k_1 - k_2)x^2 + (k_3 - k_2)z^2 = 0$ if $k = k_2$, and $(k_1 - k_3)x^2 + (k_2 - k_3)y^2 = 0$ if $k = k_3$. The only one of these three surfaces that is reducible to two *real* planes is the second. Therefore the planes cutting the ellipsoid in circular sections are those whose normals have direction numbers $\sqrt{k_2 - k_1} : 0 : \pm\sqrt{k_3 - k_2}$. Notice that the circular sections by planes through the origin can be thought of as the curve of intersection of the ellipsoid and the sphere $k_2(x^2 + y^2 + z^2) = 1$.

76. An **umbilic** of a quadric surface is a regular point of the surface at which the tangent plane is parallel to a family of circular sections of the surface. Prove the following facts about umbilics:

I. An ellipsoid or a hyperboloid of two sheets that is not a surface of revolution has four umbilics.

II. An elliptic paraboloid that is not a surface of revolution has two umbilics.

III. A spheroid or a two-sheeted hyperboloid of revolution has two umbilics, where the surface is met by its axis of revolution.

IV. A paraboloid of revolution has one umbilic, where it is met by its axis of revolution.

V. Every point of a sphere is an umbilic.

VI. No surface not enumerated above has an umbilic.

Prove that an umbilic of a quadric surface could be defined as a regular point circle section of the surface.

77. Why do such space figures as straight lines and ellipses have their characteristic appearance when represented in plane diagrams? Discuss the relationship between (1) the representation of a space figure by means of a plane diagram and (2) the projection of a space figure onto a plane by means of a system of parallel lines.

*CHAPTER IX

MATRIX ALGEBRA†

157. Introduction. Matrices have been used repeatedly in this book, but applications have been largely those that depend merely on the concept of *rank*. Chapter VII treated *rotation matrices*, which have certain distinctive properties, and Chapter VIII introduced the notion of *characteristic root*. However, an adequate treatment of these last two topics requires a part of matrix theory known as **matrix algebra,** which is concerned with operations on matrices similar to those of ordinary algebra. It is the purpose of this chapter to present a part of this theory, to indicate some of its applications to solid analytic geometry, and to point out how solid analytic geometry provides a background for interpreting certain basic concepts of matrix theory. The primary object, so far as solid analytic geometry is concerned, is the proof of Theorem I, § **153,** given in § **172.** Applications to mensuration of certain geometric figures are given in §§ **174** and **175.**

The notation (a_{ij}), as introduced in § **49,** will be used to represent a matrix A whose general element (the element in the ith row and jth column) is a_{ij}. If A is a square matrix, $|A|$ will denote its determinant. The elements a_{ii}, $i = 1, \ldots, n$, of a square matrix A are called the **diagonal elements** of A and constitute the **main** or **principal diagonal** of A. A matrix with just one row (or column) will be called a **row** (or **column**) **matrix.** All matrices considered will be assumed to have only *real* elements.

158. Multiplication of matrices. If the number of columns of a matrix A is equal to the number of rows of a matrix B, then the *product*, $C = AB$, of the matrices A and B can be defined.

DEFINITION. *If* $A = (a_{ij})$ *is a matrix of m rows and n columns and if* $B = (b_{ij})$ *is a matrix of n rows and p columns, then the **product,***

---

† In a first course in Solid Analytic Geometry, if it seems desirable to waive the proof of Theorem I, § **153,** this chapter can be omitted.

$C = AB$, of the matrices A and B is the matrix of m rows and p columns whose general element is

$$c_{ij} = \sum_{k=1}^{n} a_{ik}b_{kj} = a_{i1}b_{1j} + a_{i2}b_{2j} + \cdots + a_{in}b_{nj}.$$

Postponing the question of why matrix multiplication is defined in this way (§ **165**), let us first familiarize ourselves with the process. With practice the student will develop a scheme for running one finger across the rows of one matrix while moving another finger down the columns of the other. It is advisable to develop this mechanical association as soon as possible.

As an example, let us form the matrix product

$$\begin{pmatrix} 1 & 3 & -2 & 0 \\ 3 & 0 & 4 & -1 \end{pmatrix} \begin{pmatrix} 4 & 5 & -3 \\ 6 & 0 & -2 \\ 0 & -1 & 4 \\ 3 & 2 & 2 \end{pmatrix}.$$

The elements of the product matrix are obtained by forming the sums of the products as indicated schematically:

The result is the matrix

$$\begin{pmatrix} 22 & 7 & -17 \\ 9 & 9 & 5 \end{pmatrix}.$$

Notice that the two matrices of this example cannot be multiplied in the other order, since the number of columns of the second matrix is not equal to the number of rows of the first.

Of particular interest and importance is the special case where the matrices A and B are square and of the same order. In this case both products AB and BA are defined, and are square matrices of the same order as A and B. However, these two products are in general not the same. For example,

$$\begin{pmatrix} 1 & 2 \\ 3 & 4 \end{pmatrix} \begin{pmatrix} 5 & 6 \\ 7 & 8 \end{pmatrix} = \begin{pmatrix} 19 & 22 \\ 43 & 50 \end{pmatrix},$$

while
$$\begin{pmatrix} 5 & 6 \\ 7 & 8 \end{pmatrix} \begin{pmatrix} 1 & 2 \\ 3 & 4 \end{pmatrix} = \begin{pmatrix} 23 & 34 \\ 31 & 46 \end{pmatrix}.$$

Since equality between two matrices means that they are the *same* matrix, element for element, these two product matrices are *unequal* (even though their determinants are equal). We write
$$AB \neq BA,$$
and say that *the commutative law fails for matrix multiplication*.

If $AB = BA$ we say that the matrices **commute**, or that each commutes with the other.

It is a highly important fact that, although the commutative law for matrix multiplication fails, the **associative law**
$$(AB)C = A(BC)$$
holds whenever the products AB and BC can be formed. Proof of the associative law is obtained by writing out the general term of each product. The student is urged to do this in detail (Ex. **7**, § **164**).

The associative law permits us to write any product of matrices, such as $ABCD$, without using parentheses. If the order of the factors is not disturbed, the product is uniquely determined independently of the manner in which parentheses might be introduced.

159. Other operations on matrices. Addition and subtraction of two matrices having the same number of rows and the same number of columns and multiplication of a matrix by a constant are defined in a natural way.

DEFINITION. *If $A = (a_{ij})$ and $B = (b_{ij})$ are matrices of m rows and n columns and if k is any constant, then the **sum**, $C = A + B$, of the matrices A and B and the **product**, $D = kA$, of the constant k and the matrix A are the matrices of m rows and n columns whose general terms are $c_{ij} = a_{ij} + b_{ij}$ and $d_{ij} = ka_{ij}$, respectively. The **difference** between the two matrices A and B is the sum of A and $(-1)B$.*

To illustrate:
$$\begin{pmatrix} 1 & 2 \\ 3 & 4 \end{pmatrix} + \begin{pmatrix} 5 & 6 \\ 7 & 8 \end{pmatrix} = \begin{pmatrix} 6 & 8 \\ 10 & 12 \end{pmatrix},$$
$$5 \begin{pmatrix} 1 & 2 \\ 3 & 4 \end{pmatrix} = \begin{pmatrix} 5 & 10 \\ 15 & 20 \end{pmatrix}.$$

By writing out the general terms in each case, one can verify the following theorem (Ex. **8**, § **164**):

THEOREM I. *Whenever the products and sums can be formed the following laws hold:* (i) *the commutative and associative laws for addition,* $A + B = B + A$, $(A + B) + C = A + (B + C)$; (ii) *the distributive laws,* $k(A + B) = kA + kB$, $C(A + B) = CA + CB$, $(A + B)C = AC + BC$; (iii) *the associative laws* $(k_1 k_2)A = k_1(k_2 A)$, $(kA)B = k(AB)$; (iv) $(kA)B = A(kB)$.

Multiplication of a square matrix by a constant k does *not* correspond to multiplication of its determinant by k, since k is a factor of *each row*.

THEOREM II. *If A is a square matrix of order n and if k is any constant,* $|kA| = k^n |A|$.

160. Transpose of a matrix.

DEFINITION. *Two matrices, A and B, whose general elements satisfy the relation $a_{ij} = b_{ji}$ (each being obtained from the other by interchanging rows and columns) are said to be* **transposed matrices,** *and each is called the* **transpose** *of the other. This relation is written $A' = B$ or $A = B'$. A* **symmetric** *matrix A is one that is equal to its transpose, $A = A'$, that is, one whose elements satisfy the relation $a_{ij} = a_{ji}$.*

We record three facts, whose proofs can be supplied readily by the student (Exs. **9-11, § 164**):

THEOREM I. *The transpose of the product of two matrices is the product of their transposes in reverse order:* $(AB)' = B'A'$.

THEOREM II. *The determinant of a square matrix is equal to the determinant of its transpose.*

THEOREM III. *The rank of a matrix is equal to the rank of its transpose.*

161. Zero and the identity.
Among the square matrices of order n are two which play roles in matrix algebra similar to those of 0 and 1 in the real number system. These are defined:

$$0 = \begin{pmatrix} 0 & 0 & \cdots & 0 \\ 0 & 0 & \cdots & 0 \\ \cdot & \cdot & \cdot & \cdot \\ 0 & 0 & \cdots & 0 \end{pmatrix}, \quad I = \begin{pmatrix} 1 & 0 & \cdots & 0 \\ 0 & 1 & \cdots & 0 \\ \cdot & \cdot & \cdot & \cdot \\ 0 & 0 & \cdots & 1 \end{pmatrix},$$

and are called the **zero** and the **identity** or **unit matrix**, respectively. That is, 0 is the square matrix every element of which is zero, and

I is the square matrix whose general element is the "Kronecker delta," δ_{ij}, which is equal to 1 if i and j are equal and is equal to zero if i and j are unequal. More generally, any matrix, square or not, every element of which is zero is called a zero matrix, and is represented by the symbol O. Note that 0 and O are not the same symbol!

The student should establish the following two theorems, the first of which is concerned with square matrices only (Ex. **14, § 164**):

Theorem I. *If O and I are the zero and identity matrices of order n, respectively, if A is an arbitrary matrix of order n, and if k is an arbitrary constant, then* (i) $A + O = O + A = A;$ (ii) $AO = OA = O;$ (iii) $kO = O;$ (iv) $OA = O;$ (v) $AI = IA = A$.

Theorem II. *If O is any zero matrix, if I is any identity matrix, if A is an arbitrary matrix, and if k is an arbitrary constant, then the following relations hold whenever the sums and products involved are defined:* (i) $A + O = O + A = A;$ (ii) AO is a zero matrix; (iii) OA is a zero matrix; (iv) $kO = O;$ (v) OA is a zero matrix; (vi) $AI = A;$ (vii) $IA = A$.

162. Multiplication of determinants. The student may already have had experience in multiplying determinants by the same scheme that is used for matrices. The relation between products of determinants and products of square matrices is embodied in the theorem:

Theorem I. *The determinant of the product of two square matrices is equal to the product of their determinants:* $|AB| = |A| \cdot |B|$.

Proof will be given for third order matrices, but the method is general. Let the matrices be $A = (a_{ij})$ and $B = (b_{ij})$, and let their product AB be represented by $C = (c_{ij})$. Consider the sixth order determinant

$$D = \begin{vmatrix} a_{11} & a_{12} & a_{13} & 0 & 0 & 0 \\ a_{21} & a_{22} & a_{23} & 0 & 0 & 0 \\ a_{31} & a_{32} & a_{33} & 0 & 0 & 0 \\ -1 & 0 & 0 & b_{11} & b_{12} & b_{13} \\ 0 & -1 & 0 & b_{21} & b_{22} & b_{23} \\ 0 & 0 & -1 & b_{31} & b_{32} & b_{33} \end{vmatrix}.$$

Anybody familiar with Laplace's expansion of a determinant will recognize immediately that this determinant has the value $|A| \cdot |B|$. In order to see that this is the case, without using Laplace's development, expand D with respect to the elements of the first row. A typical term is the product of a_{11} and its cofactor, the fifth order determinant obtained by eliminating the first row and the first column. The expansion of this determinant leads quickly to its value, $A_{11} \cdot |B|$, where A_{11} is the cofactor of a_{11} in the determinant $|A|$. On factoring $|B|$ from the three non-zero terms of the expansion of D, we have the result, $|A| \cdot |B|$.

The problem now is to show that this same determinant is equal to $|C|$. This is done in three steps. (It is recommended that the student carry out the details of these steps.) First add to the fourth column the product of b_{11} and the first column, the product of b_{21} and the second column, and the product of b_{31} and the third column. (Notice the new arrangement of zeros!) Then add to the fifth column the products of b_{12}, b_{22}, and b_{32} and the first three columns in order. Finally, add to the sixth column the products of b_{13}, b_{23}, and b_{33} and the first three columns in order. With the notation for the matrix C, the result is

$$\begin{vmatrix} a_{11} & a_{12} & a_{13} & c_{11} & c_{12} & c_{13} \\ a_{21} & a_{22} & a_{23} & c_{21} & c_{22} & c_{23} \\ a_{31} & a_{32} & a_{33} & c_{31} & c_{32} & c_{33} \\ -1 & 0 & 0 & 0 & 0 & 0 \\ 0 & -1 & 0 & 0 & 0 & 0 \\ 0 & 0 & -1 & 0 & 0 & 0 \end{vmatrix}.$$

Three column interchanges (a_{1j} with c_{1j}) and multiplication of each of the last three rows by -1 give a sixth order determinant with the c's located in the upper left corner and with 1's on the main diagonal in the lower right. As with the initial evaluation of D, this can be shown to have the value $|C| \cdot 1 = |C|$. This completes the proof.

The following theorem is a corollary of Theorem I:

THEOREM II. *The product of two square matrices is non-singular if and only if both of these matrices are non-singular.*

163. Rank of the product of two matrices.

THEOREM I. *The rank of the product of two square matrices does not exceed the rank of either factor.*

Proof for matrices of the third order. Consider the two 3 by 6 matrices formed by the elements of the first three rows of the two sixth order determinants that are written out in § 162. The second matrix being obtained from the first by elementary transformations, its rank must be the same as that of the first. But this means that the rank of $C = AB$ cannot exceed the rank of A. That is, the rank of the product of two matrices cannot exceed the rank of the first factor. On the other hand, forming the transpose of each side of the equation $C = AB$, we have $C' = B'A'$. From this we can conclude that the rank of C' does not exceed the rank of B'. By Theorem III, § 160, the rank of C' is equal to the rank of AB, the rank of B' is equal to the rank of B, and the proof is complete.

Since any matrix can be filled out or completed with zero elements to form a square matrix, we can readily establish the more general theorem:

Theorem II. *The rank of the product of two matrices does not exceed the rank of either factor.*

164. Exercises.

In Exercises **1-4**, form the indicated product.

1. $\begin{pmatrix} 2 & -1 \\ -3 & 2 \end{pmatrix} \begin{pmatrix} 3 & 5 \\ 1 & 7 \end{pmatrix}.$ **2.** $\begin{pmatrix} 2 & 0 & 1 \\ -1 & 3 & 4 \end{pmatrix} \begin{pmatrix} 6 & -2 \\ 4 & 0 \\ 1 & 5 \end{pmatrix}.$

3. $\begin{pmatrix} 1 & 2 & 3 \\ 3 & 2 & 1 \end{pmatrix} \begin{pmatrix} 1 \\ 2 \\ 3 \end{pmatrix}.$ **4.** $\begin{pmatrix} 3 & 2 & 2 \\ -2 & -1 & 0 \\ 6 & 3 & 2 \end{pmatrix} \begin{pmatrix} -2 & 2 & 2 \\ 4 & -6 & -4 \\ 0 & 3 & 1 \end{pmatrix}.$

5. If $A = \begin{pmatrix} 1 & -1 & 1 \\ 0 & 3 & -2 \\ 4 & -2 & 1 \end{pmatrix}$ and $B = \begin{pmatrix} 0 & -3 & 0 \\ -1 & 4 & -2 \\ 1 & 5 & 1 \end{pmatrix},$

form the products AB, BA, $A'B'$, and $B'A'$, verify that $(AB)' = B'A'$ and $(BA)' = A'B'$, and show that the determinant of the product is in each case the product of the determinants of A and B.

6. Form the sum of the determinants A and B of Exercise **5** and show that the determinant of the sum of two matrices need not be equal to the sum of their determinants.

7. Prove the associative law for matrix multiplication.
8. Prove Theorem I, § **159**.
9. Prove Theorem I, § **160**.
10. Prove Theorem II, § **160**.
11. Prove Theorem III, § **160**.
12. Prove that the transpose of the sum of two matrices is the sum of their transposes: $(A + B)' = A' + B'$.
13. Prove that any matrix is the transpose of its transpose: $(A')' = A$.

14. Prove Theorem I and Theorem II, § 161.

15. Show by an example that it is possible to have two square matrices, neither of which is zero, whose product is zero. Conclude that the rank of the product of two matrices can be less than that of either factor.

★ 16. Let A be a column matrix with elements a_i, $i = 1, \cdots, n$, and let B be a row matrix with elements b_i, $i = 1, \cdots, n$. Prove that AB is an nth order square matrix whose general element is $a_i b_j$, and that BA is a matrix consisting of the single element $\sum_{i=1}^{n} a_i b_i$. If the ranks of A and B are 1, what is the rank of AB? of BA?

★ 17. Let R_i be the row matrix whose elements are all zero except for the element in the ith column, which is equal to 1, and let C_j be the column matrix whose elements are all zero except for the element in the jth row, which is equal to 1, the number of elements of each matrix being equal to n. Prove that if $A = (a_{ij})$ is a square matrix of order n, then $R_i A$ is the row matrix whose elements are those of the ith row of A, AC_j is the column matrix whose elements are those of the jth column of A, and $R_i A C_j$ is a matrix consisting of the single element a_{ij}.

★ 18. A square matrix (a_{ij}) whose elements that are not on the main diagonal are all zero ($a_{ij} = 0$, $i \neq j$) is called a **diagonal** matrix. Prove that the product of two diagonal matrices is a diagonal matrix and that multiplication of diagonal matrices is commutative.

★ 19. A diagonal matrix (see Ex. **18**) all of whose diagonal elements are equal is called a **scalar** matrix. Prove that a scalar matrix commutes with every square matrix of the same order and that multiplication by the scalar matrix kI multiplies every element of the other matrix by k.

★ 20. Give an example of distinct non-singular non-diagonal matrices that commute. (See Ex. **18**.)

165. Linear transformations. Consider the system of equations

(1)
$$x_1' = a_{11}x_1 + \cdots + a_{1n}x_n,$$
$$\cdots \cdots \cdots \cdots \cdots$$
$$x_n' = a_{n1}x_1 + \cdots + a_{nn}x_n,$$

with a non-singular coefficient matrix A. This system defines what is called a **linear homogeneous transformation**, or more simply a **linear transformation**, transforming the set of numbers $(x_1, \cdots, x_n)$ into the set of numbers $(x_1', \cdots, x_n')$. We say that the *point* $P(x_1, \cdots, x_n)$ is carried into the *point* $Q(x_1', \cdots, x_n')$, and again use the language and notation of transformations introduced in § **137**, writing $Q = T(P)$. Cramer's Rule guarantees that the point P is the *only* point carried into Q by the transformation T, and that we can solve for the coördinates of P in terms of the coördinates of Q to obtain the inverse transformation which carries Q into P.

Before studying this inverse transformation, however, let us see how the algebra of matrices can be used in the present situation. If we denote by X and X_1 the column matrices

$$\begin{pmatrix} x_1 \\ \cdot \\ \cdot \\ \cdot \\ x_n \end{pmatrix} \text{ and } \begin{pmatrix} x_1' \\ \cdot \\ \cdot \\ \cdot \\ x_n' \end{pmatrix},$$

respectively, we see that the original system of equations can be written in matrix form:

$$X_1 = AX.$$

The identity transformation is given (appropriately) by the identity matrix I of order n:

$$X = IX.$$

Suppose now that we have a successive application of two transformations, T and U, with matrices A and B, respectively. If $Q = T(P)$ and $R = U(Q)$ and if X, X_1, and X_2 are column matrices corresponding to P, Q, and R, respectively, then $X_1 = AX$, $X_2 = BX_1$, and the associative law for matrix multiplication gives

$$X_2 = BX_1 = B(AX) = (BA)X.$$

This corresponds to the transformation equation

$$R = U(Q) = U(T(P)) = (UT)(P).$$

The obvious correspondence between these two sets of equalities establishes the highly significant fact that products of matrices correspond to products of transformations. In fact, it is for this reason that matrix multiplication is defined as it is. Formulating this fact with more precision, we have the theorem:

THEOREM. *The product of two linear transformations is a linear transformation, whose matrix is the product of their matrices.*

166. Inverse of a linear transformation. In order to discuss the question of the inverse of a linear transformation, and therefore the inverse of a matrix, we return to Cramer's Rule and the notation of § 165. Letting A_{ij} represent the cofactor of a_{ji} (note the order of the subscripts!) in the (non-vanishing) determinant $|A|$, and solving for the variables $x_1, \cdots, x_n$, we can represent the solution for x_i by expanding the numerator determinant with respect to the elements of the ith column. The result is

$$x_i = \frac{1}{|A|}[A_{i1}x_1' + A_{i2}x_2' + \cdots + A_{in}x_n'].$$

Denoting by b_{ij} the **normalized cofactor** of a_{ji} in $|A|$,

$$b_{ij} = \frac{A_{ij}}{|A|},$$

we have a new matrix $B = (b_{ij})$, which we shall temporarily call the matrix of normalized cofactors of the elements of A, and which is the matrix of the (linear) inverse transformation T^{-1}:

(1)
$$x_1 = b_{11}x_1' + \cdots + b_{1n}x_n',$$
$$\cdots \cdots \cdots \cdots$$
$$x_n = b_{n1}x_1' + \cdots + b_{nn}x_n'.$$

Since the matrix B corresponds to the inverse transformation of T, we should expect the product of the matrices A and B to be the identity matrix. This is a fact, and the product can be formed in either order:

$$AB = BA = I.$$

Proof is obtained by application of the eighth and ninth laws of determinants given in Chapter III. The student is strongly urged to write the matrices down and carry out the multiplications in detail (Ex. **8, § 169**).

167. Inverse of a matrix.

DEFINITION. *If two matrices A and B have the property that $AB = BA = I$, each is called an **inverse** of the other.*

As was observed in **§ 166**, every non-singular matrix has a matrix of normalized cofactors. Therefore, every non-singular matrix has an inverse. If we take determinants on both sides of the matrix equation $AB = I$, obtaining the relation $|A| \cdot |B| = 1$, we see that if two matrices are inverse matrices they are both non-singular, and their determinants are reciprocals. We have thus proved part of the theorem:

THEOREM I. *A necessary and sufficient condition for a matrix A to have an inverse, written A^{-1}, such that $A^{-1}A = AA^{-1} = I$ is that A be non-singular. The inverse of a non-singular matrix is unique. The determinant of the inverse of a non-singular matrix is the reciprocal of the determinant of the matrix. A necessary and sufficient condition that two square matrices A and B be inverse matrices is that their product AB be the identity matrix. Any non-singular matrix is equal to the inverse of its inverse.*

In order to prove that a matrix cannot have more than one inverse we let $AB = AC = I$, and show that $B = C$. Since A is non-singular, there is a matrix A^{-1} such that $A^{-1}A = I$. Therefore $A^{-1}(AB) = A^{-1}(AC)$, and by the associative law $(A^{-1}A)B = (A^{-1}A)C$, or $IB = IC$, or finally, $B = C$. The remaining details of the proof are omitted.

The relation between inverse transformations and inverse matrices can now be stated formally:

THEOREM II. *If T is a linear transformation with a non-singular matrix A, its inverse T^{-1} is a linear transformation with the non-singular matrix A^{-1}.*

Since the product of two non-singular matrices is non-singular, it has an inverse. A formula for this inverse is given in the following theorem:

THEOREM III. *The inverse of the product of two non-singular matrices A and B is the product of their inverses in reverse order:*
$$(AB)^{-1} = B^{-1}A^{-1}.$$

Proof. We wish to show that $(AB)(B^{-1}A^{-1}) = I$. By the associative law this product is equal to
$$[(AB)B^{-1}]A^{-1} = [A(BB^{-1})]A^{-1} = (AI)A^{-1} = AA^{-1} = I.$$

An important consequence of Theorem II, § **163** can be established with the aid of inverse matrices.

THEOREM IV. *The rank of the product of a matrix A and a non-singular matrix B is equal to the rank of A, whether the product is AB or BA.*

Proof. We shall give a proof for the order AB, and let the student supply the proof for the product taken in the reverse order (Ex. **12**, § **169**). Denote by r the rank of A and by r' the rank of $C = AB$. Then, by Theorem II, § **163**, $r' \leq r$. By multiplying both members of the equation $C = AB$ on the right by B^{-1} we have $CB^{-1} = A$. Therefore, since the rank of CB^{-1} does not exceed that of C, $r \leq r'$. Consequently, $r = r'$.

168. Orthogonal matrices. It was observed in Chapter VII that the point transformations T and U with matrices
$$A = \begin{pmatrix} \lambda_1 & \lambda_2 & \lambda_3 \\ \mu_1 & \mu_2 & \mu_3 \\ \nu_1 & \nu_2 & \nu_3 \end{pmatrix} \quad \text{and} \quad B = \begin{pmatrix} \lambda_1 & \mu_1 & \nu_1 \\ \lambda_2 & \mu_2 & \nu_2 \\ \lambda_3 & \mu_3 & \nu_3 \end{pmatrix},$$

respectively, are inverse transformations. That A and B are inverse matrices can be verified directly, with the aid of the fact that any two rows or columns of either matrix are perpendicular directions. The student should actually expand at least one of the products AB and BA and show that the identity matrix results (Ex. **9**, § **169**).

The matrices A and B are therefore related in two ways: each is the transpose of the other, and each is the inverse of the other. Another way of expressing this fact is to say that the transpose of either of these matrices is the inverse of that matrix. Any matrix with this property is called *orthogonal* according to the definition:

DEFINITION: *An **orthogonal** matrix A is one whose transpose and inverse are equal:* $A' = A^{-1}$.

Since $|A'| = |A|$ and $|A^{-1}| = |A|^{-1}$, for any non-singular matrix A, it follows that if A is orthogonal, $|A| = |A|^{-1}$, or $|A|^2 = 1$. That is:

THEOREM I. *The determinant of any orthogonal matrix is equal to 1 or -1.*

To demonstrate how other properties of rotation matrices carry over to orthogonal matrices in general, we give a definition:

DEFINITION. *Two rows (or columns) of a matrix are **orthogonal** if and only if the sum of the products of corresponding elements is zero. The rows (or columns) of a matrix are **orthonormal** if and only if they are pairwise orthogonal and the sum of the squares of the elements in any one is equal to 1.*

We now state five orthogonality criteria for a matrix (with regard to the fifth condition, cf. Ex. **14**, § **169**):

THEOREM II. *Each of the following conditions is necessary and sufficient for a matrix A to be orthogonal, except that the fifth condition is a sufficient condition only for matrices of order greater than 2: (i) $AA' = I$; (ii) $A'A = I$; (iii) the rows of A are orthonormal; (iv) the columns of A are orthonormal; (v) A is non-singular, and either each element is equal to its cofactor in $|A|$ or each element is equal to the negative of its cofactor in $|A|$.*

Proof. The first two conditions follow immediately from Theorem I, § **167**. The third condition is equivalent to the first, and the fourth condition is equivalent to the second. For the last part, we shall prove only the following statement, and leave the remaining details to the student: "If A is non-singular, if its order n is greater than 2, and if each element is equal to its cofactor in $|A|$, then $|A| = 1$." The inverse of A is the matrix of normalized cofactors of elements of A. That is, since each element is equal to its cofactor, $A^{-1} = \frac{1}{|A|}A'$. Therefore $|A^{-1}| = \frac{|A'|}{|A|^n} = \frac{1}{|A|^{n-1}}$. But we have seen that $|A^{-1}| = \frac{1}{|A|}$. Therefore $|A|^{n-2} = 1$, and $|A| = \pm 1$. Evaluating $|A|$ by expansion with respect to the elements of any row we obtain a sum of squares. Therefore $|A| = 1$.

Two important properties of orthogonal matrices are stated in the theorem:

THEOREM III. *The product of two orthogonal matrices is orthogonal. The inverse of an orthogonal matrix is orthogonal.*

Proof. For the first part we wish to show that if A and B are orthogonal, then $(AB)(AB)' = I$. But by the associative law this product is equal to $(AB)(B'A') = [(AB)B']A' = [A(BB')]A' = (AI)A' = AA' = I$. For the second part we wish to show that $(A^{-1})(A^{-1})' = I$. This is true since $(A^{-1})(A^{-1})' = A'(A^{-1})' = (A^{-1}A)' = I' = I$.

In Chapter VII we obtained twenty-two relations involving the elements of a rotation matrix and made the statement that these were far from independent. To what extent these relations are independent and to what extent orthogonality characterizes rotation matrices are indicated in the following theorem, whose proof can be supplied easily by the student (Ex. **13**, § **169**):

THEOREM IV. *Each of the following conditions is necessary and sufficient for a third order matrix A to be a rotation matrix: (i) A is orthogonal and $|A| > 0$; (ii) the rows of A are orthonormal and $|A| > 0$; (iii) the columns of A are orthonormal and $|A| > 0$; (iv) A is non-singular and each element is equal to its cofactor in $|A|$.*

DEFINITION. *An **orthogonal transformation** is a linear transformation that has an orthogonal matrix.*

EXAMPLE. Solve for x'', y'', and z'' in terms of x, y, and z by forming the product of two matrices, if

$$7x' = 2x + 3y + 6z = -3x'' + 2y'' - 6z'',$$
$$7y' = 6x + 2y - 3z = -2x'' + 6y'' + 3z'',$$
$$7z' = -3x + 6y - 2z = 6x'' + 3y'' - 2z''.$$

220		MATRIX ALGEBRA		[Ch. IX

Solution. Let $X = \begin{pmatrix} x \\ y \\ z \end{pmatrix}$, $X_1 = \begin{pmatrix} x' \\ y' \\ z' \end{pmatrix}$, $X_2 = \begin{pmatrix} x'' \\ y'' \\ z'' \end{pmatrix}$,

$$A = \begin{pmatrix} \frac{2}{7} & \frac{3}{7} & \frac{6}{7} \\ \frac{6}{7} & \frac{2}{7} & -\frac{3}{7} \\ -\frac{3}{7} & \frac{6}{7} & -\frac{2}{7} \end{pmatrix}, \quad \text{and} \quad B = \begin{pmatrix} -\frac{3}{7} & \frac{2}{7} & \frac{6}{7} \\ -\frac{2}{7} & \frac{6}{7} & \frac{3}{7} \\ \frac{6}{7} & \frac{3}{7} & -\frac{2}{7} \end{pmatrix}.$$

Then the given systems of equations can be written $X_1 = AX = BX_2$. Furthermore, since B is orthogonal, $X_2 = (B^{-1}B)X_2 = B^{-1}(BX_2) = B^{-1}(AX) = (B'A)X$.

Finally, since $B'A = \frac{1}{49} \begin{pmatrix} -36 & 23 & -24 \\ 31 & 36 & -12 \\ 12 & -24 & -41 \end{pmatrix}$,

$$49x'' = -36x + 23y - 24z,$$
$$49y'' = 31x + 36y - 12z,$$
$$49z'' = 12x - 24y - 41z.$$

169. Exercises.

In Exercises **1-4**, find the inverse of the given matrix. Check your answer by forming the product of the given matrix and its inverse, in either order.

1. $\begin{pmatrix} 2 & 3 \\ 5 & 4 \end{pmatrix}.$		**2.** $\begin{pmatrix} 2 & 1 & 0 \\ 1 & 1 & 1 \\ 4 & 2 & 1 \end{pmatrix}.$

3. $\begin{pmatrix} 3 & 1 & -4 \\ 5 & 2 & 2 \\ 1 & -1 & -3 \end{pmatrix}.$		**4.** $\begin{pmatrix} 1 & 0 & 0 & x_0 \\ 0 & 1 & 0 & y_0 \\ 0 & 0 & 1 & z_0 \\ 0 & 0 & 0 & 1 \end{pmatrix}.$

5. If $A = \begin{pmatrix} 3 & 0 & 1 \\ 4 & 1 & -3 \\ -2 & -1 & 4 \end{pmatrix}$ and $B = \begin{pmatrix} 4 & -1 & 1 \\ 5 & 0 & 1 \\ 2 & 1 & 0 \end{pmatrix}$,

find A^{-1}, B^{-1}, AB, BA, $(AB)^{-1}$, and $(BA)^{-1}$, and check your answer by verifying that $(AB)^{-1} = B^{-1}A^{-1}$ and $(BA)^{-1} = A^{-1}B^{-1}$.

In Exercises **6** and **7**, solve for x'', y'', and z'' in terms of x, y, and z by forming the product of two matrices, if the variables are related by the given systems of equations.

6. $x'' = x' + 2y' - z',$		$x' = 2x - 5y + z,$
$y'' = 3y' + z',$		$y' = 3x + 4z,$
$z'' = 2x' - 4z';$		$z' = x + y + 2z.$
7. $x'' = 2x' - 3z',$		$x = 3x' + 2z',$
$y'' = -4y' + z',$		$y = 2x' + y',$
$z'' = -2x' + 5y';$		$z = 2x' + z'.$

8. Prove that, as stated in the last paragraph of § **166**, if B is the matrix of normalized cofactors of a non-singular matrix A, then $AB = BA = I$.

9. Verify that if A and B are the third order matrices defined in the first paragraph of § **168**, then $AB = BA = I$.

10. Prove that if B is the matrix of normalized cofactors of the elements of a non-singular matrix A, then A is the matrix of normalized cofactors of the elements of B.

11. Prove that if A is non-singular, then $AB = A$ implies $B = I$, and $BA = A$ implies $B = I$. Show that if

$$A = \begin{pmatrix} 1 & 1 \\ -1 & -1 \end{pmatrix} \quad \text{and} \quad B = \begin{pmatrix} 0 & 0 \\ 1 & 1 \end{pmatrix},$$

then $AB = A$, but $B \neq I$. Explain the apparent contradiction.

12. Prove that if B is non-singular, then the ranks of A and BA are equal.

13. Prove Theorem IV, § **168**.

14. Give an example of a non-singular non-orthogonal matrix A of order 2 each element of which is equal to its cofactor in $|A|$.

* **15.** Let (x, y, z) and (x', y', z') be the coördinates of a point P in two coördinate systems with a common origin O, the first rectangular and the second oblique. Let the directions in the xyz system of the x', y', and z' axes be $(\lambda_1, \lambda_2, \lambda_3)$, (μ_1, μ_2, μ_3), and (ν_1, ν_2, ν_3), respectively. As in Chapter VII, let the segment OP be represented as the sum of segments parallel to the primed axes, whose directed lengths are therefore x', y', and z'. By projecting this broken line segment on the x, y, and z axes, obtain the system of equations

$$x = \lambda_1 x' + \mu_1 y' + \nu_1 z',$$
$$y = \lambda_2 x' + \mu_2 y' + \nu_2 z',$$
$$z = \lambda_3 x' + \mu_3 y' + \nu_3 z',$$

where the coefficient matrix is non-singular. Hence prove that a coördinate transformation from any oblique system of axes to any oblique system of axes with the same origin is a linear transformation with a non-singular matrix. (See Ex. **14**, § **7**.)

* **16.** Prove that an orthogonal matrix of the second order must have one of the two forms

$$\begin{pmatrix} \cos\theta & \sin\theta \\ -\sin\theta & \cos\theta \end{pmatrix} \quad \text{or} \quad \begin{pmatrix} \cos\theta & \sin\theta \\ \sin\theta & -\cos\theta \end{pmatrix},$$

and that any matrix of either of these forms is orthogonal. Show that the second of these two matrices is equal to each of the products

$$\begin{pmatrix} 1 & 0 \\ 0 & -1 \end{pmatrix}\begin{pmatrix} \cos\theta & \sin\theta \\ -\sin\theta & \cos\theta \end{pmatrix} \quad \text{and} \quad \begin{pmatrix} \cos(-\theta) & \sin(-\theta) \\ -\sin(-\theta) & \cos(-\theta) \end{pmatrix}\begin{pmatrix} 1 & 0 \\ 0 & -1 \end{pmatrix}.$$

Hence prove that any (homogeneous linear) orthogonal transformation in the plane either is a rotation with fixed origin or can be represented in either the form $T_1 U$ or the form $U T_2$, where T_1 and T_2 are rotations with fixed origin and U is a reflection of the plane with respect to the x axis.

* 17. Prove that the product of any two skew-rotations with fixed origin is a rotation with fixed origin.

* 18. Prove that the inverse of any skew-rotation with fixed origin is a skew-rotation with fixed origin.

* 19. Prove that any (homogeneous linear) orthogonal transformation in space either is a rotation with fixed origin or can be represented in either the form $T_1 U$ or the form $U T_2$, where T_1 and T_2 are rotations with fixed origin and U is a reflection of space with respect to the xy plane. (Cf. Ex. **16**.)

* 20. Prove that the fourth order matrix

$$A = \begin{pmatrix} \cos \alpha & \sin \alpha & 0 & 0 \\ -\sin \alpha & \cos \alpha & 0 & 0 \\ 0 & 0 & \cos \beta & \sin \beta \\ 0 & 0 & -\sin \beta & \cos \beta \end{pmatrix}$$

is orthogonal, and show that each element is equal to its cofactor in $|A|$.

* 21. Prove that a linear homogeneous transformation T defined by system (1), § **165**, has the properties:

$$T(X + Y) = T(X) + T(Y)$$
$$T(kX) = kT(X).$$

(Think of T as transforming a column matrix into a column matrix.)

* 22. Prove that a transformation T that carries any point $(x_1, x_2, \cdots, x_n)$ into some point $(x_1', x_2', \cdots, x_n')$, that is, that transforms any column matrix into a column matrix, and that satisfies the two conditions of Exercise **21** is defined by a system of equations of the form (1), § **165**. *Suggestion:* Let C_j be defined as in Exercise **17**, § **164**. Then define the matrix $A = (a_{ij})$ in terms of the transformation T, letting the jth column of A be equal to $T(C_j)$. Then $T(X) = T(x_1 C_1 + \cdots + x_n C_n) = x_1 T(C_1) + \cdots + x_n T(C_n)$, or,

$$\begin{pmatrix} x_1' \\ \cdot \\ x_n' \end{pmatrix} = \begin{pmatrix} a_{11}x_1 + \cdots + a_{1n}x_n \\ \cdots \cdots \cdots \cdots \\ a_{n1}x_1 + \cdots + a_{nn}x_n \end{pmatrix}.$$

* 23. A **group** is a set of elements and an operation ∘ such that the following four conditions are satisfied:

(*i*) If a and b are any elements of the set (distinct or not), then $a \circ b$ is an element of the set.

(*ii*) The associative law holds:

$$(a \circ b) \circ c = a \circ (b \circ c),$$

where a, b, and c are any elements of the set.

(*iii*) There exists an identity element i of the set with the property that for any a of the set

$$a \circ i = i \circ a = a.$$

(iv) Corresponding to any element a of the set there exists an inverse element a^{-1} of the set such that

$$a^{-1} \circ a = a \circ a^{-1} = i.$$

It is not difficult to show that these axioms imply the uniqueness of the identity i and of the inverse a^{-1} of any element a.

Prove that each of the following is a group:

(a) The set of all non-singular matrices of order n, with matrix multiplication as the group operation.
(b) The set of all orthogonal matrices of order n, with matrix multiplication as the group operation.
(c) The set of all orthogonal matrices of order n with determinant equal to 1, with matrix multiplication as the group operation.
(d) The set of all rigid motions in space.
(e) The set of all rotations with fixed origin in space.
(f) The set of all translations in space.
(g) The set of all linear transformations in n-dimensional space with non-singular matrices.
(h) The set of all matrices of order n, with matrix addition as the group operation.
(i) The set of all diagonal matrices of order n, with matrix multiplication as the group operation. (See Ex. **18**, § **164**.)

* **24.** An **Abelian** or **commutative group** is a group in which the commutative law holds:

$$a \circ b = b \circ a,$$

where a and b are any elements of the group. Prove that the groups (f), (h), and (i) of Exercise **23** are Abelian, and that these are the only Abelian groups of that Exercise.

* **25.** Prove that the sum and the product of two matrices of the form

(1)
$$\begin{pmatrix} a & b \\ -b & a \end{pmatrix},$$

where a and b are real numbers, are matrices of form (1). Prove that if the matrix (1) is made to correspond to the complex number $a + bi$, then the sum and the product of two matrices that correspond to two complex numbers correspond to their sum and their product, respectively. Prove that the zero and identity matrices of the form (1) correspond to the numbers 0 and 1, respectively, and that diagonal matrices of the form (1) correspond to real numbers. Prove that the negative of a matrix of the form (1) corresponding to a complex number α and the inverse of a non-singular matrix corresponding to a non-zero number α correspond to $-\alpha$ and $1/\alpha$, respectively. Prove that the set of all matrices of the form (1) with matrix addition as the group operation and the set of all non-singular matrices of the form (1) with matrix multiplication as the group operation are Abelian groups.

The matrix (1) furnishes one method of *defining* complex numbers. (See Ex. **18**, § **164**, and Exs. **23** and **24** of this section.)

170. Direct and inverse transformations. Let $A = (a_{ij})$ and $B = (b_{ij})$ be inverse matrices and let T and U be the corresponding transformations:

$$T: \; x_i' = \sum_{j=1}^{n} a_{ij} x_j, \qquad U: \; x_i = \sum_{j=1}^{n} b_{ij} x_j'.$$

Considering these transformations as *point transformations*, we shall call T the *direct* transformation, with matrix A, and U the *inverse* transformation, with matrix B.

As in Euclidean space of three dimensions, we may wish to consider a transformation as a *coördinate transformation*. In this case we shall restrict our consideration largely to *orthogonal* transformations, which correspond to changes from one rectangular system to another, although it is possible to consider any linear transformation with a non-singular matrix as a coördinate transformation (cf. Ex. **15**, § 169). Since coördinate transformations are used principally to alter the form of an equation, we shall follow the precedent set in Chapter VII and, when regarding the transformations given above as *coördinate* transformations, think of U as the *direct* transformation, with matrix B, and T as the *inverse* transformation, with matrix A.

171. Quadratic forms.

DEFINITION. *A **quadratic form** in the variables $x_1, x_2, \cdots, x_n$ with symmetric matrix $A = (a_{ij})$ is the quadratic function* $\sum_{i,j=1}^{n} a_{ij} x_i x_j$.

THEOREM I. *If the variables $x_1, \cdots, x_n$ are subjected to a linear transformation $x_i = \sum_{j=1}^{n} c_{ij} x_j'$ with matrix $C = (c_{ij})$, a quadratic form in the variables $x_1, \cdots, x_n$ with symmetric matrix A is transformed into a quadratic form in the variables $x_1', \cdots, x_n'$ with symmetric matrix $C'AC$.*

Proof. Using notation previously adopted, we note that the transformation can be written $X = CX_1$, and that the quadratic form can be written $X'AX$. After substitution the quadratic form becomes

$$(CX_1)'A(CX_1) = X_1'(C'AC)X_1.$$

The right member of this equation is a quadratic form in the primed variables with matrix $C'AC$. This matrix is symmetric, since $(C'AC)' = (AC)'(C')' = (C'A')C = C'AC$.

DEFINITION. The **characteristic polynomial** of a square matrix A is the polynomial in the variable k defined by the determinant $|A - kI|$. The **characteristic roots** of a square matrix are the zeros of its characteristic polynomial.

THEOREM II. *If A is a square matrix and if C is an orthogonal matrix (of the same order), then the characteristic polynomials of A and $C'AC$ are identical.*

Proof. Since the determinant of an orthogonal matrix is ± 1 and since the determinant of the product of two matrices is equal to the product of their determinants,

$$|A - kI| \equiv |C'| \cdot |A - kI| \cdot |C| \equiv |C'(A - kI)C|.$$

From the laws of matrix algebra and the fact that C is orthogonal it follows that

$$|A - kI| \equiv |C'AC - C'(kI)C| \equiv |C'AC - kI|.$$

172. Proof of an earlier theorem. Sufficient machinery has now been developed to permit a simple proof of Theorem I, § 153, concerning invariants of a second degree equation $f(x, y, z) = 0$.

Since a translation leaves the matrix e unchanged, since a rotation with fixed origin has an orthogonal matrix, and since $e(x, y, z)$ is a quadratic form in the variables x, y, and z, the invariance of D, I, J, k_1, k_2, and k_3 follows from the theorems of § **171**, with the aid of the fact that corresponding coefficients of identical (characteristic) polynomials are equal. The invariance of ρ_3 is a consequence of Theorem IV, § **167**.

The invariance of ρ_4 and Δ can be established by the introduction of a fourth coördinate equal to 1, as indicated in § **117**. In this case the translation and rotation transformations are linear with the fourth order matrices

(1) $$\begin{pmatrix} 1 & 0 & 0 & x_0 \\ 0 & 1 & 0 & y_0 \\ 0 & 0 & 1 & z_0 \\ 0 & 0 & 0 & 1 \end{pmatrix} \text{ and } \begin{pmatrix} \lambda_1 & \mu_1 & \nu_1 & 0 \\ \lambda_2 & \mu_2 & \nu_2 & 0 \\ \lambda_3 & \mu_3 & \nu_3 & 0 \\ 0 & 0 & 0 & 1 \end{pmatrix},$$

respectively, the determinant of each being equal to 1 (although only the second is orthogonal). This completes the proof of the theorem.

(iv) The centers of the surface $X'AX = O$ are the points (x_1, x_2, x_3) for which AX is a column matrix the first three elements of which vanish.

(v) The singular points of the surface $X'AX = O$ are the points (x_1, x_2, x_3) for which AX is a column matrix every element of which vanishes.

(vi) The polar plane of the point (y_1, y_2, y_3) with respect to the surface $X'AX = O$ is $Y'AX = O$.

* 9. Prove that the definition of the polar plane of a point with respect to a quadric surface $f(x, y, z) = 0$ is independent of the (rectangular) coördinate system used. *Suggestion:* Using the notation of Exercise **8**, let the fixed point be (y_1, y_2, y_3), its polar plane be $Y'AX = O$, and define X_1 and Y_1 by the equations $X = CX_1$, $Y = CY_1$, C being either of the matrices (1) of § **172**. The equation of the surface $X'AX = O$ in the primed coördinate system is $X_1'(C'AC)X_1 = O$. Therefore the equation of the polar plane of (y_1', y_2', y_3') with respect to the surface, as defined in the primed system, is $Y_1'(C'AC)X_1 = O$. But this is precisely the equation $(CY_1)'A(CX_1) = O$, obtained by transforming the equation $Y'AX = O$.

174. Volume of a tetrahedron.
The vanishing of the determinant

$$T = \begin{vmatrix} x_1 & y_1 & z_1 & 1 \\ x_2 & y_2 & z_2 & 1 \\ x_3 & y_3 & z_3 & 1 \\ x_4 & y_4 & z_4 & 1 \end{vmatrix}$$

was shown in § **60** to be a necessary and sufficient condition for the four points $P_i(x_i, y_i, z_i)$, $i = 1, 2, 3, 4$, to lie in a plane. It is the purpose of this section to prove that the value of this determinant has a definite geometric significance, whether it vanishes or not.

THEOREM. *The absolute value of the determinant T is six times the volume of the tetrahedron whose vertices are the four points whose coördinates are elements of T.*

Proof. Let the variables x, y, and z be subjected to a transformation represented by either of the matrices (1) of § **172**, which we shall denote by C. In either case, it is readily shown that

$$\begin{pmatrix} x_1 & x_2 & x_3 & x_4 \\ y_1 & y_2 & y_3 & y_4 \\ z_1 & z_2 & z_3 & z_4 \\ 1 & 1 & 1 & 1 \end{pmatrix} = C \begin{pmatrix} x_1' & x_2' & x_3' & x_4' \\ y_1' & y_2' & y_3' & y_4' \\ z_1' & z_2' & z_3' & z_4' \\ 1 & 1 & 1 & 1 \end{pmatrix}.$$

Since the determinant of C is equal to 1, the value of T is independent of the coördinate system used. We choose the following coördinate system: Let the point P_4 be located at the origin, the point P_1 on the positive part of the x' axis, and the point P_2 in the half of the

§ 175]   AREA OF A TRIANGLE IN SPACE   229

$x'y'$ plane for which y' is positive. (See Fig. 50.) The determinant T then assumes the simple form:

$$T = \begin{vmatrix} x_1' & 0 & 0 & 1 \\ x_2' & y_2' & 0 & 1 \\ x_3' & y_3' & z_3' & 1 \\ 0 & 0 & 0 & 1 \end{vmatrix},$$

where x_1' and y_2' are positive. Consequently $T = x_1'y_2'z_3'$. The geometric interpretation can now be obtained immediately. The product $x_1'y_2'$ is twice the area of the triangle formed by the points P_1, P_2, and P_4. The third factor z_3' is numerically equal to the distance of the third point from the plane of the other three. Since the volume of a tetrahedron is one-third the product of the base area and the altitude, the absolute value of T must be six times the volume of the tetrahedron having the four given points as vertices.

The sign of T is the same as the sign of z_3' in the simplified form obtained in the proof just completed, and therefore depends on the orientation of the four given points. If we define right-handed and left-handed systems of three directed line segments issuing from a common point in a manner similar to that used in defining right-handed and left-handed coördinate systems in the first chapter, and furthermore, if we say that four non-coplanar points P_1, P_2, P_3, and P_4 have right-handed or left-handed orientation according as the directed segments P_4P_1, P_4P_2, and P_4P_3 form a right-handed or left-handed system, we can complete the theorem of this section with the statement:

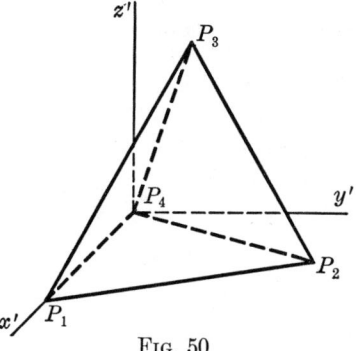

The sign of the determinant T is positive or negative according as the points P_1, P_2, P_3, and P_4 have right-handed or left-handed orientation.

175. Area of a triangle in space. It is proved in plane analytic geometry that the area of the triangle whose vertices are (x_1, y_1), (x_2, y_2), and (x_3, y_3) is

$$\pm \tfrac{1}{2} \begin{vmatrix} x_1 & y_1 & 1 \\ x_2 & y_2 & 1 \\ x_3 & y_3 & 1 \end{vmatrix},$$

the sign depending on the orientation of the points.

Consider now three non-collinear points in space, $P_1(x_1, y_1, z_1)$, $P_2(x_2, y_2, z_2)$, and $P_3(x_3, y_3, z_3)$, and the triangle Δ having these points as vertices. The triangle referred to above is the projection of Δ on the xy plane. Using the notation

$$a = \tfrac{1}{2}\begin{vmatrix} y_1 & z_1 & 1 \\ y_2 & z_2 & 1 \\ y_3 & z_3 & 1 \end{vmatrix}, \quad b = \tfrac{1}{2}\begin{vmatrix} z_1 & x_1 & 1 \\ z_2 & x_2 & 1 \\ z_3 & x_3 & 1 \end{vmatrix}, \quad c = \tfrac{1}{2}\begin{vmatrix} x_1 & y_1 & 1 \\ x_2 & y_2 & 1 \\ x_3 & y_3 & 1 \end{vmatrix},$$

and letting A be the area of Δ, we wish to prove the following theorem:

THEOREM. *The square of the area of any triangle is equal to the sum of the squares of the areas of its projections on the coördinate planes:*

$$A^2 = a^2 + b^2 + c^2.$$

Proof. The equation of the plane through the three points P_1, P_2, P_3 can be written

$$T(x, y, z) \equiv \tfrac{1}{2}\begin{vmatrix} x & y & z & 1 \\ x_1 & y_1 & z_1 & 1 \\ x_2 & y_2 & z_2 & 1 \\ x_3 & y_3 & z_3 & 1 \end{vmatrix} = 0,$$

or
$$ax + by + cz + d = 0,$$

where a, b, and c are defined above. If $P_4(x_4, y_4, z_4)$ is a point one unit from the plane of P_1, P_2, and P_3, then A is equal to the absolute value of $T(x_4, y_4, z_4)$, by the theorem of § **174**. But the formula for the distance between a plane and a point gives the relation

$$\frac{|ax_4 + by_4 + cz_4 + d|}{\sqrt{a^2 + b^2 + c^2}} = 1.$$

Therefore $a^2 + b^2 + c^2 = [T(x_4, y_4, z_4)]^2 = A^2$.

176. Transforming a transformation. Consider simultaneously a linear point transformation T with matrix A and a linear coördinate transformation Γ with non-singular matrix C:

$$T: y_i = \sum_{j=1}^{n} a_{ij} x_j, \qquad \Gamma: x_i = \sum_{j=1}^{n} c_{ij} x_j'.$$

The principal concern of this section will be to answer the following question: "What is the effect on the point transformation T of the coördinate transformation Γ? In other words, if $y_i = \sum_{j=1}^{n} c_{ij} y_j'$, how are the coördinates $y_1', y_2', \cdots, y_n'$ expressed in terms of the coördinates $x_1', x_2', \cdots, x_n'$?"

§ 176] TRANSFORMING A TRANSFORMATION

To answer this question, let X, X_1, Y, and Y_1 be the column matrices corresponding to the points $(x_1, \cdots, x_n)$, $(x_1', \cdots, x_n')$, $(y_1, \cdots, y_n)$, and $(y_1', \cdots, y_n')$, respectively. Then $Y = AX$, $X = CX_1$, and $Y = CY_1$. Therefore $Y_1 = (C^{-1}C)Y_1 = C^{-1}(CY_1) = C^{-1}Y = C^{-1}AX = (C^{-1}AC)X_1$. We therefore have the theorem:

THEOREM I. *If a linear coördinate transformation with non-singular matrix C is applied to the points $(x_1, \cdots, x_n)$, any linear point transformation with matrix A is transformed into a linear point transformation with matrix $C^{-1}AC$. If the coördinate transformation is orthogonal, the matrix A is transformed into the matrix $C'AC$.*

Since any rotation in three-dimensional Euclidean space, different from the identity, has a definite axis and a definite angle of rotation, a new coördinate system can be chosen so that the axis of the rotation is the z' axis. The rotation matrix assumes a particularly simple form, as described in § 143. This result is expressed in the theorem:

THEOREM II. *If A is any rotation matrix, there exists a rotation matrix C that reduces A to the canonical form:*

$$(1) \quad C'AC = \begin{pmatrix} \cos\theta & \sin\theta & 0 \\ -\sin\theta & \cos\theta & 0 \\ 0 & 0 & 1 \end{pmatrix},$$

where θ is the angle of the rotation with matrix A, the sense of the rotation angle θ being related to the direction of the z' axis in accordance with the right-handed convention: a right-handed screw would advance in the direction of the positive or negative z' axis according as the angle θ is positive or negative.

EXAMPLE. Solve the Example of § **144** by the method of this section.

Solution. As in the Example of § **143**, we find the equations of the axis: $\frac{x}{3} = \frac{y}{3} = z$. Let the direction of the z' axis be $\left(\frac{3}{\sqrt{19}}, \frac{3}{\sqrt{19}}, \frac{1}{\sqrt{19}}\right)$. Two directions that are perpendicular to this direction and to each other can be chosen with direction numbers $1, -1, 0$ and $1, 1, -6$. Accordingly, define the matrix

$$C = \begin{pmatrix} \frac{1}{\sqrt{2}} & \frac{1}{\sqrt{38}} & \frac{3}{\sqrt{19}} \\ -\frac{1}{\sqrt{2}} & \frac{1}{\sqrt{38}} & \frac{3}{\sqrt{19}} \\ 0 & -\frac{6}{\sqrt{38}} & \frac{1}{\sqrt{19}} \end{pmatrix}.$$

Denoting by A the given rotation matrix, we find that

$$C'AC = \begin{pmatrix} -\frac{5}{14} & \frac{3}{14}\sqrt{19} & 0 \\ -\frac{3}{14}\sqrt{19} & -\frac{5}{14} & 0 \\ 0 & 0 & 1 \end{pmatrix}.$$

Therefore $\cos\theta = -\frac{5}{14}$ and $\sin\theta > 0$. Therefore θ is the obtuse angle whose cosine is $-\frac{5}{14}$, which would cause a right-handed screw to advance in the direction chosen for the z' axis.

177. Exercises.

1. In Exercises **1-4**, find the volume of the tetrahedron whose vertices are the four given points, and state whether the four points have right-handed or left-handed orientation.

  1. (1, 0, 0), (1, 1, 0), (8, −7, 18), (0, 0, 0).
  2. (0, 1, 2), (2, −1, 0), (−2, 0, 1), (0, −1, −2).
  3. (0, 0, 0), (2, −1, 3), (4, 1, 0), (1, 2, −2).
  4. (3, 1, 2), (4, 2, −2), (1, −2, −1), (5, 1, 1).

5. Two opposite edges of a tetrahedron have equal lengths a and are perpendicular both to each other and to the segment of length b that joins their midpoints. Express the volume of the tetrahedron in terms of a and b.

In Exercises **6** and **7**, find the area of the triangle whose vertices are the three given points.

  6. (0, 0, 0), (2, −1, −3), (1, 5, −1).
  7. (0, 2, −4), (6, 2, −2), (−4, 0, 4).

8 and 9. For each of Exercises **29** and **30**, § **147**, determine an orthogonal matrix C that reduces the given matrix to the canonical form (1) of § **176**. In this way find the angle of each rotation.

10. Prove that the three points $P_1(x_1, y_1, z_1)$, $P_2(x_2, y_2, z_2)$, and $P_3(x_3, y_3, z_3)$ are collinear if and only if the three determinants

$$\begin{vmatrix} y_1 & z_1 & 1 \\ y_2 & z_2 & 1 \\ y_3 & z_3 & 1 \end{vmatrix}, \quad \begin{vmatrix} z_1 & x_1 & 1 \\ z_2 & x_2 & 1 \\ z_3 & x_3 & 1 \end{vmatrix}, \quad \text{and} \quad \begin{vmatrix} x_1 & y_1 & 1 \\ x_2 & y_2 & 1 \\ x_3 & y_3 & 1 \end{vmatrix}$$

vanish.

11. Prove that the characteristic roots of any symmetric rotation matrix A (corresponding to an involutory transformation) that is not the identity are $-1, -1,$ and 1. *Suggestion:* Let C be an orthogonal matrix that reduces A to the canonical form (1) of § **176**. Then $C'AC$ is symmetric, $\sin\theta = 0$, and θ is a multiple of 180°. (See Ex. **26**, § **147**.)

12. Prove that the angle of a rotation with fixed origin is 180° if and only if the trace of its matrix is -1. (See Ex. **11**.)

13. Find an orthogonal matrix C that reduces the matrix of Exercise **26**, § **147**, to diagonal form.

* 14. Prove that the characteristic roots of any rotation matrix have the form $\cos\theta + i\sin\theta$, $\cos\theta - i\sin\theta$, and 1, where θ is the angle of the rotation, and therefore that the absolute value of each characteristic root of a rotation matrix is 1. (More generally, the absolute value of each characteristic root of any real orthogonal matrix is 1.)

* 15. Let T be a linear point transformation in Euclidean three-dimensional space, with matrix A, and let P_1, P_2, P_3, and P_4 be four non-coplanar points, which are carried by the transformation T into the points Q_1, Q_2, Q_3, and Q_4, respectively. Prove that the volume of the tetrahedron (collapsed or not) whose vertices are the transformed points is equal to the volume of the tetrahedron whose vertices are the original points, multiplied by the absolute value of $|A|$, the determinant of the matrix of the transformation. If the matrix A is non-singular, prove that orientation is preserved or reversed according as the determinant $|A|$ is positive or negative. Prove that the transformation T carries the points of space into the points of space, the points of a plane through the origin, the points of a line through the origin, or the origin itself according as the rank of A is 3, 2, 1, or 0.

* 16. Prove that if a skew-rotation has any fixed points other than the origin, its fixed points constitute a plane through the origin. *Suggestion:* Assume that there is a line of fixed points, and choose axes so that this line is the z axis. The coördinates of the fixed points must satisfy a system of equations of the form (1), § **143**, where ν_3 is equal to 1, since each point on the z axis is fixed. Consequently, $\nu_1 = \nu_2 = \lambda_3 = \mu_3 = 0$, and the transformation must have a matrix of the form

$$\begin{pmatrix} \cos\theta & \sin\theta & 0 \\ \sin\theta & -\cos\theta & 0 \\ 0 & 0 & 1 \end{pmatrix}.$$

But any transformation with a matrix of this form must have a plane of fixed points. (Cf. Ex. **16**, § **169**.)

* 17. Let $(\lambda_i, \mu_i, \nu_i, \xi_i)$, $i = 1, \cdots, 4$, be four mutually perpendicular directions in E_4 (see § **62**). Explain why it is meaningless to say that a system of lines or axes having these directions is either right-handed or left-handed. What would you call the lines through the origin with directions $(1, 0, 0, 0), \cdots, (0, 0, 0, 1)$? Show that the value of the determinant whose rows are the four directions that were given first is ± 1. Formulate a definition that will tell whether two systems are similarly or oppositely oriented.

* 18. Develop a theory of invariants for the general equation of the second degree in two variables, x and y, similar to that of Chapters VIII and IX, and apply it to the theory of conic sections in plane analytic geometry.

* 19. Consider the following two definitions of *equivalence* between two matrices A and B (in the first of which A and B are not necessarily square):

 (*i*) $A \sim B$ if and only if there exist non-singular matrices C and D such that $CAD = B$;

 (*ii*) $A \sim B$ if and only if there exists a non-singular matrix C such that $C^{-1}AC = B$.

Prove that each is an equivalence relation, that the first is logically equivalent to the definition of § **50**, and that the second is not equivalent to the earlier definition. *Suggestion:* If A can be transformed into B by means of an elementary transformation (§ **50**), show that there exists either a non-singular matrix C such that $CA = B$ or a non-singular matrix D such that $AD = B$. (See Ex. **36**, § **13**.)

ANSWERS

Page 6, § 7

3. (a) A, D; C, G; B, F; E, H.   (d) A, E; B, C; D, H; F, G.
(g) A, H; B, G; C, F; D, E.
4. (a) D_2; (b) C_2; (c) OB_2; (d) C_2; (e) E_1E_2; (f) B_1B_2.
6. $(0, 4, -3)$, $(5, 0, -3)$, $(5, 4, 0)$, $(5, 0, 0)$, $(0, 4, 0)$, $(0, 0, -3)$.
7. $(2, 3, 7)$, $(2, 5, 5)$, $(2, 5, 7)$, $(8, 3, 5)$, $(8, 3, 7)$, $(8, 5, 5)$.   **8.** $(7, 4, 1)$.
9. B, C, E; the xz plane.   **10.** 6 and 2.   **11.** 4, 4, 2.
12. Partial answer: (a) $x_1 = -x_2$; (d) $y_1 = -y_2$, $z_1 = -z_2$;
(g) $x_1 = -x_2$, $y_1 = -y_2$, $z_1 = -z_2$.   **13.** (a) $(0, y, z)$; (d) $(x, 0, 0)$.

Page 12, § 13

1. 7.   **2.** 9.   **3.** 13.   **4.** $\sqrt{62}$.   **5.** 5.
6. 1.   **7.** 11.   **8.** 9.   **9.** $\sqrt{29}$.   **10.** 3.
13. $6\sqrt{2}$. **16.** $x^2 + y^2 + z^2 = 16$.   **17.** $(x-5)^2 + (y+1)^2 + (z-3)^2 = 4$.
18. $(x-2)^2 + (y+6)^2 + (z-3)^2 = (x-4)^2 + (y-2)^2 + (z+5)^2$,   or
$x + 4y - 4z + 1 = 0$.
19. $y = 0$.   **20.** $z = 2$.   **21.** $x = y$.
22. $(2, 4, 4)$.   **23.** $(1, 3, 2)$.   **24.** $(5, -2, \tfrac{5}{2})$.
25. $(0, 0, 0)$.   **26.** $(1, 6, 8\tfrac{1}{2})$.   **27.** $(18, -11, 17)$.
28. $(-6, 13, 5)$.   **29.** $(22, -15, 19)$.   **30.** $(4, 2, -1)$, $(5, 6, -3)$.
31. $(\tfrac{37}{5}, 8, 1)$, $(\tfrac{42}{5}, 5, 5)$, $(\tfrac{61}{5}, 2, 9)$, $(\tfrac{73}{5}, -1, 13)$.   **34.** $(2, 4, 3)$.
36. Relations (i), (iii), (iv), (vii), (ix), and (x) are equivalence relations.
Relations (ii) and $(viii)$ violate the reflexive and transitive laws, and relations (v) and (vi) violate the symmetric law.

Page 16, § 15

1. (a), (d), (e), and (f).   **2.** (b), (d), and (f).   **3.** $45°$ or $135°$.
4. 1, 0, 0.   **5.** 0, 1, 0.   **6.** 0, 0, 1.
7. $-\dfrac{1}{\sqrt{2}}, -\dfrac{1}{\sqrt{2}}, 0$.   **8.** $\dfrac{1}{\sqrt{3}}, \dfrac{1}{\sqrt{3}}, \dfrac{1}{\sqrt{3}}$.   **9.** $-\dfrac{1}{\sqrt{3}}, -\dfrac{1}{\sqrt{3}}, -\dfrac{1}{\sqrt{3}}$.
10. $54° 44'$.
11. $\tfrac{4}{9}, \tfrac{7}{9}, -\tfrac{4}{9}$.   **12.** $\dfrac{1}{\sqrt{14}}, \dfrac{2}{\sqrt{14}}, \dfrac{3}{\sqrt{14}}$.   **13.** $-\tfrac{2}{11}, -\tfrac{6}{11}, \tfrac{9}{11}$.
14. $\tfrac{2}{7}, \tfrac{3}{7}, -\tfrac{6}{7}$.   **15.** $-\tfrac{4}{21}, -\tfrac{8}{21}, \tfrac{19}{21}$.   **16.** $-\tfrac{2}{15}, \tfrac{2}{3}, \tfrac{11}{15}$.
17. The line makes an acute angle with the z axis and is therefore directed upward.
18. The line is directed downward.
19. The line lies in a horizontal plane.
20. The line makes an acute angle with the x axis and is therefore directed toward the observer.

ANSWERS

21. The line is parallel to the yz plane.
22. The line is directed to the left.
23. The line is parallel to the x axis and similarly directed.
24. The line is parallel to the y axis and oppositely directed.
25. The line is vertical, directed upward or downward.
26. 2, −4, −1.   27. −2, −5, 1.   28. (24, 12, 8).
29. (c) $\sin^2 \theta + \cos^2 \theta = 1$.   (d) Directed lines.   (e) Slope = μ/λ.
    (f) Undirected lines.

Page 18, § 17

1. $\frac{9}{11}, -\frac{2}{11}, \frac{6}{11}$.   2. $\frac{11}{15}, \frac{2}{3}, \frac{2}{15}$.   3. $\frac{3}{\sqrt{35}}, \frac{5}{\sqrt{35}}, \frac{1}{\sqrt{35}}$.
4. $\frac{2}{3}, -\frac{1}{3}, \frac{2}{3}$.   5. 0, $\frac{3}{5}$, $-\frac{4}{5}$.   6. 0, 0, 1.
7. 3, 5, −1.   8. 1, 1, 1.   9. −2, 3, 4.
10. 1, 2, 1.   11. 1, −1, 1.   12. 0, 1, 0.
13. (b) $b, -a$;   (c) a, b;   (d) Slope = m/l.

Page 20, § 20

1. $-\frac{5}{77}$.   2. $\frac{2}{27}$.   3. $\frac{169}{171}$.   4. $\frac{3}{13}$.
5. $\frac{3}{13}\sqrt{2}$.   6. $\frac{2}{3}$.   7. $\frac{1}{6}\sqrt{2}$.   8. $\frac{1}{2}$.
9. Yes.   10. Yes.   11. No.   12. Yes.
13. $-\frac{57}{11}$.   14. 3.
17. (a) $\cos \theta = \lambda_1 \lambda_2 + \mu_1 \mu_2$.   (b) $l_1 l_2 + m_1 m_2 = 0$.

Page 22, § 22

2. $\frac{1}{21}\sqrt{41}$.

Page 23, § 24

1. 1, 1, 1.   2. 1, 0, 0.   3. 2, 3, 0.
4. 1, 2, −3.   5. 3, 5, 1.   6. 3, −2, 4.
7. $(\pm\frac{2}{3}, \mp\frac{2}{3}, \pm\frac{1}{3})$.   8. $(\pm\frac{6}{7}, \pm\frac{2}{7}, \mp\frac{3}{7})$.

Page 24, § 26

7. 45°, 45°, 90°.   8. 30°, 60°, 90°.   9. 21° 47′, 38° 13′, 120°.
10. (a) $l = m = 0$;   (b) $n = 0$.   11. $k = -3$.
13. 1 : 1;   (4, 3, 0).   14. (7, 0, 1).   17. 0, 2, −1.
18. 3, 4, −5.   19. 12, 5, 9.   20. 32, 1, 23.

Page 29, § 29

4. $4x + y - 3z + 10 = 0$.   5. $x + 7y + 3z + 29 = 0$.
6. $x + 2y - 4z - 21 = 0$.   7. $5x - y - 2z = 0$.
8. $2x + 5y - z + 9 = 0$.
9. $7x - 5y + 4z - 36 = 0$ and $7x - 5y + 4z + 54 = 0$.
10. $2x - 2y + 7z + 4 = 0$.   11. $z = -1$.   12. $x = 5$.
13. $y = 3$.   14. $x - z - 6 = 0$.   15. $4x + 3y - 7 = 0$.
16. $7x - 3z - 22 = 0$.   17. $k = 2$.

§§ 17–37]   ANSWERS   237

18. $3x - y + 2z = 0$.
20. $2x - 3y - 6z + 6 = 0$.
22. $3x + 5y - z - 2 = 0$.
24. $5x - y - 2z + 1 = 0$.
19. $x + 2y - 4z - 8 = 0$.
21. $x + y + z - 7 = 0$.
23. $2x - 7y + 3z - 6 = 0$.
25. $k = 3$; $7x + 3y - z - 2 = 0$.

Page 33, § 33

1. $4x + 12y - 3z - 12 = 0$.   2. -3; 5; $-\tfrac{15}{2}$.   3. $x + y + z - 8 = 0$.
4. $6x + 3y + 2z - 15 = 0$.   5. -6; 2; 6.
6. Planes through the origin or parallel to at least one axis.
7. Divide the coefficient of each term (including the constant term) as follows:
   (a) by ± 3; (b) by ± 9; (c) by ± 11;
   (d) by ± 7; (e) by $\pm \sqrt{3}$; (f) by ± 9.
8. (a) 0; (b) $\tfrac{8}{9}$; (c) 3; (d) 4; (e) $2\sqrt{3}$; (f) 8.
9. (b) $(-\tfrac{7}{9}, \tfrac{4}{9}, -\tfrac{4}{9})$;   (c) $(\tfrac{6}{11}, -\tfrac{9}{11}, -\tfrac{2}{11})$;
   (d) $(-\tfrac{2}{7}, -\tfrac{3}{7}, \tfrac{6}{7})$;   (e) $(1/\sqrt{3}, 1/\sqrt{3}, 1/\sqrt{3})$;
   (f) $(-\tfrac{8}{9}, \tfrac{4}{9}, \tfrac{1}{9})$;   (d), (e), and (f).
10. (a) 1; (b) $\tfrac{1}{3}$; (c) $\tfrac{15}{11}$; (d) $\tfrac{38}{7}$; (e) $4\sqrt{3}$; (f) $\tfrac{26}{3}$;
    (b), (d), (e), and (f); (c), (d), (e), and (f).
11. $4x + 7y - 4z \pm 63 = 0$.   12. $6x + 3y + 2z \pm 14 = 0$.
13. (a) 3; (b) 1; (c) $\tfrac{13}{2}$; (d) $\tfrac{81}{4}$.
14. (a) and (d).   16. (b) and (c) same side.
17. $x - 2y - 2z + 8 = 0$, $x - 2y - 2z - 22 = 0$.
18. $9x + 2y - 6z + 98 = 0$, $9x + 2y - 6z - 34 = 0$.
19. $7x - 9y + 2z - 2 = 0$.   20. $\tfrac{2}{7}\sqrt{21}$.

Page 39, § 37

1. $\tfrac{4}{21}$.   2. $\tfrac{2}{3}$.   3. $\tfrac{2}{7}$.   4. $1/\sqrt{2}$.
7. (a) $2, 1, 4$; (b) $2x + y + 4z = 0$.   8. (a) $2, 3, 1$; (b) $2x + 3y + z = 0$.
9. (a) $3, -2, 7$; (b) $3x - 2y + 7z = 0$.
10. (a) $4, 6, -5$; (b) $4x + 6y - 5z = 0$.
11. $11x + 14y - 3z = 0$.   12. $12x + 13y - z + 5 = 0$.
13. $5x + 5y + 3 = 0$.   14. $4x + y + 3z + 9 = 0$.
15. $5x + 5z + 14 = 0$.   16. $10x + 5y + 5z + 17 = 0$.
17. $\sqrt{7}(x - y + 2z + 5) \pm \sqrt{3}(2x + 3y - z - 1) = 0$.
18. Answer to Ex. 17 with the minus sign.
19. $3x + 5y + 4z + 5 = 0$.   20. The acute angle.
21. The obtuse angle.   22. $3y + 2z + 5 = 0$.
23. $10x - 3y - 7z - 45 = 0$.   24. $x + 2y - z + 3 = 0$.
25. $x + 4y - 1 = 0$.   26. $2x + 3y = 0$.
27. $2x + 3y + 10 = 0$.   28. $3x - 2y + 17z - 24 = 0$.
29. $3x - 2y - 15z = 0$.   30. $3x = 2y$.
31. $x + 2y + 2z - 6 = 0$; $x = 2$.   32. $7x + 6y + 6z - 26 = 0$; $x = 2$.
33. $k_1(x - 3) + k_2(z - 1) = 0$.   34. $k_1(2x + y - 8) + k_2(z - 1) = 0$.
35. $k_1(x - y) + k_2(x - z) = 0$.
36. $k_1(2x + y - 7) + k_2(2x + z - 10) = 0$.

Note: Each of the answers to Exercises **33–36** is one of infinitely many possibilities.

ANSWERS

Page 46, § 40

1. (a) $\arccos(-\frac{1}{3})$; (b) $\arccos\frac{1}{3}$. 2. (a) $60°$; (b) $120°$.
3. (a) $\arccos\frac{31}{35}$; (b) $\arccos\frac{31}{35}$. 4. (a) $120°$; (b) $60°$.
11. $\dfrac{x-1}{2} = \dfrac{y-3}{1} = \dfrac{z-1}{6}$. 12. $\dfrac{x-1}{3} = \dfrac{y+2}{-2} = \dfrac{z+4}{1}$.
13. $\dfrac{x-5}{5} = \dfrac{y-1}{0} = \dfrac{z}{-3}$. 14. $\dfrac{x+2}{0} = \dfrac{y}{1} = \dfrac{z-7}{0}$.
15. $\dfrac{x-5}{2} = \dfrac{y-1}{3} = \dfrac{z-8}{-3}$. 16. $\dfrac{x+3}{4} = \dfrac{y-2}{-1} = \dfrac{z}{1}$.
17. $\dfrac{x-3}{0} = \dfrac{y-4}{1} = \dfrac{z-1}{2}$. 18. $\dfrac{x-7}{1} = \dfrac{y+9}{0} = \dfrac{z+2}{0}$.
19. $x = 1 + 2t$, $y = 3 + t$, $z = 1 + 6t$; $x = 1 + 3t$, $y = -2 - 2t$, $z = -4 + t$;
    $x = 5 + 5t$, $y = 1$, $z = -3t$; $x = -2$, $y = t$, $z = 7$.
20. $(0, 9, 4)$; $(-27, 0, 22)$; $(6, 11, 0)$.
21. $(0, 39, -45)$; $(26, 0, 20)$; $(18, 12, 0)$.
22. $(0, 11, 8)$; $(6, 11, 0)$. 23. $(0, -12, 14)$; $(-12, 0, -10)$; $(-7, -5, 0)$.
24. $\dfrac{x-3}{5} = \dfrac{y-5}{8} = \dfrac{z-6}{-1}$. 25. $\dfrac{x+2}{1} = \dfrac{y-1}{-3} = \dfrac{z}{-2}$.
26. $\dfrac{x}{3} = \dfrac{y-5}{1} = \dfrac{z-1}{0}$. 27. $\dfrac{x-1}{0} = \dfrac{y+3}{1} = \dfrac{z+4}{0}$.
28. $\dfrac{x}{3} = \dfrac{y-2}{-20} = \dfrac{z-\frac{5}{2}}{-5}$. 29. $\dfrac{x-2}{3} = \dfrac{y+5}{1} = \dfrac{z}{-4}$.
30. $\dfrac{x+1}{5} = \dfrac{y-3}{2} = \dfrac{z+4}{-3}$. 31. $\dfrac{x-3}{0} = \dfrac{y+1}{3} = \dfrac{z-5}{2}$.
32. $\dfrac{x}{1} = \dfrac{y}{1} = \dfrac{z}{1}$. 33. $\dfrac{x+4}{1} = \dfrac{y+7}{0} = \dfrac{z-2}{0}$.
34. $\dfrac{x+5}{3} = \dfrac{y-1}{1} = \dfrac{z}{7}$. 35. $\dfrac{x}{5} = \dfrac{y}{-2} = \dfrac{z}{-2}$.
40. $3x - 2y - 6z - 17 = 0$. 41. $4x - 5z - 20 = 0$.
42. $5x - 3y + 6z - 28 = 0$.
43. $6x + 5y - 8 = 0$; $x + 5z - 3 = 0$; $y - 6z + 2 = 0$.
44. $x - 2 = 0$; $3y + 2z + 23 = 0$.
45. $2x + 3y + 7 = 0$; $5x - 3z - 17 = 0$; $5y + 2z + 23 = 0$.
46. $4x + y - 6 = 0$; $5x + z - 7 = 0$; $5y - 4z - 2 = 0$.
47. $\dfrac{x - x_1}{l} = \dfrac{y - y_1}{m} = \dfrac{z}{0}$.
48. $(3, 5, -2)$. 49. $(3, -1, 4)$. 50. $(-2, -3, 1)$. 51. $(4, 2, -7)$.
52. $(1, -1, 0)$. 53. $(3, 1, 5)$. 54. $(-2, -4, 1)$. 55. $(3, 7, -2)$.
56. $(1, 0, 1)$. 57. $(2, 3, -1)$. 58. $\arcsin\frac{7}{11}$. 59. $30°$.
60. $x - y - z - 2 = 0$. 61. $3x - 4y + z + 8 = 0$.
62. $89x + 167y - 22z + 29 = 0$.
63. $2x - y - 3z = 0$. 64. $2x - y - 3 = 0$.
65. $\dfrac{x-1}{1} = \dfrac{y+2}{-2} = \dfrac{z+3}{-1}$.
66. (a) $\dfrac{x-1}{2} = \dfrac{y-2}{3} = \dfrac{z-3}{-3}$; (b) $\dfrac{x-1}{0} = \dfrac{y-2}{1} = \dfrac{z-3}{1}$;
    (c) $\dfrac{x-1}{3} = \dfrac{y-2}{-1} = \dfrac{z-3}{1}$.

67. $x = 5 + \dfrac{4t}{9}$, $y = 1 - \dfrac{t}{9}$, $z = -6 + \dfrac{8t}{9}$.

68. $x = -\dfrac{t}{3}$, $y = \dfrac{2t}{3}$, $z = \dfrac{2t}{3}$.

69. $x = 1 - \dfrac{6t}{7}$, $y = -3 - \dfrac{2t}{7}$, $z = \dfrac{3t}{7}$.

70. $x = -\dfrac{3t}{5}$, $y = 0$, $z = \dfrac{4t}{5}$.

Page 52, § 43

1. $x = 5t_1 + 7t_2,\quad y = t_1 + 4t_2,\quad z = 8t_1 + 5t_2;$
$x = -3t_1 + t_2,\quad y = 2t_1 + t_2,\quad z = t_2;$
$x = 3t_1 + 3t_2,\quad y = 4t_1 + 7t_2,\quad z = t_1 + 7t_2;$
$x = 7t_1 + 20t_2,\quad y = -9t_1 - 9t_2,\quad z = -2t_1 - 2t_2.$

3. $(2, -2, -21)$. **4.** $t_1 = 2$, $t_2 = -\tfrac{1}{3}$, $t_3 = -\tfrac{2}{3}$.

Page 54, § 46

1. $5\sqrt{5}$. **2.** $3\sqrt{2}$. **3.** 9. **4.** 7. **5.** $\tfrac{1}{3}\sqrt{61}$.
6. $\sqrt{3}$. **7.** 8. **8.** 4. **9.** 2. **10.** $\sqrt{14}$.
11. $\tfrac{1}{2}\sqrt{14}$. **12.** $5/\sqrt{14}$. **13.** $\tfrac{1}{3}\sqrt{6}\,a$. **14.** $a/\sqrt{2}$.
15. (a) $60°$; (b) $a/\sqrt{3}$. **16.** $(2, 5, -6)$.
17. $(\tfrac{1}{3}, \tfrac{1}{3}, \tfrac{1}{3})$. **18.** $(2, -1, 0)$. **19.** $(1, 1, 0)$.
20. $x - 6 = y - 9 = z + 2;\quad x - 1 = -2y = -2z;$
$x - 7 = y - 4 = -z - 5;\quad x - 1 = -y + 1 = z/k.$
21. $2x = 2y = z$. **22.** $\dfrac{x-1}{1} = \dfrac{y-5}{-3} = \dfrac{z+2}{-4}$.
23. $(5, 6, -2)$ and $(-3, 4, 8)$; $2\sqrt{42}$.

Page 60, § 51

1. 76. **2.** 0. **3.** 120.
4. -23. **5.** 1. **6.** 1.
7. 3. **8.** 2. **9.** 2.

Page 64, § 53

1. Consistent; $x = z - 2$, $y = -z$.

2. Consistent; $x = 2w + 3$, $y = \dfrac{w-1}{2}$, $z = \dfrac{4w+2}{3}$.

3. Inconsistent.

4. Consistent; $x = z - w + 2$, $y = 2z - w + 3$.

Page 68, § 55

1. $r = 3$, $R = 3$; one point, $(1, 1, 1)$.
2. $r = 2$, $R = 3$; planes intersect pairwise in parallel lines.
3. $r = 2$, $R = 2$; line in common, $x - 2 = y - 1 = z$.
4. $r = 2$, $R = 3$; planes intersect pairwise in parallel lines.

5. $r = 2$, $R = 2$; line in common, $\dfrac{x}{1} = \dfrac{y+7}{5} = \dfrac{z+3}{3}$.
6. $r = 3$, $R = 3$; one point, $(1, -2, 0)$.
7. $r = 2$, $R = 2$; line in common, $\dfrac{x-1}{3} = \dfrac{y+1}{1} = \dfrac{z+1}{-2}$.
8. $r = 3$, $R = 4$; no point in common.

Page 70, § 57

1. 1; first row is three-fifths of the second.
2. 3; first row is 3 times 2nd $-$ 5 times 3rd $+$ 2 times 4th.
3. 2; first row is 4 times 2nd $-$ 3 times 3rd.
4. $(x - z - 2) = (x - y - 1) + (y - z - 1)$.
5. $(x + 4y - 7z + 7) = 3(x + y - 2z + 1) - (2x - y + z - 4)$.
6. $(3x + y + 5z + 3) = (x - y + z - 1) + 2(x + y + 2z + 2)$;
$(4x - 2y + 5z - 1) = 3(x - y + z - 1) + (x + y + 2z + 2)$.
7. 2; first row is twice 2nd $-$ 3 times 3rd $= -$3rd $+$ twice 4th $= (k + 2)$ times 2nd $- (k + 3)$ times 3rd $- k$ times 4th, where k is any number.

Page 72, § 59

1. Collinear; $\dfrac{x-5}{2} = \dfrac{y+1}{-5} = \dfrac{z+2}{1}$. **2.** Not coplanar.
3. Coplanar; $x + z = 2$. **4.** Coplanar; $2x - y + z + 1 = 0$.
5. Collinear; $\dfrac{x+3}{1} = \dfrac{y+2}{2} = \dfrac{z+1}{3}$. **6.** Not coplanar.

Page 74, § 61

1. $\begin{vmatrix} x & y & z & 1 \\ 1 & 0 & 0 & 1 \\ 0 & 1 & 0 & 1 \\ 0 & 0 & 1 & 1 \end{vmatrix} = 0$. **2.** $\begin{vmatrix} x & y & z & 1 \\ 3 & 8 & 1 & 1 \\ 0 & -3 & 2 & 1 \\ 2 & 4 & -5 & 1 \end{vmatrix} = 0$.

Page 91, § 70

1. $x^2 + y^2 - 4 = 0$; circle.
2. $xz = 1$; hyperbola.
3. $y + z + 3 = 0$; straight line.
4. $x^2 + z^2 + 1 = 0$; imaginary circle.
5. $(x + 2y - 1)^2 = 0$; two coincident straight lines.
6. Vacuous.
7. $3, -1$. **8.** 2 (once).
9. $\pm i$ (imaginary). **10.** Vacuous.
11. $x = 0$, $x + 3y - 5z + 2 = 0$; multiplicity 1.
12. $x - y = 0$, $x + z = 0$; multiplicity 1.
13. $x + 2y + z = 0$, $x + 2y + 1 = 0$; multiplicity 1.
14. $x + 3y - 2z + 1 = 0$; multiplicity 2.

§§ 57–80]   ANSWERS   241

15. z axis.   **16.** y axis.   **17.** x axis.
18. z axis.   **19.** y axis.   **20.** x axis.
21. No.   **22.** $y^2 + z^2 = 9$.   **23.** $x^2 + y^2 = 25$.
24. $x^2 + z^2 = 4z$.   **25.** $(x-3)^2 + (y-4)^2 = 25$.   **34.** $(1, 1, 0), (-11, -3, 0)$.
35. $(0, 2, -1), (0, -2, 1), (0, \sqrt{2}, -\sqrt{2}), (0, -\sqrt{2}, \sqrt{2})$.

Page 95, § 72

1. $y^2 - x^2 = 5, \ z = 0$; hyperbola.
2. $(y - 2z)(2y + 4z + 7) = 0, \ x = 0$; two straight lines.
3. $(x^2 + y^2)^2 + x^2 = 4, \ z = 0$.
4. $x(3xz + 5z^2 + 4) = 0, \ y = 0$; a straight line and a hyperbola.
5. $2x = z^2, \ y = 0$.   **6.** $2x^2 + 2xy + y^2 = 16, \ z = 0$.
9. $7x - y - 5z - 15 = 0$.
10. $x - y = 4, \ z = x^2 - 2x + 3, \ z = y^2 + 6y + 11$.

Page 96, § 74

1. (a) A chair; (b) a two-bladed propeller, or two right hands in a hand-clasp; (c) a hand-clasp with a right hand and a left hand, palms together, fingers oppositely directed.

Page 98, § 76

1. xy plane; z axis; yes.   **2.** None; y axis; no.
3. None; none; yes.   **4.** All; all; yes.
5. yz plane; none; no.   **6.** None; none; yes.

Page 101, § 78

2. $f(x, y, z) = 0$ and $f(2k - x, y, z) = 0$ are equivalent equations.
3. $f(x, y, z) = 0$ and $f(y, x, z) = 0$ are equivalent equations.
4. $f(x, y, z) = 0$ and $f\left(\dfrac{-x + 2y + 2z}{3}, \dfrac{2x - y + 2z}{3}, \dfrac{2x + 2y - z}{3}\right) = 0$ are equivalent equations.
5. $f(x, y, z) = 0$ and $f(2x_0 - x, 2y_0 - y, 2z_0 - z) = 0$ are equivalent equations.

Page 103, § 80

1. (a) $x^2 + 2y^2 + 2z^2 = 8$; (b) $x^2 + 2y^2 + z^2 = 8$.
2. (a) $4x^2 - 9y^2 - 9z^2 = 5$; (b) $4x^2 + 4y^2 - 9z^2 = 5$.
3. (a) and (b) $6x^2 + 6y^2 + 6z^2 = 7$.
4. (a) $4x^2 - 9y^2 - 9z^2 - 24x + 36 = 0$; (b) $4x^2 - 9y^2 + 4z^2 + 36y - 36 = 0$.
5. (a) $y = 2$; (b) $x^2 + y^2 = 4$.
6. (a) $x^{\frac{2}{3}} + (y^2 + z^2)^{\frac{1}{3}} = 1$; (b) $(x^2 + y^2)^{\frac{1}{3}} + z^{\frac{2}{3}} = 1$.
7. z axis; $x^2 + z = 2, \ y = 0$.   **8.** x axis; $x^2 - 4y^2 = 8, \ z$
9. x axis; $x^2 - 4z^2 = 0$ (or $x - 2z = 0$ or $x + 2z = 0$), $y = 0$.
10. y axis; $z = 2y^2 - y + 1, \ x = 0$.
11. $4a^2(x^2 + z^2) = (x^2 + y^2 + z^2 + a^2 - b^2)^2$.

Page 107, § 85

1. $(x - 2)^2 + (y + 5)^2 + (z - 1)^2 = 36$.
2. $(x - 3)^2 + y^2 + (z + 4)^2 = 4$.
3. $(x - 4)^2 + (y - 1)^2 + (z - 3)^2 = 9$.
4. $(x + 2)^2 + (y + 1)^2 + (z - 5)^2 = 140$.
5. $(3, -1, -2)$; real; 5.
6. $(-1, -7, 5)$; imaginary.
7. $(-\frac{3}{2}, 2, \frac{1}{2})$; point sphere.
8. $(\frac{1}{3}, -\frac{5}{3}, 1)$; real; $\frac{2}{3}$.
9. $x^2 + y^2 + z^2 - x - y - z = 0$.
10. $x^2 + y^2 + z^2 - 3x + 27y - 13z + 62 = 0$.
12. $2x - y + 4z = 34$.
13. $2x - y + 4z = 13$.
14. Circle, center $(-3, -3, 2)$, radius $\sqrt{19}$, direction numbers of normal 7, 4, -4.
15. Circle, center $(0, 2, 5)$, radius 5, dir. nos. of normal 1, 3, 1.
16. Point $(-\frac{8}{11}, \frac{3}{11}, \frac{42}{11})$.
17. $x^2 + y^2 + z^2 - 4x - 3z - 10 = 0$.
18. $3x^2 + 3y^2 + 3z^2 - 2x + 8y + 23z + 3 = 0$.
19. $13x - 7y + z + 18 = 0$.
20. $24x - 56y + 4z + 35 = 0$.
21. $x^2 + y^2 + z^2 - 4x + 6y - 10z + 28 = 0$.
22. $x^2 + y^2 + z^2 - 6x - 2y + 8z + 1 = 0$,
    $81x^2 + 81y^2 + 81z^2 + 74x + 398y - 332z - 1179 = 0$.
23. $x^2 + y^2 + z^2 - 6x + 2y - 12z + 30 = 0$.
24. $x^2 + y^2 + z^2 - 6x + 2y - 12z - 10 = 0$,
    $x^2 + y^2 + z^2 - 6x + 2y - 12z - 178 = 0$.

Page 112, § 87

1. $(x + 7y - 14z - 7)(7x - 5y - 2z - 1) = 0$.
2. $5x - 3y - z + 4 = 0$.
3. $(5x + 8y - 22z - 11)(11x - 4y - 10z - 5) = 0$.
4. $(5x - 3y - z - 17)(15x - 9y - 3z + 5) = 0$.
5. xy plane, $z = 0$.
6. Cylinder, $x^2 + y^2 = a^2$.
7. Sphere $x^2 + y^2 + z^2 - 8x - 4y + 2z + 1 = 0$.
8. Sphere $12[(x - 1)^2 + y^2 + z^2] = -4x - 8y - 8z + 11$.
9. Two spheres $12[(x - 1)^2 + y^2 + z^2] = \pm(4x + 8y + 8z - 3)$.
10. Sphere $12[(x - 1)^2 + y^2 + z^2] = -4x - 8y - 8z + 7$ and point $(\frac{7}{6}, \frac{1}{3}, \frac{1}{3})$.
11. Line $\dfrac{x - 3}{1} = \dfrac{y - 6}{1} = \dfrac{z - 2}{-2}$.
12. The z axis.
13. Four lines $\dfrac{x}{1} = \dfrac{y}{2} = \dfrac{z}{3}, \dfrac{x}{1} = \dfrac{y}{2} = \dfrac{z}{-3}, \dfrac{x}{1} = \dfrac{y}{-2} = \dfrac{z}{3}, \dfrac{x}{-1} = \dfrac{y}{2} = \dfrac{z}{3}$.
14. Circle $x = y$, $3(x^2 + y^2 + z^2) + 2x - 1 = 0$; center $(-\frac{1}{6}, -\frac{1}{6}, 0)$; radius $\frac{1}{6}\sqrt{14}$.
15. Circle $x = -\frac{19}{4}$, $y^2 + z^2 = \frac{119}{16}$; center $(-\frac{19}{4}, 0, 0)$; radius $\frac{1}{4}\sqrt{119}$.
16. Sphere $x^2 + y^2 + z^2 = 1$.
17. Hyperbolic cylinder $8x^2 - y^2 = 8$.
18. Sphere or plane.
19. Line perpendicular to plane of points.
20. Sphere; center at midpoint of segment joining given points.
21. Right circular cylinder; axis is line of intersection of planes.

Page 115, § 89

1. (3, 0, 2).
2. (0, −2, 3).
3. (3, 3√3, −1).
4. (3, 4, 0).
5. (4, 180°, 1).
6. (5, 90°, −2).
7. (4√3, 30°, 0).
8. (0, any θ, 0).
9. (−7, 0, 0).
10. (0, 4√2, 4√2).
11. (−3, 3√3, −6√3).
12. (2, 2, 2√2).
13. (5, 90°, 270°).
14. (5, 53° 8′, 60°).
15. (3, 180°, any θ).
16. (√3, 54° 44′, 45°).
17. Cylinder.
18. Half-plane.
19. Cylinder.
21. Sphere.
22. Half-cone.
23. Half-plane.
24. Surface of revolution.
25. $x^2 + y^2 - z^2 = 0$.
26. $x + y = \sqrt{2}$.
27. $\rho \cos \theta + 2\rho \sin \theta + 3z = 3$.
28. $\rho = \cos \theta$.
29. $x^2 + y^2 + z^2 = 9$.
30. $(x^2 + y^2 + z^2)^2 = x^2 + y^2$.
31. $r = 3 \sec \phi$.
32. $\phi = 60°, 120°$.
33. Circle.
34. Helix.
35. Half-line.
36. Circle.
37. $z^2 = 9\rho^2$; right circular cone.
38. $r^2 = \cos^2 \phi$; two spheres tangent to the plane $\phi = 90°$ at the origin; **radius** of each $= \frac{1}{2}$.

Page 118, § 92

3. 3, 6, 2.
4. √6, √3, √2.
5. Oblate; y axis.
6. Prolate; z axis.
7. Oblate; x axis.
8. Prolate; x axis.

Page 120, § 97

3. 1 sheet; transverse: 3, 6; conjugate: 2.
4. 2 sheets; transverse: 3; conjugate: 6, 2.
5. 1 sheet; all = 1.
6. 2 sheets; all = 1.

Page 128, § 116

5. Imaginary cone.
6. Real elliptic cylinder.
7. Hyperbolic cylinder.
8. Imaginary parallel planes.
9. 1-sheeted hyp. of rev.; y axis.
10. Elliptic paraboloid.
11. Real parallel planes.
12. Right. circ. cone; x axis.
13. Imag. elliptic cylinder.
14. Ell. parab. of rev.; y axis.
15. Prolate spheroid; z axis.
16. Hyperbolic paraboloid.
17. Imaginary ellipsoid.
18. Coincident planes.
19. Imag. intersecting planes.
20. Sphere.
21. 2-sheeted hyp. of rev.; x axis.
22. Parabolic cylinder.
23. Real intersecting planes.
24. Hyperbolic paraboloid.
25. Right circular cylinder; x axis.
26. Sphere.
27. Real parallel planes.
28. Parabolic cylinder.
29. $\dfrac{x^2}{4} + \dfrac{y^2}{9} + \dfrac{z^2}{16} = 1$.
30. $\dfrac{x^2}{25} + \dfrac{y^2}{25} - \dfrac{z^2}{4} = 1$.
31. $\dfrac{x^2}{9} + \dfrac{y^2}{9} - \dfrac{z^2}{9} = -1$.
32. $\dfrac{x^2}{2} + \dfrac{y^2}{2} - \dfrac{z^2}{1} = 0$.

33. $\dfrac{x^2}{4} + \dfrac{y^2}{36} + 2z = 0.$ **34.** $\dfrac{x^2}{100} - \dfrac{y^2}{25} + 2z = 0.$
35. $x^2 + y^2 - z^2 = 1.$ **36.** $x^2 + 6z = 0.$
37. (a) $y^2 + z^2 = 9;$ (b) $9x^2 - 4y^2 - 4z^2 = 0;$ (c) $9x - 2y^2 - 2z^2 = 0;$
  (d) $x^2 + y^2 + z^2 + kx = 13 + 2k;$ k a parameter.
38. Paraboloid of revolution; $x^2 + y^2 + 4z = 0.$
39. Right circular cone; $x^2 + y^2 - z^2 = 0.$
40. Right circular cone; $x^2 + y^2 - 3z^2 = 0.$
41. Oblate spheroid; $x^2 + y^2 + 2z^2 = 4.$
42. Prolate spheroid. **43.** Sphere.
44. Prolate spheroid; $4x^2 + 4y^2 + z^2 = 4.$
45. 2-sheeted hyp. of rev.; $8x^2 - y^2 - z^2 = 8.$
46. $\dfrac{x^2}{9} + \dfrac{y^2}{9} - \dfrac{z^2}{16} = 0.$
50. The locus is that of a point whose distances from a fixed point and a fixed plane are in a constant ratio.
  (i) (If the point is in the plane): if $a^2 + b^2 + c^2 = 1$, a line (or two imaginary intersecting planes); if $a^2 + b^2 + c^2 > 1$, a right circular cone; if $a^2 + b^2 + c^2 < 1$, an imaginary cone.
  (ii) (If the point is not in the plane): if $a^2 + b^2 + c^2 = 1$, a paraboloid of revolution; if $a^2 + b^2 + c^2 > 1$, a two-sheeted hyperboloid of revolution; if $a^2 + b^2 + c^2 < 1$, a prolate spheroid.

Page 135, § 120

1. $(-11, -8, 19)$ and $(5, 0, -5)$. **2.** $(11, 3, -14)$.
3. Imaginary $(1 \pm 2i, -2 \pm i, 1 \mp 3i)$. **4.** No points.
5. $(1, 3, 3)$ and $(7, -15, -9)$. **6.** $(2, 0, 1)$ twice.
7. Ruling. **8.** No points.

Page 138, § 123

1. $3x - 2y - 6z = 0.$ **2.** $5x - 4y + 1 = 0.$
3. $x + y + z = 0.$ **4.** $x - 3y - 2z + 3 = 0.$
5. Point $(0, 6, 1)$. **6.** Line $\dfrac{x-1}{2} = \dfrac{y}{-1} = \dfrac{z-1}{0}.$
7. Plane $x + 2y + 2 = 0.$ **8.** No center. **9.** No.

Page 143, § 128

1. $5x - 6y + 4z - 49 = 0;$ $\dfrac{x-1}{5} = \dfrac{y+6}{-6} = \dfrac{z-2}{4}.$
2. $9x + 8y + 4z - 21 = 0;$ $\dfrac{x-1}{9} = \dfrac{y-2}{8} = \dfrac{z+1}{4}.$
3. $x + 2z = 4;$ $\dfrac{x-2}{1} = \dfrac{y}{0} = \dfrac{z-1}{2}.$
4. $2x - y + 7z = 0;$ $\dfrac{x}{2} = \dfrac{y}{-1} = \dfrac{z}{7}.$
5. $(1, 5, -3)$. **6.** $(2, -3, 1)$.

§§ 120–140]   ANSWERS   245

8. $2x - 4y + 3z + 3 = 0$; $\dfrac{x-1}{2} = \dfrac{y-2}{-4} = \dfrac{z-1}{3}$.
9. $13x + 10y + 34z = 294$; $\dfrac{x-2}{13} = \dfrac{y-3}{10} = \dfrac{z-7}{34}$.
11. $x + y + z = 0$; $x + 1 = y - 2 = z + 1$.
16. $2x + 5y + z = \pm 7$; $(\pm \tfrac{10}{7}, \pm \tfrac{5}{7}, \pm \tfrac{1}{7})$.
17. $x + y - z = \pm 4$; $(\pm 3, \pm 2, \pm 1)$. **18.** No real tangent plane.
20. $5x - 2y + z - 31 = 0$; $(10, -6, -31)$.
21. $4x + 3y - 2z + 6 = 0$; $(-12, 12, -3)$.

Page 148, § 130

1. $2x - 3y - 12z + 4 = 0$. **2.** $4x - 5y - 3z = 10$.
3. $3x + 7y + 6z = 0$. **4.** No polar plane.
5. $(2, -8, 1)$. **6.** $(4, 6, -2)$. **7.** $\dfrac{x}{2} = \dfrac{y-2}{2} = \dfrac{z}{-1}$.

Page 156, § 133

1. $\dfrac{x-1}{1} = \dfrac{y-1}{0} = \dfrac{z-1}{1}$, $\dfrac{x-1}{0} = \dfrac{y-1}{1} = \dfrac{z-1}{1}$.
2. $\dfrac{x-12}{9} = \dfrac{y+14}{-8} = \dfrac{z-8}{5}$, $\dfrac{x-12}{15} = \dfrac{y+14}{-24} = \dfrac{z-8}{13}$.
3. $\dfrac{x+6}{2} = \dfrac{y-15}{3} = \dfrac{z-8}{8}$, $\dfrac{x+6}{2} = \dfrac{y-15}{-3} = \dfrac{z-8}{-2}$.
4. $\dfrac{x-1}{0} = \dfrac{y+1}{0} = \dfrac{z-1}{1}$, $\dfrac{x-1}{35} = \dfrac{y+1}{-7} = \dfrac{z-1}{5}$.
5. $\dfrac{x \pm 4}{16} = \dfrac{y \mp 4}{-6} = \dfrac{z \pm 2}{5}$. **6.** $\dfrac{x-15}{5} = \dfrac{y-3}{-1} = \dfrac{z}{-6}$.
7. Cone: $x^2 - 9y^2 - z^2 = 0$. **8.** Hyp. I: $x^2 - 9y^2 + 4z^2 = 1$.
9. Hyp. Par.: $x^2 - 9y^2 = 4z$. **10.** $x^2 - y^2 + 2z + 1 = 0$.
11. $4x^2 + 4y^2 - z^2 = 36$. **12.** $9x^2 - y^2 - z^2 + 4 = 0$.

Page 170, § 140

1. One set: $x = x' + 8$, $y = y' - 3$, $z = z' + 9$.
2. $(3, -1, -5)$; $x'^2 + y'^2 + z'^2 = 16$.
3. $(-2, -6, 0)$; $4x'^2 - y'^2 + 2z'^2 = 5$.
4. $(1, -4, -3)$; $x'^2 + 3z'^2 - 6x'y' + 11 = 0$.
5. $(-3 + 2t, 1 - t, -t)$; $2y'^2 - 2z'^2 + 2x'y' - 2x'z' + 15 = 0$.
6. $(9 + t, 4 + t, -t)$; $2x'^2 + 3y'^2 + 3z'^2 - 2x'y' + 2x'z' + 4y'z' = 6$.
7. Plane $2x - y + 3z = 2$; $4x'^2 + y'^2 + 9z'^2 - 4x'y' + 12x'z' - 6y'z' = 12$.
8. $\dfrac{x'^2}{2} + \dfrac{y'^2}{4} + 2z' = 0$; elliptic paraboloid, vertex $(2, 11, 17)$.
9. $x'^2 + 7y' = 0$; parabolic cylinder.
10. $2x'^2 + 3y'^2 + z'^2 + 2 = 0$; imaginary ellipsoid.
11. $5x'^2 - y'^2 + 6z'^2 = 0$; real cone.

Page 179, § 147

1. $\frac{2}{7}, \frac{3}{7}, \frac{6}{7}$.

2. $\frac{10}{19}, -\frac{15}{19}, \frac{6}{19}, -\frac{10}{19}$.

3. $-\frac{2}{3}, -\frac{2}{3}, -\frac{2}{3}, -\frac{2}{3}, -\frac{1}{3}$.

4. $0, \pm\frac{1}{\sqrt{3}}, \frac{1}{\sqrt{3}}, \frac{2}{\sqrt{6}}, \mp\frac{1}{\sqrt{6}}, \mp\frac{1}{\sqrt{6}}$.

5. One set: $x' = \frac{1}{\sqrt{3}}(x+y+z);\ y' = \frac{1}{\sqrt{6}}(x-2y+z);\ z' = \frac{1}{\sqrt{2}}(x-z)$.

6. One set: $x' = \frac{1}{\sqrt{70}}(3x+6y-5z);\ y' = \frac{1}{\sqrt{14}}(x+2y+3z);$

$$z' = \frac{1}{\sqrt{5}}(2x-y).$$

7. $x' + y' = 8$.
8. $(\sqrt{6}+4\sqrt{2}+3\sqrt{3})x' + (-\sqrt{6}+4\sqrt{2}-3\sqrt{3})y' + (2\sqrt{6}-3\sqrt{3})z' + 5 = 0$
9. $2x'^2 - 4y'^2 - z'^2 + 4 = 0$; hyperboloid of one sheet.
10. $x'^2 - y'^2 = 2\sqrt{3}$; hyperbolic cylinder.

12. $\begin{pmatrix} \frac{1}{3} & \mp\frac{2}{3} & \pm\frac{2}{3} \\ \frac{2}{3} & \mp\frac{1}{3} & \mp\frac{2}{3} \\ \frac{2}{3} & \pm\frac{2}{3} & \pm\frac{1}{3} \end{pmatrix}, \begin{pmatrix} -\frac{1}{3} & \pm\frac{2}{3} & \pm\frac{2}{3} \\ -\frac{2}{3} & \pm\frac{1}{3} & \mp\frac{2}{3} \\ -\frac{2}{3} & \mp\frac{2}{3} & \pm\frac{1}{3} \end{pmatrix}$.

13. $x'^2 - z'^2 = 1$.
14. $x'^2 + y'^2 + z'^2 = 25$

19. $\begin{pmatrix} 1 & 0 & 0 \\ 0 & -1 & 0 \\ 0 & 0 & -1 \end{pmatrix}$.
20. $\begin{pmatrix} 0 & 0 & -1 \\ 0 & 1 & 0 \\ 1 & 0 & 0 \end{pmatrix}$.

22. $\begin{pmatrix} 0 & 0 & 1 \\ \pm 1 & 0 & 0 \\ 0 & \pm 1 & 0 \end{pmatrix}$ or $\begin{pmatrix} 0 & 0 & -1 \\ \pm 1 & 0 & 0 \\ 0 & \mp 1 & 0 \end{pmatrix}$.

23. $\begin{pmatrix} 0 & 0 & -1 \\ 0 & \pm 1 & 0 \\ \pm 1 & 0 & 0 \end{pmatrix}$ or $\begin{pmatrix} 0 & 0 & 1 \\ \pm 1 & 0 & 0 \\ 0 & \pm 1 & 0 \end{pmatrix}$.

24. $x = \frac{2}{3}x' - \frac{2}{3}y' - \frac{1}{3}z' + 4$ $x' = \frac{1}{3}(2x+y+2z-7)$
    $y = \frac{1}{3}x' + \frac{2}{3}y' - \frac{2}{3}z' - 3$ $y' = \frac{1}{3}(-2x+2y+z+13)$
    $z = \frac{2}{3}x' + \frac{1}{3}y' + \frac{2}{3}z' + 1$ $z' = \frac{1}{3}(-x-2y+2z-4)$.

27. Circle, radius = 1.
28. Parabola.

29. $\frac{x}{0} = \frac{y}{2} = \frac{z}{-1}$.
30. $x = -y = -z$.

31. arc cos $(-\frac{2}{3})$.
32. arc cos $\frac{1}{7}$.

Page 188, § 150

1. $x + 3 = 0,\ y + 2z = 0,\ 2y - z = 0$.
2. $x - y - 3 = 0,\ x + y - 4z - 9 = 0$.
3. $4x - 2y + 3z + 5 = 0$.
4. $x + y + z - 4 = 0,\ x + (1 \mp \sqrt{3})y + (-2 \pm \sqrt{3})z + 2 \mp \sqrt{3} = 0$.
5. $2x + y + z + 3 = 0,\ y - z - 1 = 0$.
6. $x - 4y - z = 0.\ x + z + 2 = 0$.

§§ 147–156]   ANSWERS   247

7. $y + z = 0$, $x - y + z = 0$, $2x + y - z = 0$.
8. $2x + y + z - 2 = 0$, $x - y - z + 5 = 0$, $y - z + 4 = 0$.
9. $2y - z - 4 = 0$, $y + 2z + 3 = 0$.
10. $x - 2y + 2z = 0$, $2x - y - 2z = 0$.

Page 193, § 154

1. Hyperboloid of two sheets.
2. Hyperbolic cylinder.
3. Coincident planes.
4. Real quadric cone.
5. Real intersecting planes.
6. Hyperbolic paraboloid.
7. Real ellipsoid.
8. Hyperboloid of one sheet.
9. Hyperbolic paraboloid.
10. Elliptic paraboloid.
11. Elliptic paraboloid.
12. Imaginary ellipsoid.
13. Parabolic cylinder.
14. Hyperbolic paraboloid.
15. Hyperboloid of one sheet.
16. Real ellipsoid.
17. Imaginary intersecting planes.
18. Hyperbolic cylinder.
19. Real quadric cone.
20. Hyperboloid of two sheets.

Page 198, § 156

1. (a) $\dfrac{x'^2}{20} + \dfrac{y'^2}{4} - \dfrac{z'^2}{4} = -1$; (b) center $(-3, 0, 0)$.

2. (a) $\dfrac{x'^2}{9} - \dfrac{y'^2}{3} = -1$; (b) centers $(6 + 2t, 3 + 2t, t)$.

3. (a) $x'^2 = 0$; (b) planes $(4x - 2y + 3z + 5)^2 = 0$.

4. (a) $2x'^2 + (-1 + \sqrt{3})y'^2 - (1 + \sqrt{3})z'^2 = 0$; (b) center $(1, 1, 2)$.

5. (a) $x'^2 - y'^2 = 0$; (b) planes $2x + y + z + 3 = \pm\sqrt{3}(y - z - 1)$; centers $(-2 - t, 1 + t, t)$.

6. (a) $3x'^2 - \tfrac{1}{3}y'^2 + 2z' = 0$; (b) vertex $(1, 1, -3)$; tangent at vertex $2x + y - 2z - 9 = 0$.

7. (a) $\dfrac{x'^2}{6} + \dfrac{y'^2}{4} + \dfrac{z'^2}{2} = 1$; (b) center $(0, 0, 0)$.

8. (a) $x'^2 + \dfrac{y'^2}{4} - \dfrac{z'^2}{2} = 1$; (b) center $(-1, 0, 4)$.

9. (a) $\tfrac{3}{2}x'^2 - y'^2 + 2z' = 0$; (b) vertex $(3, 1, -2)$; tangent at vertex $x - 3 = 0$.

10. (a) $x'^2 + 2y'^2 + 2z' = 0$; (b) vertex $(4, 4, 2)$; tangent at vertex $2x + 2y + z - 18 = 0$.

11. Imaginary elliptic cylinder.
12. Real elliptic cylinder.

13. (a) Real elliptic cylinder; $\dfrac{x'^2}{9} + \dfrac{y'^2}{6} = 1$; (b) $y + z + 2 = 0$, $x + y - z + 1 = 0$; (c) centers $(1 + 2t, -2 - t, t)$.

14. (a) Real parallel planes; $x'^2 = \tfrac{16}{11}$; (b) $3x - y + z + 3 = 0$; (c) centers $3x - y + z + 3 = 0$; planes $3x - y + z + 7 = 0$, $3x - y + z - 1 = 0$.

15. (a) Right circular cylinder; $\dfrac{x'^2}{4} + \dfrac{y'^2}{4} = 1$;
   (b) axis $x = 1 + t$, $y = 4 + 2t$, $z = t$; also line of centers.

16. (a) Hyperboloid of two sheets; $\dfrac{x'^2}{4} + \dfrac{y'^2}{4} - \dfrac{z'^2}{2} = -1$;
   (b) axis $x = 3 + t$, $y = 1 + t$, $z = 1 + 2t$; center $(3, 1, 1)$.

17. (a) Prolate spheroid; $\dfrac{x'^2}{24} + \dfrac{y'^2}{4} + \dfrac{z'^2}{4} = 1$;
(b) axis $x = t, y = 1 + 2t, z = 0$; center $(0, 1, 0)$.
18. (a) Elliptic paraboloid; $\dfrac{x'^2}{\frac{2}{3}} + \dfrac{y'^2}{\frac{2}{3}} + 2z' = 0$;
(b) axis $x = 1 + 2t, y = 1 + 2t, z = 1 + t$; vertex $(1, 1, 1)$.
19. (a) Hyperboloid of one sheet; $\dfrac{x'^2}{2} + \dfrac{y'^2}{2} - z'^2 = 1$;
(b) axis $x = y = z$; center $(0, 0, 0)$.
20. (a) Line cylinder or imaginary intersecting planes, normal sections point circles; $x'^2 + y'^2 = 0$;
(b) axis $x = -2 - t, y = -2 + t, z = t$; also line of centers.
21. (a) Oblate spheroid; $\dfrac{x'^2}{4} + \dfrac{y'^2}{4} + z'^2 = 1$;
(b) axis $x = -2 + t, y = 2 + t, z = t$; center $(-2, 2, 0)$.
22. (a) Right circular cone; $x'^2 + y'^2 - 6z'^2 = 0$;
(b) axis $x = 2 + 2t, y = -5 + t, z = 1 + 3t$; center $(2, -5, 1)$.
23. (a) Hyperbolic cylinder; $1, 0, -1$;
(b) principal planes $x - 2y + z - 1 = 0, x + y + z - 1 = 0$;
centers $(1 + t, 0, -t)$; canonical form $\dfrac{x'^2}{4} - \dfrac{y'^2}{2} = -1$.
24. (a) Parabolic cylinder; $7, -4, -1$;
(b) principal plane $x + y + 3z + 1 = 0$; line of vertices $(8 + 7t, -4t, -3 - t)$;
tangent at vertices $x + 2y - z - 11 = 0$; canonical form $x'^2 + \dfrac{2\sqrt{6}}{11} z' = 0$.
25. (a) Imaginary elliptic cylinder; $1, -1, 2$;
(b) principal planes $2x - z = 0, x + 5y + 2z = 0$;
centers $(t, -t, 2t)$; canonical form $\dfrac{x'^2}{10} + \dfrac{y'^2}{20} = -1$.
26. (a) Parabolic cylinder; $6, 2, -3$;
(b) principal plane $3x - 6y + 2z = 0$; line of vertices $(2 + 6t, 1 + 2t, -3t)$;
tangent at vertices $2x + 3y + 6z - 7 = 0$; canonical form $x'^2 + \frac{8}{7}z' = 0$.
27. $x^2 + y^2 + z^2 - 2xy - 2xz - 2yz = 0$; right circular cone; axis $x = y = z$; generating angle arc cot $\sqrt{2}$.
28. If $a < 0$ or $0 < a < \frac{1}{2}$, hyperboloid of one sheet; if $a > \frac{1}{2}$, hyperboloid of two sheets; if $a = \frac{1}{2}$, real cone; if $a = 0$, hyperbolic paraboloid.
29. (i) $d > 9$: if $a < 4$, hyperbolic cylinder; if $a > 4$, imaginary elliptic cylinder; if $a = 4$, imaginary parallel planes. (ii) $d < 9$: if $a < 4$, hyperbolic cylinder; if $a > 4$, real elliptic cylinder; if $a = 4$, real parallel planes. (iii) $d = 9$: if $a < 4$, real intersecting planes; if $a > 4$, imaginary intersecting planes; if $a = 4$, coincident planes.
30. $a = 1$, hyperboloid of two sheets, axis $(0, t, -t)$; $a = -1$, hyperboloid of one sheet, axis $(0, t, t)$.
31. (i) $d > 0$: if $a > 1$, imaginary ellipsoid; if $a = 1$, imaginary parallel planes; if $-2 < a < 1$, hyperboloid of one sheet; if $a = -2$, right circular cylinder; if $a < -2$, prolate spheroid. (ii) $d < 0$: if $a > 1$, oblate spheroid; if $a = 1$, real parallel planes; if $-2 < a < 1$, hyperboloid of two sheets; if $a = -2$, imaginary elliptic cylinder; if $a < -2$, imaginary ellipsoid. (iii) $d = 0$:

§§ 156–173] ANSWERS 249

if $a > 1$, imaginary cone; if $a = 1$, coincident planes; if $-2 < a < 1$, right circular cone; if $a = -2$, imaginary intersecting planes; if $a < -2$, imaginary cone.

33. (i) $d = 1$; (ii) $d < 1$; (iii) $d > 1$.
35. $x^2 + y^2 + z^2 - R^2 = (\lambda x + \mu y + \nu z)^2$.
37. $(a^2 + b^2 + c^2 - R^2)(x^2 + y^2 + z^2) = (ax + by + cz)^2$.
38. $e(x, y, z) = (\alpha x + \beta y + \gamma z)^2$. **39.** $e(x, y, z) = k(x^2 + y^2 + z^2)$.
40. Cone $xy + xz + yz = 0$ (except for points on the coördinate axes).
42. $x^2 + y^2 + 2z^2 - 2xz - 2yz - 1 = 0$.
44. $x^2 + y^2 + z^2 - 2xz - 2yz + 2z - 1 = 0$.
49. Hyperbolic paraboloid. **50.** Ellipse.
51. Hyperboloid of one sheet.
56. Sphere, center at the fixed point on the z axis, passing through the origin.
61. $xy + xz + yz = 0$.

Page 213, § 164

1. $\begin{pmatrix} 5 & 3 \\ -7 & -1 \end{pmatrix}$. **2.** $\begin{pmatrix} 13 & 1 \\ 10 & 22 \end{pmatrix}$. **3.** $\begin{pmatrix} 14 \\ 10 \end{pmatrix}$. **4.** $\begin{pmatrix} 2 & 0 & 0 \\ 0 & 2 & 0 \\ 0 & 0 & 2 \end{pmatrix}$.

6. $60 \neq -5 + 3$. **15.** $\begin{pmatrix} 1 & 0 \\ 0 & 0 \end{pmatrix}$, $\begin{pmatrix} 0 & 0 \\ 0 & 1 \end{pmatrix}$. **16.** 1; 0 or 1.

20. $\begin{pmatrix} 2 & 1 \\ 2 & 3 \end{pmatrix}$, $\begin{pmatrix} 3 & 1 \\ 2 & 4 \end{pmatrix}$.

Page 220, § 169

1. $\begin{pmatrix} -\frac{4}{7} & \frac{3}{7} \\ \frac{5}{7} & -\frac{2}{7} \end{pmatrix}$. **2.** $\begin{pmatrix} -1 & -1 & 1 \\ 3 & 2 & -2 \\ -2 & 0 & 1 \end{pmatrix}$.

3. $\begin{pmatrix} -\frac{4}{33} & \frac{7}{33} & \frac{10}{33} \\ \frac{17}{33} & -\frac{5}{33} & -\frac{26}{33} \\ -\frac{7}{33} & \frac{4}{33} & \frac{1}{33} \end{pmatrix}$. **4.** $\begin{pmatrix} 1 & 0 & 0 & -x_0 \\ 0 & 1 & 0 & -y_0 \\ 0 & 0 & 1 & -z_0 \\ 0 & 0 & 0 & 1 \end{pmatrix}$.

6. $x'' = 7x - 6y + 7z$, **7.** $x'' = -8x + 13z$,
$y'' = 10x + y + 14z$, $y'' = -6x - 4y + 13z$,
$z'' = -14y - 6z$. $z'' = 12x + 5y - 24z$.

14. Any matrix of the form $\begin{pmatrix} a & b \\ -b & a \end{pmatrix}$, where $a^2 + b^2 \neq 0, 1$.

Page 227, § 173

1. $\begin{pmatrix} 0 & \frac{2}{\sqrt{6}} & -\frac{1}{\sqrt{3}} \\ \frac{1}{\sqrt{2}} & \frac{1}{\sqrt{6}} & \frac{1}{\sqrt{3}} \\ \frac{1}{\sqrt{2}} & -\frac{1}{\sqrt{6}} & -\frac{1}{\sqrt{3}} \end{pmatrix}$. **2.** $\begin{pmatrix} \frac{1}{\sqrt{3}} & \frac{1}{\sqrt{6}} & -\frac{1}{\sqrt{2}} \\ -\frac{1}{\sqrt{3}} & \frac{2}{\sqrt{6}} & 0 \\ \frac{1}{\sqrt{3}} & \frac{1}{\sqrt{6}} & \frac{1}{\sqrt{2}} \end{pmatrix}$.

3. $\begin{pmatrix} \frac{1}{\sqrt{6}} & \frac{1}{\sqrt{3}} & \frac{1}{\sqrt{2}} \\ \frac{2}{\sqrt{6}} & -\frac{1}{\sqrt{3}} & 0 \\ \frac{1}{\sqrt{6}} & \frac{1}{\sqrt{3}} & -\frac{1}{\sqrt{2}} \end{pmatrix}.$
4. $\begin{pmatrix} \frac{2}{\sqrt{14}} & \frac{1}{\sqrt{5}} & \frac{6}{\sqrt{70}} \\ -\frac{3}{\sqrt{14}} & 0 & \frac{5}{\sqrt{70}} \\ \frac{1}{\sqrt{14}} & -\frac{2}{\sqrt{5}} & \frac{3}{\sqrt{70}} \end{pmatrix}.$

Page 232, § 177

1. 3; right-handed.
2. 2; left-handed.
3. $\frac{3}{2}$; left-handed.
4. $\frac{29}{6}$; right-handed.
5. $\frac{1}{6}a^2b$.
6. $\frac{3}{2}\sqrt{42}$.
7. $2\sqrt{206}$.
8. One answer:

$\begin{pmatrix} 1 & 0 & 0 \\ 0 & \frac{1}{\sqrt{5}} & -\frac{2}{\sqrt{5}} \\ 0 & \frac{2}{\sqrt{5}} & \frac{1}{\sqrt{5}} \end{pmatrix}.$

9. and 13. One answer:

$\begin{pmatrix} \frac{1}{\sqrt{6}} & \frac{1}{\sqrt{2}} & -\frac{1}{\sqrt{3}} \\ -\frac{1}{\sqrt{6}} & \frac{1}{\sqrt{2}} & \frac{1}{\sqrt{3}} \\ \frac{2}{\sqrt{6}} & 0 & \frac{1}{\sqrt{3}} \end{pmatrix}.$

INDEX

(The numbers refer to pages.)

Abelian group, 223 (Ex. 24)
Addition of matrices, 209
Algebraic curve, 87, 178
Algebraic equation, 84
Algebraic geometry, 83
Algebraic surface, 84
Angle between two lines, 9
  cosine of the, 19
  sine of the, 21
Angle between two planes, 35
  cosine of the, 35
Angle of a rotation, 177, 231
Angles, direction, 15
Associative law, 164, 209, 210
Asymptotic cone, 121, 121 (Ex. 5), 154, 199 (Ex. 32)
Augmented matrix, 61
Axes, coördinate, 2
  of an ellipsoid, 118
  of a hyperboloid, 119, 120
  rotation of, 160, 172
  translation of, 160, 161
Axis, conjugate, 119, 120
  of a pencil of planes, 37
  of a rotation, 175, 176
  of revolution, 101
  of symmetry, 96
  transverse, 119, 120

Bisector, of the angle between two lines, 25 (Exs. 15, 16)
  of the angle between two planes, 40 (Exs. 17, 18)
Bundle, of planes, 77 (Ex. 22)
  of spheres, 109 (Ex. 35)

Canonical equation for a quadric surface, 117
  reduction to, 188
Canonical form for a matrix, 226, 231

Center, of a quadric surface, 136
  of a sphere, 104
  of symmetry, 96
Central quadric, 137
Centroid, of a tetrahedron, 14 (Ex. 35)
  of a triangle, 13 (Ex. 33)
Characteristic direction, 185
Characteristic equation, 185
Characteristic polynomial, 185, 225
Characteristic root, 185, 225, 232 (Ex. 14)
Characteristic vector, 185
Chord of a quadric surface, 133
Circle, 106
Circular sections, 101, 106, 205 (Exs. 74, 75)
Classification, of families of points, 71, 72
  of quadric surfaces, 117–126, 132, 192
  of systems of planes, 67
Coefficient matrix, 61
Cofactor, 57
  normalized, 216
Coincident planes, 126
Collinear points, 7, 71, 72
Column matrix, 207
Commutative group, 223 (Ex. 24)
Commutative law, 164, 209, 210, 223 (Ex. 24)
Complete analysis of a quadric surface, 194
Complex numbers, definition of, 223 (Ex. 25)
Complex of spheres, 110 (Ex. 36)
Complex space, 79
Cone, asymptotic, 121, 121 (Ex. 5), 154, 199 (Ex. 32)
  circular sections of a, 205 (Ex. 75)
  imaginary quadric, 122
  projecting, 201 (Ex. 43)
  quadric, 121, 122, 203 (Exs. 57, 61)
  real quadric, 121, 157 (Ex. 17)
  tangent, 200 (Ex. 36)

INDEX

Confocal quadrics, 145 (Ex. 24)
Conicoid, 86
Conjugate axis, 119, 120
Conjugate diameters, 139 (Ex. 24)
Conjugate diametral planes, 139 (Exs. 23, 24)
Conjugate plane, 136
Conjugate polar lines, 149 (Exs. 25–27)
Consistent system of equations, 60
Coördinate axes, 2
Coördinate planes, 1
Coördinate system, 1, 7 (Ex. 14), 113, 114
Coördinate transformation, 113–115, 159, 162, 224
Coördinates, cylindrical, 113
  oblique, 7 (Ex. 14), 221 (Ex. 15)
  rectangular, 1
  spherical, 114
Coplanar lines, 7
Coplanar points, 7, 71, 72
Cosine of the angle, between two lines, 19
  between two planes, 35
Cosines, direction, 15, 17 (Ex. 29)
Cramer's Rule, 61
Curve, 87
  algebraic, 87
Cylinder, 89
  hyperbolic, 124
  imaginary elliptic, 124
  line, 125
  parabolic, 125
  projecting, 92, 200 (Ex. 41)
  quadric, 126
  real elliptic, 124
  tangent, 200 (Ex. 34)
Cylindrical coördinates, 113

Degenerate sections of quadrics, 134
Degree, of an algebraic equation, 84
  of an algebraic surface, 85
Dependence, linear, 68
Determinant, of a matrix, 57, 174
  of a rotation matrix, 174
Determinants, multiplication of, 211
  properties of, 56
Diagonal, main, 207
Diagonal element, 207
Diagonal matrix, 214 (Ex. 18)
Diameter of a quadric surface, 139 (Ex. 23), 201 (Ex. 45)

Diameters, conjugate, 139 (Ex. 24)
  mutually perpendicular, 201 (Ex. 45)
Diametral plane, 135, 139 (Exs. 15–22), 144 (Ex. 13)
Diametral planes, conjugate, 139 (Exs. 23, 24)
Directed distance, 1
  from a plane to a plane, 32
  from a plane to a point, 1, 32
  on a line, 1
Direction, 1, 16
  characteristic, 185
  negative, 1
  of the line of intersection of two planes, 36
  perpendicular to two directions, 22
  positive, 1
Direction angles, 15
Direction cosines, 15
  in plane analytic geometry, 17 (Ex. 29)
Direction numbers, 17
  in plane analytic geometry, 19 (Ex. 13)
Director sphere, 201 (Ex. 46)
Directrix of a cylinder, 89
Distance, 1
  between a plane and a point, 32
  between a point and a line, 52
  between two planes, 32, 35 (Ex. 21)
  between two points, 7, 8
  between two skew lines, 53, 75 (Ex. 10)
  directed, 1
Distributive law, 210
Doubly ruled surface, 153

Element of a matrix, 57
Elementary transformation, of an equation, 26
  of a matrix, 58
Elimination, 38, 92, 94
  by addition or subtraction, 38, 94
  by substitution, 94
Ellipsoid, 117, 118
  circular sections of an, 205 (Ex. 75)
  imaginary, 118
  perpendicular diameters of an, 201 (Ex. 45)
  point, 122
  real, 117
  tangent planes of an, with given normal, 144 (Ex. 15)

INDEX

Elliptic cylinder, 124
  circular sections of an, 205 (Ex. 75)
Elliptic paraboloid, 122
  circular sections of an, 205 (Ex. 75)
  tangent plane of an, with given normal, 145 (Ex. 19)
Equation, algebraic, 84
  characteristic, 185
  graph of an, 26
  homogeneous, 60
  linear, 26, 60
  of the first degree, 26, 60
  of the second degree, 86, 131
Equation of a plane, 27
  in intercept form, 30
  in normal form, 31
  through three points, 29, 72, 74 (Ex. 5)
Equations, equivalent, 26
  system of homogeneous, 60
  system of linear, 60
Equations of a line, 41
  in parametric form, 44, 50
  in symmetric form, 42
Equations of a plane in parametric form, 50
Equivalence, of equations, 26
  of first degree equations, 28
  of matrices, 58, 233 (Ex. 19)
  of second degree equations, 190
  of systems of linear equations, 60
Equivalence relation, 14 (Ex. 36), 30 (Ex. 27), 60 (Ex. 10), 64 (Ex. 5), 65 (Ex. 6), 233 (Ex. 19)
Equivalent equations, 26
Equivalent matrices, 58, 233 (Ex. 19)
Equivalent systems of linear equations, 60

Factor of an algebraic surface, 84
Families, of lines, 76 (Ex. 12)
  of planes, 65, 67
  of points, 71, 72
Field, 83
Fixed point of a transformation, 175
Form, quadratic, 132, 224
Four-dimensional space, 77

General element of a matrix, 57
General equation of the second degree, 86, 131
  complete analysis of the, 194
Generator of a cylinder, 89
Generatrix of a surface of revolution, 101
Graph, of an equation, 26
  of a function, 26
  of an inequality, 34 (Ex. 15)
  of a system of equations, 26
Group, 222 (Ex. 23), 223 (Ex. 24)
  Abelian, 223 (Ex. 24)

Homogeneous equation, 60
Hyperbolic cylinder, 124
Hyperbolic paraboloid, 123, 152, 157 (Exs. 13–16, 19, 20), 158 (Ex. 22), 202 (Exs. 53, 54), 203 (Ex. 55)
  tangent plane of a, with given normal, 145 (Ex. 19)
Hyperboloid of one sheet, 119, 130 (Ex. 49), 151, 154, 157 (Exs. 16, 18, 21), 158 (Ex. 22), 202 (Ex. 52)
  circular sections of a, 205 (Ex. 75)
  tangent planes of a, with given normal, 144 (Ex. 15)
Hyperboloid of two sheets, 120, 130 (Ex. 48)
  circular sections of a, 205 (Ex. 75)
  tangent planes of a, with given normal, 144 (Ex. 15)

Identity matrix, 210
Identity transformation, 163, 173, 215
Imaginary circle, 106
Imaginary ellipsoid, 118
Imaginary elliptic cylinder, 124
Imaginary intersecting planes, 125
Imaginary parallel planes, 126
Imaginary quadric cone, 122
Imaginary sphere, 104
Inequality, graph of an, 34 (Ex. 15)
Infinity, point at, 88
Intercept, of a plane, 5
  of a surface, 89
Intercept form of the equation of a plane, 30
Intersections of quadrics and lines, 133
Invariance, 165
  of degree, 177
  of Δ, 169, 191, 225
  of the notion of polar plane, 228 (Ex. 9)

INDEX

Invariant property or quantity, 165
Invariants of a quadric surface, 191, 225
Inverse of a matrix, 216
Inverse transformation, 164, 215
Involution, 181 (Ex. 26), 232 (Ex. 11)
Irreducible algebraic surface, 84
Irreducible polynomial, 84

Left-handed coödinate system, 3
Left-handed orientation, 229
Line, equations of a, 41
  equations of a, in parametric form, 44, 50
  equations of a, in symmetric form, 42
  of intersection of two planes, 36
  of symmetry, 96
  of vertices, 126
  polar, 149 (Exs. 24–27)
Line cylinder, 125
Linear combination, 38, 69, 93
  of planes, 38
  of surfaces, 93
Linear dependence, 68
Linear equation, 26, 60
Linear transformation, 214, 222 (Exs. 21, 22), 233 (Ex. 15)
  between oblique systems, 221 (Ex. 15)
Lines, angle between two, 9
  bisector of the angle between two, 25 (Exs. 15, 16)
  collection of, 76 (Ex. 12)
  conjugate polar, 149 (Exs. 25–27)
  coplanar, 7
  cosine of the angle between two, 19
  distance between two, 53, 75 (Ex. 10)
  parallel, 7, 9
  perpendicular, 20
  plane parallel to two, 75 (Ex. 7)
  sine of the angle between two, 21
  skew, 7
Locus problems, 110

Main diagonal, 207
Mapping, 162
Matrices, addition of, 209
  equivalence of, 58, 233 (Ex. 19)
  multiplication of, 207
Matrix, 23, 57
  augmented, 61
  coefficient, 61
  column, 207
  diagonal, 214 (Ex. 18)
  e, E, 131
  elementary transformations of a, 58
  general element of a , 57
  identity, 210
  inverse, 216
  order of a square, 57
  orthogonal, 217, 218
  rank of a, 57
  rotation, 173
  row, 207
  scalar, 214 (Ex. 19)
  singular, 58
  square, 57
  symmetric, 131, 210
  transposed, 210
  unit, 210
  zero, 210
Matrix algebra, 207
Median of a triangle, 13 (Ex. 33)
Midpoint of a line segment, 12
Motion, rigid, 162, 165, 171 (Ex. 18), 181 (Ex. 25)
Multiplication, of determinants, 211
  of matrices, 207
  of transformations, 163
Multiplicity of a factor, 84
Multiplying transformation, 191

Negative direction, 1
Non-singular matrix, 58
Non-singular quadric surface, 131
Non-trivial solution, 61
Normal form for the equation of a plane, 31
Normal line of a quadric, 140
Normal plane of a curve, 142
Normal to a plane, 27
Normal to two directions, 22
Normalized cofactor, 216

Oblate spheroid, 118
Oblique coördinate system, 7 (Ex. 14), 221 (Ex. 15)
Octants, 2
One-to-one transformation, 170 (Ex. 16)
Order of a square matrix, 57
Orientation, right-handed, 229

INDEX

Origin of coördinates, 2
Orthogonal matrix, 217, 218
Orthogonal spheres, 106 (Ex. 27)
Orthogonal transformation, 219, 227 (Exs. 5, 6)
Orthonormal rows (columns), 218

Parabolic cylinder, 125
Paraboloid, elliptic, 122, 205 (Ex. 75)
  hyperbolic, 123, 152, 157 (Exs. 13–16, 19, 20) 158 (Ex. 22), 202 (Exs. 53, 54), 203 (Ex. 55)
  tangent plane of a, with given normal, 145 (Ex. 19)
Parallel circles, 205 (Exs. 74, 75)
Parallel lines, 7, 9
Parallel planes, 7, 32, 126
Parallel sections, 204 (Ex. 72), 205 (Ex. 73)
Parallelopiped, coördinate, 4
Parametric equations, of a line, 44, 50, 76 (Ex. 15)
  of a plane, 50, 76 (Ex. 17), 77 (Ex. 24)
  of a surface, 86
Peano space-filling curve, 87
Pencil, of planes, 37
  of spheres, 106
  of surfaces, 93
Perpendicular lines, 20
Perpendicular planes, 36
Perpendicular to two directions, 22
Plane, conjugate to a given direction, 136
  coördinate, 1
  diametral, 135, 139 (Exs. 15–22), 144 (Ex. 13)
  distance from a, 32
  equation of a, 27
  equation of a, in intercept form, 30
  equation of a, in normal form, 31
  equation of a, through three points, 29, 72, 74 (Ex. 5)
  intercept of a, 5
  normal, 142
  normal to a, 27
  of symmetry, 96
  parallel to a coördinate axis, 28
  parallel to a coördinate plane, 28
  parallel to two lines, 75 (Ex. 7)
  parametric equations of a, 50
  polar, 146, 149 (Exs. 22–24), 171

(Ex. 28), 228 (Ex. 9)
  principal, 184
  projecting, 39, 42, 43
  section by a, 89, 134, 178
  tangent, 140
  through the origin, 28
  trace of a, 5
Plane section, of an algebraic surface, 89, 178
  of a quadric surface, 134, 178
Planes, angle between two, 35
  bundle of, 77 (Ex. 22)
  intersecting, 125
  line of intersection of two, 36
  parallel, 7, 32, 126
  pencil of, 37
  perpendicular, 36
  projecting, 39, 42, 43
  systems of, 65, 67
Point, distance from a plane to a, 32
  of division, 10
  of symmetry, 96
  regular, 140
  singular, 140, 140 (Exs. 1, 2)
Point circle, 106
Point ellipsoid, 122
Point sphere, 104
Point transformation, 162, 224
Points, collinear, 7, 71, 72
  coplanar, 7, 71, 72
  families of, 71
Polar line, 149 (Exs. 24–27)
Polar plane, 146, 149 (Exs. 22–24), 171 (Ex. 28), 228 (Ex. 9)
Pole, 146, 149 (Exs. 16–21)
Polynomial, characteristic, 185, 225
  reducible, 84
  variable, 84
Positive direction, 1
Principal diagonal, 207
Principal plane, 184
Product, of algebraic surfaces, 85
  of determinants, 211
  of matrices, 207
  rank of the, 212, 217
  of transformations, 163
Projecting cone, 201 (Ex. 43)
Projecting cylinder, 92, 200 (Ex. 41)
Projecting planes of a line, 39, 42, 43
Projection, 5
  of a broken line segment, 24
  of a curve, 92
  on a directed line, 9

Prolate spheroid, 118, 130 (Ex. 47)
Properties of determinants, 56

Quadratic form, 132, 224
Quadric cone, 121, 122, 203 (Exs. 57, 61)
  circular sections of a, 205 (Ex. 75)
Quadric cylinder, 126, 200 (Exs. 34, 41)
Quadric surface, 86, 131
  a line and a, 133
  central, 137
  invariants of a, 191, 225
  plane section of a, 134, 178
  reducible, 125, 126, 203 (Exs. 58–60)
  ruled, 150
  singular, 131
Quadric surfaces, classification of, 117–126, 132, 192
  confocal family of, 145 (Ex. 24)

Radical center, 110 (Ex. 36)
Radical plane, 106
Radius vector, 7
Rank, of a matrix, 57
  of the product of matrices, 212, 217
Real ellipsoid, 117
  director sphere of a, 201 (Ex. 46)
  perpendicular diameters of a, 201 (Ex. 45)
  tangent plane of a, with given normal, 144 (Ex. 15)
Real elliptic cylinder, 124
Real intersecting planes, 125
Real parallel planes, 126
Real quadric cone, 121, 157 (Ex. 17)
  asymptotic, 121, 121 (Ex. 5), 154, 199 (Ex. 32)
  circular sections of a, 205 (Ex. 75)
  projecting, 201 (Ex. 43)
  tangent, 200 (Ex. 36)
Rectangular coördinates, 1
Reducible polynomial, 84
Reducible quadric surface, 125, 126, 203 (Exs. 58–60)
Regular point of a quadric surface, 140
Relation, equivalence, 14 (Ex. 36), 30 (Ex. 27), 60 (Ex. 10), 64 (Ex. 5), 65 (Ex. 6), 233 (Ex. 19)
Revolution, surface of, 101
Right-handed coördinate system, 3, 7 (Ex. 14), 78, 233 (Ex. 17)

Right-handed orientation, 229
Rigid motion, 162, 165, 171 (Ex. 18), 181 (Ex. 25)
Root, characteristic, 185, 225, 232 (Ex. 14)
Rotation, 160, 172
  angle of, 177, 231
  axis of, 175, 176
Rotation matrix, 173
Row-equivalence of matrices, 65 (Ex. 6)
Row matrix, 207
Ruled surface, 150
Ruling, of a cylinder, 89
  of a quadric surface, 134

Scalar matrix, 214 (Ex. 19)
Section of a surface, 89, 178
  circular, 101, 106, 205 (Exs. 74, 75)
Sections by parallel planes, 204 (Ex. 72), 205 (Ex. 73)
Self-polar tetrahedron, 149 (Ex. 27)
Semi-axes, 118, 119, 120
Simultaneous linear equations, 60
Sine of the angle between two lines, 21
Singular matrix, 58
Singular point of a quadric surface, 140, 140 (Exs. 1, 2)
Singular quadric surface, 131
Skew lines, 7
Skew-rotation, 174, 222 (Exs. 17, 18), 233 (Ex. 16)
Solution of a system of linear equations, 60
  trivial, 61
Sphere, 104
  through four points, 105
Spheres, bundle of, 109 (Ex. 35)
  complex of, 110 (Ex. 36)
  pencil of, 106
Spherical coördinates, 114
Spheroid, oblate, 118
  prolate, 118, 130 (Ex. 47)
Spur of a matrix, 185
Square matrix, 57
Sum of matrices, 209
Surface, 83
  algebraic, 84
  degree of an algebraic, 85
  intercept of a, 89
  of revolution, 101
  plane section of a, 89, 178

INDEX

quadric, 86, 131
ruled, 150
trace of a, 89
Surfaces, pencil of, 93
Symbolic notation for transformations, 162
Symmetric form for the equations of a line, 42
Symmetric matrix, 131, 210
Symmetry, 4, 95
  for algebraic surfaces, 98
  for graphs, 97
  line of, 96
  plane of, 96
  point of, 96
System of linear equations, 60
  consistent, 60
  equivalence of one, with another, 60
  homogeneous, 60
  inconsistent, 60
  solution of a, 60
Systems of planes, 65, 67

Tangent cone, 200 (Ex. 36)
Tangent cylinder, 200 (Ex. 34)
Tangent line, 140, 142
Tangent plane, 140
Tangent planes with given normal direction, 144 (Ex. 15), 145 (Ex. 19)
Tangent to a sphere, 107 (Ex. 11), 108 (Ex. 26)
Tetrahedron, centroid of a, 14 (Ex. 35)
  perpendicular bisectors of the edges of a, 75 (Ex. 8)
  points interior to a, 77 (Ex. 20)
  self-polar, 149 (Ex. 27)
  volume of a, 228
Torus, 104 (Ex. 11)
Trace, of a matrix, 185
  of a plane, 5

  of a surface, 89
Transformation, identity, 163, 173, 215
  inverse, 164, 215
  involutory, 181 (Ex. 26), 232 (Ex. 11)
  linear, 214, 221 (Ex. 15), 222 (Exs. 21, 22), 233 (Ex. 15)
  multiplying, 191
  of an equation, 26, 166, 194
  of a matrix, elementary, 58
  of coördinates, 113–115, 159, 162, 224
  of points, 162, 224
  one-to-one, 170 (Ex. 16)
  orthogonal, 219, 227 (Exs. 5, 6)
Transforming a transformation, 230
Translation, 160, 161
  simplification of an equation by a, 166
Transpose of a matrix, 210
Transverse axis, 119, 120
Triangle, area of a, 229
  centroid of a, 13 (Ex. 33)
  points interior to a, 52 (Ex. 6)
Trivial solution, 61

Umbilic, 206 (Ex. 76)
Unique factorization, 85, 203 (Ex. 60)
Unit matrix, 210

Variable polynomial, 84
Vector, characteristic, 185
  radius, 7
Vertex, of a cone, 121
  of an elliptic paraboloid,
  of a hyperbolic paraboloid
  of a parabolic cylinder,

Zero matrix, 210